Further Milestones In Biochemistry

FOUNDATIONS OF MODERN BIOCHEMISTRY
A Multi-Volume Treatise, Volume 3

Editors: **MARGERY G. ORD and LLOYD A. STOCKEN,** *Department of Biochemistry, University of Oxford, Oxford, England*

Further Milestones In Biochemistry

Edited by: **MARGERY G. ORD**
LLOYD A. STOCKEN
Department of Biochemistry
University of Oxford
Oxford, England

JAI PRESS INC.

Greenwich, Connecticut *London, England*

Library of Congress Cataloging-in-Publication Data

Foundations of modern biochemistry / editors, Margery G. Ord and Lloyd
A. Stocken
p. cm.
Includes bibliographical references and indexes.
ISBN 1-55938-960-5 (v. 1)
1. Biochemistry—History. I. Ord, Margery G. II. Stocken, Lloyd
A.
QD415.F68 1995
574.19'2'09—dc20 95-17048
CIP

55 Old Post Road No. 2
Greenwich, Connecticut 06836

JAI PRESS LTD.
38 Tavistock Street
Covent Garden
London WC2E 7PB
England

ISBN: 0-7623-0078-7

Library of Congress Catalog Number 95-17048

Transferred to digital printing 2006

CONTENTS

LIST OF CONTRIBUTORS

Judy P. Armitage	Department of Biochemistry University of Oxford Oxford, England
Patricia H. Clarke	Department of Chemical and Biochemical Engineering University College London London, England
Stuart J. Ferguson	Department of Biochemistry University of Oxford Oxford, England
Robin F. Irvine	Department of Pharmacology University of Cambridge Cambridge, England
Margery G. Ord	Department of Biochemistry University of Oxford Oxford, England
S. V. Perry	Department of Physiology University of Birmingham Birmingham, England
Philip J. Randle	Nuffield Department of Clinical Biochemistry University of Oxford Oxford, England

Lisa A. Steiner — Department of Biology
Massachusetts Institute of Technology
Cambridge, Massachusetts

Lloyd A. Stocken — Department of Biochemistry
University of Oxford
Oxford, England

F. R. Whatley — Department of Plant Sciences
University of Oxford
Oxford, England

ACKNOWLEDGEMENTS*

Once more we are very grateful to Professor Radda for allowing us space in the Department and to our colleagues for help and advice. Most of the photographs and some of the figures were copied for us by Ms A.Morgan. We are indebted to her for her skill and patience. We also thank the departmental librarian, Mr. Brian Taylor, and the staff of the Radcliffe Science Library, especially Miss Yvonne Hibbott, for their patience in obtaining and checking references for us.

We are very grateful to Professors Cohen, Andrew Huxley and Milstein and the Drs. Hokin for their photos. Those of Kenneth Bailey, Jean Hanson, A.Szent-Gyorgi and H.H.Weber were kindly given to us by Professor Perry, those of Julius Adler, Engelmann and Pfeffer by Dr.Armitage, and that of Daniel Arnon by Professor Whatley. We are grateful to his family for the photograph of Robin Hill and to the Department of Biochemistry, University of Cambridge, England, for that of Marjory Stephenson. The History of Science Museum, Oxford, allowed us to reproduce their portrait of Leeuwenhoek. The Nobel Foundation gave us permission to reproduce photographs of MacFarlane Burnet, Gerald Edelman, Rodney Porter, Susumu Tonegawa, Earl Sutherland, Andre Lwoff, Francois Jacob and Jacques Monod. Photographs of Emil von Behring and Karl Landsteiner are reproduced with permission from "The History of Immunology" (Silverstein, A.M., 1989), Academic Press, and those of Hugh Huxley and Elvin Kabat by permission of Annual Reviews, Inc.

We are also grateful to the American Society for Biochemistry and Molecular Biology, the American Society for Microbiology, the Biochemical Society, Academic Press, Blackwells Scientific Publishers, Cambridge University Press, and W.H.Freeman & Co. for permission to reproduce their copyright material.

Margery G. Ord
Lloyd A. Stocken
Editors

* Superscript numbers next to surnames throughout the volume refer to photographs, pages 165–171. Superscript numbers in **bold** refer to earlier or planned volumes of this series where photographs can be found.

Chapter 1

INTRODUCTION

The first volume of this series was an attempt to draw the attention of younger scientists to the remarkable achievements of early biochemists. These early biochemists with poor equipment, working alone or in small groups, made fundamental advances in our understanding of intermediary metabolism and enzymology.

We were then encouraged to carry the story forward from the mid-1950s. The international expansion in biochemistry from this period was greatly promoted by the prevailing optimistic assessment of the importance of biomedical research. There was an increased availability of resources, more research workers, and more sophisticated, commercially available, apparatus. Communications between scientists also accelerated, not as dramatically as at present, but there was a noticeable expansion in the numbers of journals, meetings and interlaboratory collaborations.

We therefore invited distinguished biochemists who had worked during this period to focus on what seemed to them to be the "quantum leaps" or milestones in the development of their topics. Up-to-the-minute details are available elsewhere in journals or comprehensive reviews. We are very grateful to these authors for their efforts to make their subjects intelligible to other scientists not expert in the field. Their personal involvement has resulted in authorative, sometimes quite individualistic, accounts of the way the subjects have developed.

Volume 2 recounted the elucidation of the mechanisms of information storage and transfer from DNA, and these topics are now taken further by an account of the control of information transfer in microorganisms, and the consequences of Jacob and Monod's concept of gene regulation.

New methods for the analysis of protein structures and the shift in emphasis since 1960, from preoccupation with the catalytic properties of proteins to the analysis of their roles in regulation, were described in Volume 2. The properties of a special class of proteins—the glycoproteins—were also considered. This has now been followed by an article on a further highly specialized class of proteins—the immunoglobulins—and the genetic basis of their diversity. Muscle contraction is also more fully characterized in molecular terms.

Consideration of cytoplasmic reactions in Volume 1 focused on intracellular energy considerations and the formation and utilization of ATP, and the concept of "the high-energy phosphate bond". Photophosphorylation was not then included but is considered now. In 1960 ATP was seen as the immediate driving force for endergonic reactions in cells while its mechanism of synthesis remained an enigma. The chemiosmotic hypothesis of Mitchell not only provided the key to understanding the mechanisms of ATP synthesis but also led to an appreciation of how a transmembrane electrochemical gradient of protons or sodium ions, rather than ATP, was the immediate driving force for some endergonic processes in cells. Progress in our understanding of bacterial motility, driven by the proton motive force, is also presented.

The importance of phosphate was first recognized by Wroblewski at the start of this century (see Volume 1). Phosphorylation of sugars provides a charge center for the attachment of otherwise neutral substrates to proteins—a key step in their subsequent transformations. Similarly, the introduction or removal of phosphate residues on OH groups of serine, threonine, or tyrosine in proteins with consequent conformational changes has emerged as an almost ubiquitous mechanism for the control of metabolic pathways. This, together with the transduction of information from the environment across the cell membrane and its onward transmission to the nucleus to regulate protein synthesis and/or cell division is now considered in articles on cell signaling and intracellular regulation by reversible protein phosphorylation.

Chapter 2

BIOENERGETICS AFTER 1960:
FROM THE CHEMIOSMOTIC HYPOTHESIS TO STRUCTURE-BASED MOLECULAR MECHANISMS

Stuart J. Ferguson

INTRODUCTION*

Most investigators of oxidative phosphorylation and photophosphorylation must in the early 1960s have been peacefully unaware of the formulation of Mitchell's[1] chemiosmotic hypothesis (see Mitchell, 1961a,b,c, 1966) that was, following its announcement in 1961, to revolutionize the way in which biochemists thought about both oxidative phosphorylation and related processes. Indeed some of these related processes, e.g. bacterial flagella motion,

* This account is presented conventionally, in the sense that the development of the subject is treated as deriving from Mitchell's papers. It is recognized that the origin of the ideas (cf. Williams, 1961) of the use of the proton in energy coupling is disputed. Brief reference is made to this in the concluding section.

had scarcely been recognized (see Armitage, Chapter 5), let alone connected to the same fundamental mechanism that drives ATP synthesis in oxidative phosphorylation. Furthermore, there was no hint of the challenge to the prevalent idea, which was actually to persist for another 20 years, that ATP was the fundamental minimal unit of energy currency in the cell. Photophosphorylation by both thylakoids of green plants and a cell free-membrane fraction (chromatophores) from certain photosynthetic bacteria, had been identified only six or so years before 1960 (see Whatley, Chapter 3). Consequently, the main effort in the field that has come to be known as bioenergetics was oxidative phosphorylation, and it is with this that we shall first be concerned here, returning later to see how studies on photophosphorylation by both thylakoids and chromatophores provided seminal contributions to our general understanding of bioenergetics.

The study of oxidative phosphorylation throughout the 1950s had mainly been concerned with the process in mitochondria. Investigations of the bacterial process had been a minor focus, with the result that the field did not benefit from a molecular genetic approach at an early stage. One reason for this is that during the 1950s methods had been developed, especially by David Green's laboratory, for the large scale preparation of bovine heart mitochondria which provided the raw material for the investigation of oxidative phosphorylation. Such large scale preparation of bacterial material was not so easy and furthermore, unlike intact mitochondria, intact bacterial cells cannot be supplied with the substrates for oxidative phosphorylation because external ADP is not taken up; thus it is difficult to separate oxidative phosphorylation from other concommitant cellular metabolism. Seemingly it was not fully realized in 1960 that the bacterial cytoplasmic membrane would be the site of ATP synthesis, and even if it had been, development of methods for preparation of inside-out membranes capable of oxidative phosphorylation with added substrates were, with the exception of the chromatophores mentioned above, still some way in the future.

By 1960 the consensus view was that mitochondria contained a respiratory chain that conducted electrons, originating principally from NADH or succinate, to oxygen. The composition of the respiratory chain was by no means fully characterized, but most progress had been made in identifying the cytochromes and the order in which they occurred in the sequence. Electrons originating from NADH were believed to generate three molecules of ATP, whereas those coming from succinate gave two molecules. Thus it was widely envisaged that there were three sites intimately associated with the respiratory chain where ATP was made.

Naturally, the hot topic was the mechanism (or mechanisms—there could have been different chemical processes at the three sites) whereby the energy released from electron transfer through these sites was coupled to the condensation of phosphate with ADP to give ATP. The expectation was that the mechanism would be akin to that used in glycolysis where the key step is the coupling of glyceraldehyde-3-phosphate oxidation to the incorporation of inorganic phosphate into the carbon skeleton of the reaction product, 1,3-bisphosphoglycerate. Such an idea, further developed to allow for known uncoupling of electron transport from ATP synthesis in the presence of molecules such as dinitrophenol, was the basis of the chemical hypothesis for the mechanism of ATP synthesis which is associated with Slater's name (Slater, 1953). The mitochondrial membrane was not envisaged as playing a key role in the process; indeed there was the belief that it should be possible to observe oxidative phosphorylation in a membrane-free system. However, as has been rehearsed many times, no evidence for any kind of chemical intermediate emerged. Against this background, Mitchell proposed his revolutionary chemiosmotic hypothesis (we return to the history of its origins later):

> The equilibrium constant, K', for the hydrolysis of glucose-6-phosphate to glucose and inorganic phosphate is approximately 250 in homogeneous aqueous solution at pH 7..., and the concentration of glucose-6-phosphate in equilibrium with 10^{-2} M glucose and 10^{-2} M phosphate would be only 4×10^{-7} M. If glucose-6-phosphatase were located in the anisotropic membrane complex..., a pH difference of only 3 units between phases I and II, or a potential difference of 60 mV and a pH difference of 2 units, would lower the dissociation constant by a factor of 1000 and would raise the concentration of glucose-6-phosphate in equilibrium with 10^{-2} M glucose and 10^{-2} M phosphate to 4×10^{-4} M.... I need hardly point out that a similar, but greater asymmetry of electrochemical hydrogen ion activity to that considered in the above example, could be responsible for converting the ATPases of the particulate systems of photosynthetic and oxidative phosphorylation into the ATP-synthesizing catalysts. I hope to develop this interesting and important aspect of translocation catalysis on another occasion (Mitchell, 1961a).

In retrospect this idea was simple, although outrageous because there was not really a scrap of evidence to support it. Moreover, it was frequently misunderstood which led to many years of intense argument. What Mitchell proposed from the beginning (see e.g. Mitchell, 1961a,b,c, 1966, 1968) (Figure 1) was that electron transport along the mitochondrial respiratory chain caused the displacement of protons from one side of the intact inner mitochondrial membrane to the other, with the consequence that a proton electrochemical gradient would be set up. Such a gradient would have two components—a pH gradient and a membrane potential—and would in turn drive protons through an ATP synthesizing enzyme. Mitchell offered proposals for how the electron transport chain and ATP synthase might function in this context at a simple molecular level but these proposals can now be seen as supplementary to the main, novel, hypothesis.

Substantiation of the chemiosmotic hypothesis relied very significantly on the experimental work of Mitchell and Moyle (see e.g. Mitchell, 1976, 1979) who produced evidence that the mitochondrial respiratory chain did indeed

The author proposes, therefore, to define explicitly a chemiosmotic hypothesis of electron-transport phosphorylation (Mitchell, 1960), as a basis for extension or disproof.

(i) Electron transfer, driven by oxido-reduction or photons, occurs vectorially across a membrane, separating aqueous phases A and B.

(ii) Process (i) effectively generates H^+ in A and OH^- in B.

(iii) The membrane is relatively impermeable to ions, but may allow exchange (Ussing, 1947) of H^+ and/or OH^- against ions of equivalent and like charge. The skew of $[H^+]$ ([] denoting electrochemical activity) therefore shows as a pH difference (pH_{B-A}) plus a membrane potential (mv_{A-B}). Approximately,

$$[H^+]_A/[H^+]_B = 10^{pH_{B-A}} \times 10^{mv_{A-B}/60} \leqslant e^{-(\text{work per electron translocated}/kT)}$$

(iv) The membrane contains an anisotropic adenosine triphosphatase system (phosphate-accepting active centre, E) catalysing the reaction: phosphate + ADP $\rightleftharpoons$ ATP + H^+ + OH^-.

(v) E communicates rapidly with OH^- of A and H^+ of B, but slowly with H^+ of A, OH^- of B, and H_2O of A and B. Consequently,

$$[H_2O]_{A \text{ or } B} \geqslant [H_2O]_E \geqslant ([H_2O]_{A \text{ or } B} \times [H^+]_B)/[H^+]_A,$$

when $$[H^+]_B/[H^+]_A < 1.$$

The inequalities of (iii) and (v) depend upon 'leakiness' and show as uncoupling. The H^+ differential, generated by electron translocation, dehydrates phosphate + ADP (or other acidic acceptor) by withdrawing OH^- and H^+ from phosphorylium and acidic acceptor respectively along different, chemically specific, translocation paths in the adenosine triphosphatase system. Using (iii), (v), and equilibrium constant data (Atkinson, Johnson & Morton, 1959): at 10 mM-inorganic phosphate, [ATP]/[ADP] would be about unity if, for example, A were 2 pH units below and 300 mv above B, and the system were well coupled.

Figure 1. Mitchell's early definition of the chemiosmotic hypothesis of electron-transport phosphorylation. (Reproduced with permission from Mitchell, 1961b).

translocate protons and establish a proton electrochemical gradient, or, in Mitchell's own terms, *a proton motive force* (pmf).

Meanwhile, however, a very important development was the identification by Racker and colleagues of the mitochondrial ATPase (see Racker, 1977). It proved possible to release from mitochondrial membranes a water-soluble protein that had ATPase activity which was called F_1. The removal of this species caused the capacity of the membranes for oxidative phosphorylation to decline and the disappearance of knob-like projections from the matrix surface of the inner mitochondrial membrane. These projections, and ATP synthesis, could be restored by mixing the soluble ATPase with depleted membranes under appropriate conditions. This ATPase activity has proved to be the locus of all the ATP synthesis catalyzed by oxidative phosphorylation and thus the concept of sites of ATP synthesis associated with discrete components of the respiratory chain proved to be fallacious. Rather today we understand there to be three sites of proton translocation within the electron transport system.

The establishment of the existence of a discrete ATPase was a significant step forward but, like the chemiosmotic hypothesis, it was contested. The main objection to a role for the soluble ATPase in oxidative phosphorylation was that it was insensitive to oligomycin which had been established as a specific inhibitor of the ATP synthesis and ATP hydrolysis by mitochondrial membranes. The resolution of this apparent quandary came when it was shown that F_1 was associated with a component known as F_0 located in the inner mitochondrial membrane. This component proved to be the proton conducting channel and to be the locus of oligomycin binding. Thus when F_1 was attached to the membrane ATP hydrolysis was linked to the driving of protons through the channel so that oligomycin could inhibit.

The 1960s were also the times when the idea that the mitochondrial electron transport chain could be split into four complexes was established. This development gave credence to the idea that ATP synthesis occurred at a spatially quite distinct machinery from the electron transport chain (Hatefi et al., 1962; and see Nicholls and Ferguson, 1992).

Following from these developments an important series of experiments was performed, mainly in Racker's laboratory, in which purified electron transport complexes were reconstituted, with appropriate topological considerations, together with the ATP synthase to ensure that a chemiosmotic proton circuit could be achieved. The complexes then catalyzed oxidative phosphorylation (Racker, 1977).

These observations implied there was neither a direct structural link between electron transport and ATP synthesis, nor an unrecognized set of proteinaceous components that were needed for coupling. At much the same time, bacteriorhodopsin was found to be a light-driven proton pump in the archaebacterium *Halobacterium halobium*. Reconstitution of this protein with the ATPase synthase from bovine mitochondria allowed light-driven ATP synthesis. This was considered to be further evidence in favor of the chemiosmotic theory.

STUMBLING BLOCKS TO THE ACCEPTANCE OF THE CHEMIOSMOTIC THEORY

As Mitchell himself remarked on many occasions, his theory proposed something that was completely outside the experience of almost all investigators of oxidative and photophosphorylation. Their backgrounds were almost exclusively in chemical biology and thus they were strongly conditioned to look for chemical intermediates and not to consider membrane phenomena. Looking back to 1960 one sees that membrane potentials were probably best understood in terms of the nerve. Thus the resting membrane potential of approximately 60 mV was known to arise from the diffusion of potassium ions down the concentration gradient; the resting potential is a reflection of an electrochemical equilibrium for potassium in which the concentration gradient, generated by an ATP-dependent pump, leads to the potential. The chemiosmotic theory envisaged something quite different. The movement of protons driven by an electron transfer chain would itself be electrogenic and could generate both a concentration gradient and a membrane potential. In other words, the proton would most definitely not be at electrochemical equilibrium. The idea that you could have both a membrane potential and a pH gradient definitely caused confusion. Slater once famously remarked in this context (e.g. Slater, 1994) that Mitchell wanted to have his cake and eat it! The word chemiosmosis conjured up thoughts of only concentration gradients and this too contributed to lack of understanding. Contrary to what has often been written elsewhere, Mitchell's (1961b) paper clearly describes both a pH gradient and a membrane potential (see Figure 1).

Another of the stumbling blocks to an understanding was that appreciation of how little charge needed to be moved across a biological membrane to generate a sizeable membrane potential, although comprehended in the nerve field, was not widely recognized by those studying oxidative phosphorylation. It turns out that the number of protons that must move across the mitochondrial membrane to establish a potential of 200 mV is so small that, given reasonable buffering power in the two aqueous phases on either side of the membrane, a negligible pH gradient may develop. If on the other hand the buffering capacity is very low, part of the proton electrochemical gradient may be expressed as a pH gradient. An important point is that there will be an upper limit to the total electrochemical gradient, set by the thermodynamics of the proton translocation reactions, but that the relative contributions of the membrane potential and pH gradient can vary considerably. Indeed, at the extreme the pH gradient can be the dominant term. This is exactly what is found for the thylakoids of green plants. The reason is now known to be the considerable flux of Mg^{2+} and Cl^- ions that can occur across this membrane. Such charge movements dissipate the membrane potential and result in the proton electrochemical gradient being fully expressed as a pH gradient.

Understandably, this difference between thylakoids on the one hand, and mitochondria or bacteria on the other, caused great confusion. In particular, it is the reason why the famous acid bath experiment of Jagendorf and Uribe (1966), which considerably advanced acceptance of the chemiosmotic hypothesis, works for thylakoids but not for mitochondria. Actually, it is noteworthy that Mitchell himself (1968) seemingly did not at first fully understand the acid bath experiment in which thylakoids were first soaked in a weak acid and then "jumped" to a higher external pH. The resultant ATP synthesis seen when the experiment is done in the dark, and which is abolished by uncouplers, can only be explained if proton movement out of the thylakoid, through the ATP synthase, is linked to the synthesis of ATP (Figure 2). The problem that Mitchell presumably had with this experiment is that the outward movement of a tiny number of protons, down their pH gradient, might have been expected to build up a membrane potential, positive outside, that would impede further proton movement. As mentioned earlier, only very small amounts of proton movement would be necessary for this state to be reached—so small that the ATP formed would be undetectable. Hence Mitchell's reservations. Fortunately this reservation was not apparently entertained by the experimentalists or else the acid bath experiment might not have been performed. Of course, the reason why ATP synthesis was observed was that the then unsuspected movement of chloride and magnesium, which does not occur across the mitochondrial membrane, prevented the build up of the otherwise anticipated membrane potential.

A further impediment to the acceptance of the chemiosmotic theory was the doubt that the maximum membrane potential envisaged, of approximately 200 mV, could be sustained by a biological membrane. In the 1960s there was a belief from studies with nerve that potentials exceeding approximately 80 mV would cause dielectric breakdown. This view has subsequently been discounted although exact quantitation of the membrane potential remains a problem to this day. Important electrophysiological work by Slayman on the cytoplasmic membrane of *Neurospora crassa* showed that cytoplasmic membrane potentials can reach 250 mV. Nevertheless the misleading idea that there is usually a large pH gradient rather than a membrane potential across the mitochondrial inner membrane persisted through the 1970s. The present writer can well remember being told during that time that the chemiosmotic hypothesis had to be wrong because it required, assuming a cytoplasmic pH of 7.0, a matrix pH of 9.5, incompatible with the function of mitochondrial enzymes. On the other hand, some investigators were convinced by measurements of membrane potentials from a light-dependent carotenoid

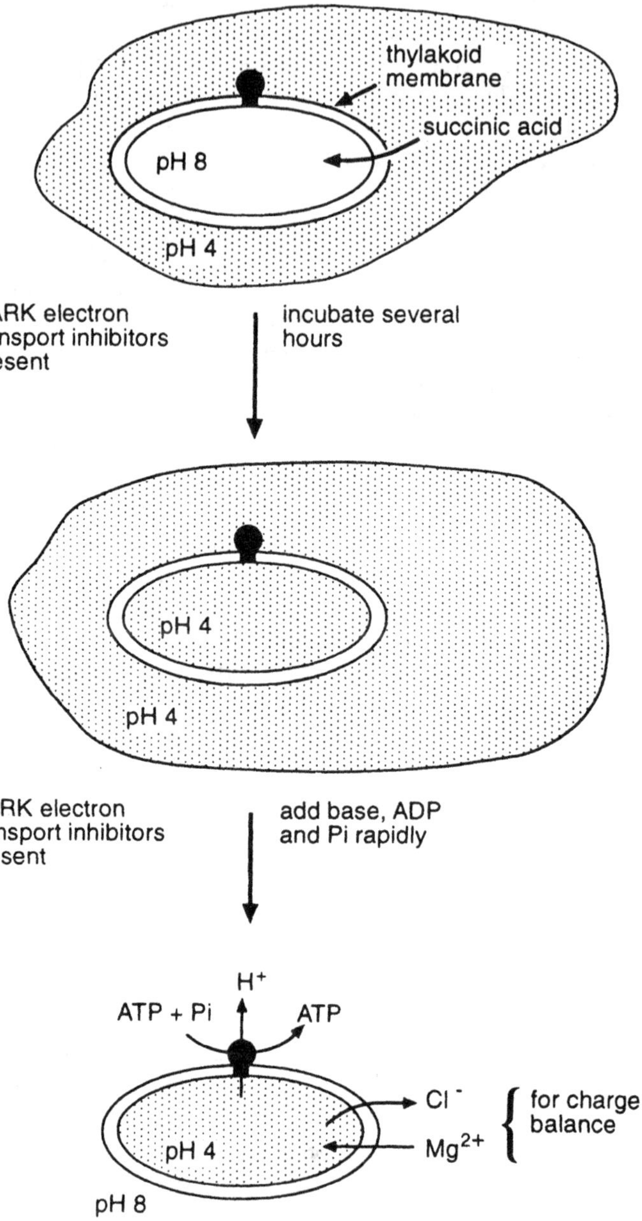

Figure 2. The Jagendorf acid-bath experiment. For fuller explanation, see text. (Reproduced with permission from Nicholls and Ferguson, 1992).

shifts in chromatophores from photosynthetic bacteria. However, it took a long time for it to be accepted that mitochondrial, thylakoid, and bacterial respiratory or photosynthetic membranes catalyzed energy transduction by the same mechanism.

NONINTEGRAL P/O RATIOS

The gradual acceptance of a chemiosmotic mechanism for oxidative and photophosphorylation was not at first accompanied by any reconsideration of the idea that the stoichiometry of ATP synthesis by mitochondria, the P/O ratio, was 3 if electrons passed from NADH to oxygen, and 2 if they originated with succinate. Establishment of these stoichiometries had been a hard fought issue, but one that had been addressed in the context that whole number stoichiometries were expected because ATP synthesis in mitochondrial oxidative phosphorylation was anticipated to be the consequence of discrete chemically coupled reactions analogous to the combination of the glyceraldehyde 3-phosphate dehydrogenase and 3-phosphoglycerate kinase reactions. Gradually, however, investigators' thoughts turned to the possibility that the stoichiometry of ATP synthesis might be nonintegral. Although Mitchell himself apparently did not champion this idea, the chemiosmotic theory provided the very framework on which such nonintegral stoichiometries might be hung. Thus if, say, x protons are translocated across the membrane per pair of electrons passing from NADH to oxygen and if y protons must pass back through the ATP synthase, then the stoichiometry of ATP synthesis is x/y, which evidently need not be integral. It is the fact that oxidative phosphorylation occurs via a delocalized mechanism with many electron transfer chains and ATP synthases communicating with the same two aqueous phases at the two sides of the membrane that allows such nonintegral stoichiometry to be possible.

The above is a theoretical argument; what were the experimental observations that led to increasing acceptance of nonintegral stoichiometry? Consistent with a P/O ratio of 3 for NADH, Mitchell argued that six protons are translocated per pair of electrons passing from NADH to oxygen and two through the ATP synthase for each ATP synthesized. This description became inadequate when it was realized that the exchange of adenine nucleotides across the inner mitochondrial membrane is electrogenic. Specifically, ATP^{4-} exits in exchange for ADP^{3-} entry. Add this to the evidence that phosphate enters mitochondria as a univalent anionic species in exchange for a hydroxyl ion (or in symport with a proton) and one realizes that one net proton moves into the mitochondrion in association with the combination of these two transport processes. Thus to generate ATP external to mitochondria from cytoplasmic ADP and P_i it became the view that three protons must move into the mitochondrial matrix, two via the ATP synthase and one via the combined transport processes (Figure 3). If

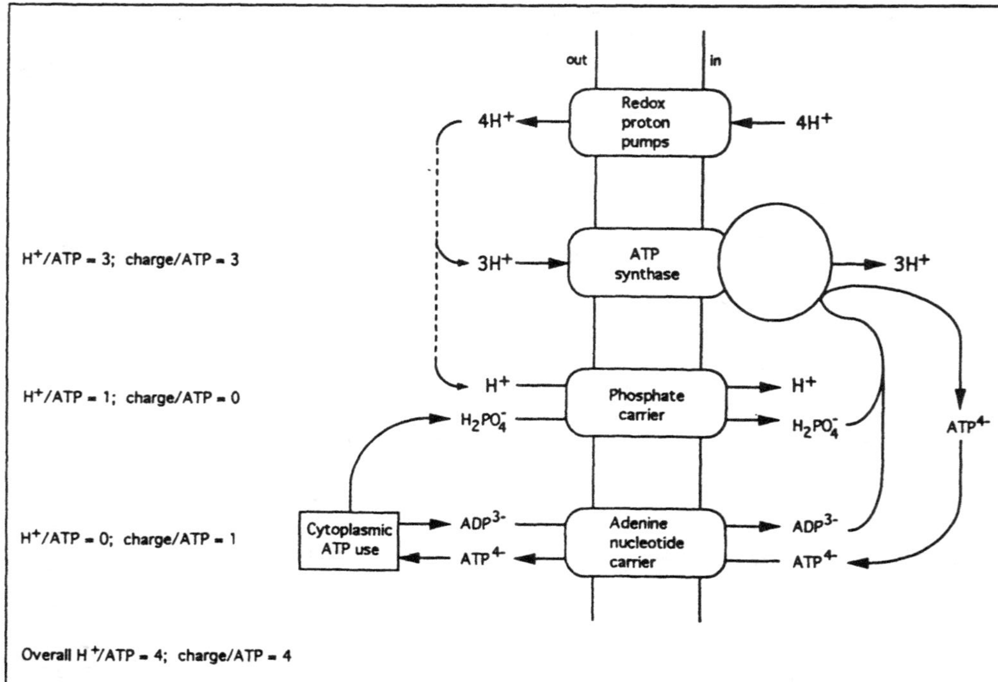

Figure 3. Proton circuits in the mitochondria. ATP synthesis requires return to the matrix of four H^+ for each ATP exported. Three H^+ are translocated through the F_0 carrier of the ATP synthase, and one H^+ is translocated by the turnover of the phosphate and adenine nucleotide carriers. Thus 25% of the free energy input into oxidative phosphorylation is injected at the level of these substrate transporters. As explained in the text, cytochrome oxidase effectively translocates four positive charges per pair of electrons, as shown for the redox proton pump. As discussed in the text, the cytochrome bc_1 complex translocates only two positive charges per pair of electrons. (Reproduced with permission from Brand, 1994).

six protons per NADH and four per succinate were translocated, then the respective P/O ratios had to be 2 and 1.5, respectively.

This realization was one of several factors that caused investigators from approximately the mid-1970s onwards to reconsider both the direct experimentally determined P/O ratios and the proton translocation stoichiometries. One of the first of these, Peter Hinkle, who had previously worked with Mitchell at Glyn as a post-doctoral fellow, was, with his group, especially prominent. They concluded that the P/O ratio was approximately 2.5 for NADH and 1.5 for succinate (Hinkle and Yu, 1979; Hinkle et al., 1991). Careful scrutiny of the values for the proton translocation stoichiometries in the laboratories of Brand and Lehninger among others eventually led to a consensus that 10 protons are translocated per NADH oxidized and 6 per succinate (Brand et al., 1976). The value of 6 can be considered a currently agreed value but that 10 for NADH may yet prove in need of revision. Meanwhile study of the stoichiometry of proton translocation through the ATP synthase has also called for revision of this figure upwards from two to three. When this latter value is taken together with the requirement, described above, for one net proton to move with the ATP/ADP exchange plus phosphate transport, then P/O ratios are predicted, from consideration of proton translocation stoichiometries, to be 2.5 (10/4) and 1.5 (6/4), respectively, for NADH and succinate oxidation. These values are close to those determined by direct measurement of ATP synthesis and oxygen consumption.

The bald stoichiometries of proton translocation by the mitochondrial electron transport chain disguise another fundamental revision to the original concept of oxidative phosphorylation occurring at three distinct, but equivalent, sites in the respiratory chain. It has transpired that as electrons pass from succinate to oxygen they indeed pass through two segments of the electron transport chain where net proton translocation occurs but that the stoichiometry is not equal. The first of these segments is the cytochrome bc_1 complex (otherwise known as complex III). Here, the well established Q-cycle mechanism shows how two positive changes are moved outwards per pair of electrons transferred (see e.g. Nicholls and Ferguson, 1992). In contrast cytochrome oxidase moves four positive charges per pair of electrons. Hence, the bc_1 complex makes only half the contribution to ATP synthesis that the oxidase does. Thus, contributions to ATP synthesis of these two proton translocating regions of the chain are unequal. Figure 3 shows a proton circuit that includes a redox proton pump such as cytochrome oxidase, which effectively moves four protons per pair of electrons together with the ATP synthase and transporter systems.

The proposal that the stoichiometry of ATP synthesis by mitochondria is lower than was widely accepted during the period 1960 to 1990 means that the standard textbook statements about the stoichiometry of ATP synthesis from full oxidation of a fatty acid or glucose need to be revised downwards.

The reader may at this point be wondering if the P/O ratio for matrix ATP synthesis may be larger, because of the lack of requirement for adenine nucleotide and phosphate transport than for cytoplasmic ATP synthesis. The expectation indeed is that it should be, with values of $10/3 = 3.33$ for NADH and $6/3 = 2$ for succinate, but definitive experimental verification has proved very difficult.

This discussion of nonintegral stoichiometries brings us to address the question of whether ATP should still be regarded as the fundamental energy currency of the cell. The answer is no. A dramatic illustration of this can be seen in the bacterium *Propionigenium modestum* which derives energy for growth from the decarboxylation of succinate to propionate. Classically one would have expected such an organism to make at least one ATP molecule, catalyzed by a cytoplasmic water-soluble enzyme, for each succinate utilized. However, consideration of some cellular thermodynamics shows that this is not possible because the free energy for succinate decarboxylation is less than that needed for synthesis of one ATP under realistic conditions of succinate, propionate, ADP, P_i, and ATP concentration. In fact, in classical pre-chemiosmotic days it would have to have been concluded that such an organism could not exist. However, we now know how this organism organizes its energy metabolism. The key decarboxylation step is linked to the translocation of sodium ions across the cytoplasmic membrane, thus setting up a sodium electrochemical gradient. Subsequent passage of sodium ions back down the electrochemical gradient, and through an ATP synthase, drives the ATP synthesis. The stoichiometries of sodium ion movement by these two systems are not known with absolute certainty but the decarboxylation reaction is thought to drive movement of two sodiums while the ATP synthase is believed to require three or four. Either way, the overall stoichiometry of ATP synthesis per succinate utilized is less than one and thereby compatible with thermodynamic constraints. The fundamental unit of energy currency in the cell is thus not ATP but rather the proton or Na^+ electrochemical gradients.

CHANGES IN PERSPECTIVE ON ADP AND P_i CONDENSATION

It was axiomatic in the chemical mechanism for oxidative phosphorylation that phosphate would be transferred from a covalent phosphorylated species, not necessarily a protein, to ADP in the final step of ATP synthesis. The notion that a phosphorylated state of the ATP synthase might be formed en route to ATP synthesis was not incompatible with the chemiosmotic theory and was seriously considered by many workers in the field, although not by Mitchell himself who had other ideas (see later). Another reason for anticipating formation of a phosphoenzyme intermediate in the mechanism of the ATP synthase was that such phosphoenzyme intermediates were identified for the

Na^+/K^+ ATPases of plasma membranes and the Ca^{2+} ATPase of the sarcoplasmic reticulum. However, for the ATP synthases of oxidative phosphorylation no evidence of such an intermediate has ever been found and several observations now positively argue against covalent attachment of phosphate to an amino acid side chain prior to transfer to ADP (see e.g. Nicholls and Ferguson, 1992).

Right from the first formulation of his theory, Mitchell sought to provide a mechanistic basis for the coupling of proton translocation to ATP synthesis by the ATP synthase. The essence of his proposal was that protons would flow directly into the active site of the ATP synthase where they would protonate key oxygen atoms and thus "drive" the condensation of phosphate with ADP to give ATP. The precise mechanistic details proposed by Mitchell were hotly contested. The idea that protons entered the site where ATP is synthesized and drove the dehydration of ADP plus P_i to give ATP was also proposed by Williams (see later). However, subsequent discoveries have made it very unlikely that protons passing through the membrane directly pass through catalytic sites of the ATP synthase.

One of the earliest pointers in this direction was the work of Boyer and colleagues in the 1970s which involved isotope exchange reactions (see e.g. Boyer, 1977). The idea can be illustrated by considering the consequences of hydrolyzing ATP in ^{18}O enriched water. Since the bond split in ATP hydrolysis is that between the γ-phosphorus atom and the bridging oxygen the product phosphate should be enriched with ^{18}O to the same extent as the water. However, if ATP splitting at the active site can be followed by resynthesis of ATP, without ADP and/or P_i leaving the active site, then provided the P_i and/or β-phosphate of ADP can reorientate to some extent, the oxygen eliminated as water need not be the ^{18}O previously incorporated into the P_i from water. Hence a second hydrolysis of ATP at the active site will result in the P_i carrying excess ^{18}O, a manifestation of an exchange reaction.

From experiments of this kind two important conclusions emerged. First, such exchanges could be observed, albeit under certain limited conditions, with the soluble F_1 fragment of the ATP synthase to which protons could not be delivered from the membrane. Second, such studies indicated that the free energy change for ATP hydrolysis at the active site of the enzyme is very much less negative than in bulk aqueous solution. Thus, while traditionally it had been postulated that the energy coupling step would be associated with the chemical reaction of condensing P_i with ADP, the notion that this step might be close to equilibrium at the active site shifted attention to the idea that the binding of ADP and/or P_i together with release of ATP might be the steps most intimately involved with energy transduction. In other words, conformational changes would be linked to the proton motive force driving protons through the ATP synthase. As we shall see later, further information about the structure of the ATP synthase and a sodium translocating ATP synthase substantiated this point.

REVERSED ELECTRON TRANSPORT

In the 1960s the energetics of certain groups of bacteria were an enigma. Organisms of the genus *Nitrobacter* provide an example. They gain energy for growth from the aerobic oxidation of nitrite (NO_2^-) to nitrate (NO_3^-). The puzzle they presented was that the reaction,

$$NO_2^- + H_2O + NAD(P)^+ \rightarrow NO_3^- + NAD(P)H$$

lies very very far to the left and thus it is not immediately obvious how these organisms acquire the NAD(P)H required for reductive biosynthesis. The reaction,

$$NO_2^- + 1/2O_2 \rightarrow NO_3^-$$

is sufficiently exergonic to drive cellular ATP synthesis with a maximum stoichiometry of approaching one ATP per nitrite oxidized. Accordingly, *Nitrobacter* was initially envisaged as synthesizing ATP by a soluble enzyme system in the cytoplasm. NAD(P)H was then thought to be formed in a nebulous ATP-dependent reaction whereby the energy of ATP hydrolysis would be harnessed to drive electrons uphill from nitrite to $NAD(P)^+$. This unsatisfactory picture has now been supplemented by a chemiosmotic mechanism that illustrates its unifying features.

The current view of the energetics of *Nitrobacter* is that electrons enter an electron transport chain at the level of *c*-type cytochrome (Figure 4). The flow of electrons to oxygen generates a proton motive force relying substantially, it is proposed, on the proton pumping activity of a cytochrome aa_3 oxidase that has recently been shown to be structurally closely related to the counterpart enzyme in mitochondria and other bacteria (Berber, 1996). This proton motive force can drive ATP synthesis as in other respiratory organisms. But the proton motive force also drives several other processes among which reversed electron transport from nitrite to NAD^+ is very important. What this means is that a minority of electrons originating from nitrite is driven uphill along an electron transport chain at the expense of the proton motive force generated by the dominant electron flow pathway from nitrite to oxygen. Thus again we see that ATP should not be regarded as the fundamental unit of cell energy currency. This unit is the proton electrochemical gradient.

ACTIVE TRANSPORT

Mitchell's main interest through the 1950s had been active transport. He recognized that his chemiosmotic hypothesis could also account for coupling

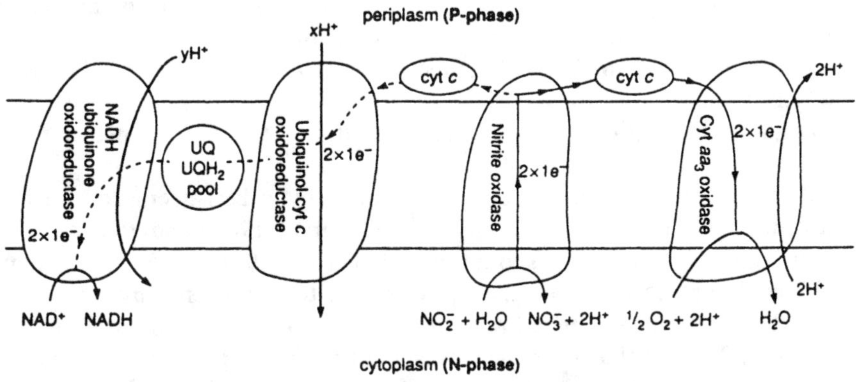

Figure 4. Protonmotive force generation and reversed electron transport in *Nitrobacter*. See text. (Reproduced with permission from Nicholls and Ferguson, 1992).

between electron and solute transport. Thus, in the case of *E. coli*, the uptake of lactose could occur in symport with one or more protons. The active uptake of the uncharged lactose could be driven by the proton motive force without any involvement of ATP. The first experimental support of this proposition was provided by West (1970) (see also Henderson, 1991). Subsequently many transport processes have proved to be linked to proton movements, not only in bacteria but also, for example, in the uptake of adrenaline into chromaffin granules (Nicholls and Ferguson, 1992).

A CONTROVERSY

The foregoing is a personal selection of some of the significant changes that occurred in the field of bioenergetics between 1960 and 1996. Although others would doubtless choose to highlight other specific developments, and Beinert (1992) provides a valuable complementary account, it is expected that the general theme will be widely recognized. However, in a brief historical survey such as this, one must note that the origins of the chemiosmotic hypothesis are disputed.

In a paper published in 1961, but submitted in August, 1960, Williams (1961) proposed that protons were involved in the coupling of electron transport to ATP synthesis. The hypothesis was that upon oxidation of a hydrogen carrier in the electron transport chain, electrons and protons separated with the latter going to a site where they drove the condensation of ADP and P_i to give ATP. This was undoubtedly a novel idea and was clearly submitted before any formal description of the chemiosmotic idea by Mitchell. A clear hint "...in its oxidized state hydrogen as the proton can bring about condensation polymerization such as polyphosphate formation...." of Williams' idea was also presented in an article published in 1959 (Williams, 1959). It was advanced at a time when the bilayer structure of a membrane was not always accepted and certainly ideas about the relative organization and sidedness of membrane proteins were scant.

Mitchell's first truly explicit description of his hypothesis seems to have been presented to the Biochemical Society in March, 1961, and the text was, according to Weber (1991), submitted in February, 1961. In contrast to many of his contemporaries, Mitchell already appears to have been convinced at that time that a biological membrane presents a barrier to ions and to have sidedness. Thus at that time there was a considerable difference between the style of writings of Williams and Mitchell. However, to put it plainly, Williams (1993) believes that there are grounds for suspecting that the idea of using the proton as the coupling agent may have come to Mitchell through reading Williams' contributions, either in *The Enzymes* (Williams, 1959) or in Williams' 1961 paper, perhaps as a referee.

It is doubtful whether we will ever now know if there is substance to this claim, although Slater (1994) has remarked that it is curious that Mitchell never cited Williams' papers, despite having been sufficiently interested by them to engage in a correspondence with Williams. In these letters he asked Williams how he thought protons could "make" ATP and Williams replied that local or delocalized (the latter not being favored for energetic reasons) proton movements could be involved. We can, however, ask whether we, having read Williams' papers, could have made the intellectual jump to Mitchell's position in which the key, in the the present writer's opinion, was the concept of electrochemical differences in proton potential between bulk aqueous phases across a membrane. The term "chemiosomosis" deflects attention from the electrical component of this term and in retrospect the attendant mechanistic proposals can be seen not to be the central issue. The present writer has to confess that he doubts if he would have done so but cannot be certain. He might have grasped the idea of a proton flow between the sites for electron transport and ATP synthesis but, as proved the case for many investigations for nearly 20 years from 1960, the concept of the electrochemical potential of the proton would have probably been difficult to recognize. However, it is important that other scientists examine the written records and consider this matter for themselves.

We are left with two possibilities concerning the relationship between Mithell's and Williams' ideas. First, Mitchell had read only Williams' 1959 article and did not consider that the phrase reproduced above sparked any thoughts in his own mind towards the chemiosmotic hypothesis. Second, that reading *The Enzymes* article and/or the *Journal of Theoretical Biology* article of Williams (possibly as a referee, see Williams, 1993) did at least focus

Mitchell's mind on a coupling role for protons in ATP synthesis. In this latter case it would have been appropriate to have cited Williams' papers. Williams (1993) also argues that other papers by Davies and Krebs and by Crane (for refs. see Williams, 1993) ought to have been cited. However, one must stress that there is no evidence, only Williams' suspicion, that Mitchell saw Williams' paper as a referee. If only the Journal had kept the list of the referees it used, this dispute may have proven unfounded. There is here a lesson. Journals ought to keep for long periods records of referees used. Referees may be tempted to use information gained while reviewing manuscripts (and grants). Knowing that records would be kept for many years might place a restraint on such activities.

CONCLUSIONS

What Mitchell got fundamentally correct in 1961 was the idea that coupling between exergonic electron tranpsort and endergonic ATP synthesis was achieved via a proton electrochemical gradient across an ion-impermeable membrane. Later the idea was developed to include coupling to other processes, e.g. active transport of lactose into bacteria by symport with protons, and bacterial movement via the flagella (see Armitage, Chapter 5). The chemiosmotic paradigm also provided a unifying link, probably widely unsuspected in 1960, between mitochondrial oxidative phosphorylation and both the light reactions of photosynthesis and diverse modes of bacterial respiration (Figure 5).

It is not surprising that some of Mitchell's original proposals for mechanisms have have not stood the test of time. This is most notably the case for the ATP synthase where, for reasons explained earlier, translocated protons are not believed to enter the active site. Remarkably, we now have a high resolution structure of the ATP synthase (Abrahams et al., 1994) and this shows that the catalytic sites are separated from the membrane bilayer by a very thin and long stalk which is most unlikely to provide a proton conducting pathway. Furthermore, as described elsewhere (Thauer and Morris, 1984; Nicholls and Ferguson, 1992) it is now known that some bacteria achieve chemiosmotic coupling by a sodium circuit. Included on the circuit is a sodium translocating ATP synthase which is remarkably similar in structure to the proton translocating enzyme. As the chemistry of sodium ions is rather different from that of the proton, mechanisms for involving translocated protons at active sites cannot be easily interchanged for those involving sodium. All the evidence available now points to the translocated ions driving conformational changes in the globular head group part of the F_1 ATPase.

Despite the structural information now available, we remain some way from a detailed understanding of exactly how the ATP synthase functions. It is

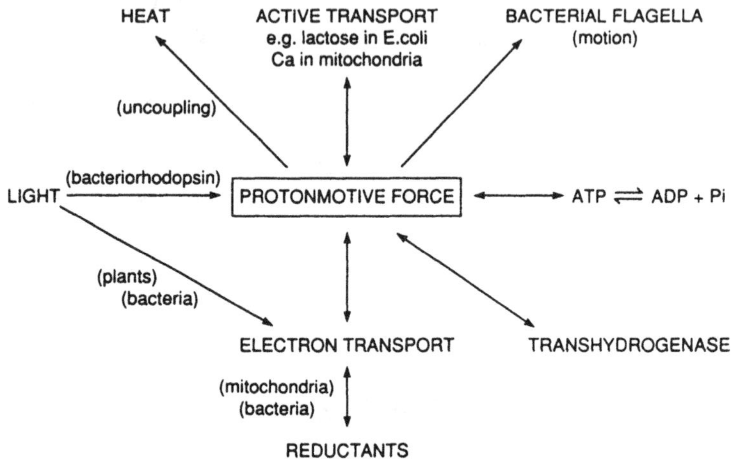

Figure 5. Pathways of energy transduction that involve the protonmotive force, or proton electrochemical gradient, as intermediate. (Reproduced with permission from Nicholls and Ferguson, 1992).

generally accepted that there are three catalytic sites on the ATP synthase and that any one time each is at a different stage of the catalytic process. It is thought that the subunits of the enzyme interchange position during the reaction (Boyer, 1977), a concept that has very recently been supported by an elegant labeling experiment (Duncan et al., 1995).

On the other hand, Mitchell's early proposals for how electron transport chains might translocate protons have to a significant extent stood the test of time. The original idea was that hydrogen would be split into protons and electrons at one side of the membrane, followed by deposition of the protons into the aqueous phase and movement of electrons to the opposite side of the membrane. Subsequent combination on a hydrogen carrier of the electrons with protons taken from the adjacent aqueous phase would lead to net proton translocation across the membrane and thus generation of proton electrochemical gradient. This is the so-called redox-loop mechanism, in which charge separation across the membrane is substantially achieved by electron movement. We now know that this description applies to some systems. For example, the first high resolution crystal structure of a membrane protein was of the photosynthetic reaction center and showed how a series of prosthetic groups carried electrons across the membrane. In contrast, the very recent structures of cytochrome oxidase (Iwata et al., 1995; Tsukihara et al., 1996) show that such electron movement is only a small part of the mechanism. In the oxidase, protons are taken from the aqueous phase at one side of the membrane and channeled through the protein either to a site where they meet oxygen and electrons coming from the opposite side and water is formed, or to a gate which can be opened by the action of the proton, using the available redox energy, from which they can proceed to the other side of the membrane. Both processes generate a proton electrochemical gradient.

The ubiquinol–cytochrome *c* oxidoreductase part of the electron transfer chain functions according to a Q-cycle mechanism (e.g. Nicholls and Ferguson, 1992) which was itself proposed by Mitchell. Although not a simple redox-loop mechanism it retains elements of Mitchell's original idea. The area of bioenergetics is continuing to yield protein structures; that for a ubiquinol–cytochrome *c* oxidoreductase is forthcoming (Yu et al., 1996). A further example of proton translocation now being given a structural framework is bacteriorhodopsin, a light driven proton pump. Its function, first suggested around 1971, would not have been evident without the chemiosmotic hypothesis.

Overall, we see that from 1960 to 1996 we have proceeded from a mystery—the coupling mechanism in oxidative phosphorylation and related processes—via the experimental vindication of the solution to the mystery, that is the chemiosmotic principle, to being on the brink of understanding the molecular enzymology of oxidative phosphorylation at a structural level.

REFERENCES

Abrahams, J.P., Leslie, A.G.W., Lutter, R., & Walters, J.E. (1994). Structure at 2.8Å resolution of F_1-ATPase from bovine heart mitochondria. Nature (London) 370, 621-628.

Beinert, H. (1992). Trails of inquiry and thought leading towards today's bioenergetics. Biochim. Biophys. Acta 1101, 125-133.

Berber, G. (1996). *Nitrobacter winogradski* cytochrome oxidase genes are organized in a repeated gene cluster. Antonie von Leeuwenhoek 69, 305-315.

Boyer, P.D. (1977). Coupling mechanisms in capture, transmission and use of energy. Annu. Rev. Biochem. 46, 957-966.

Brand, M.D. (1994). The stoichiometry of proton pumping and ATP synthesis in the mitochondria. The Biochemist Aug/Sept., 22.

Brand, M.D., Reynafarje, B., & Lehninger, A.L. (1976). Stoichiometric relationship between energy-dependent proton ejection and electron transport in mitochondria. Proc. Natl. Acad. Sci. USA, 74, 1955-1959.

Duncan, T.M., Bulygin, V.V., Zhou, Y., Hutcheon, M.L., & Gross, R.C. (1995). Rotation of subunits during catalysis by *Escherichia coli* F_1 ATPase. Proc. Natl. Acad. Sci. USA 93, 10964-10968.

Hatefi, Y., Haavik, A.G., Fowler, L.R., & Griffiths, D.E. (1962). Studies on the electron transfer system. XLII, Reconstitution of the electron transfer system. J. Biol. Chem. 237, 2661-2669.

Henderson, P.J.F. (1991). Studies of translocation catalysis. Bioscience Reports 11, 477-458.

Hinkle, P.C., Kumar, M.A., Resetar, A., & Harris, D.L. (1991). Mechanistic stoichiometry of mitochondrial phosphorylation. Biochemistry 30, 3576-3582.

Hinkle, P.C. & Yu, M.L. (1979). The P/O ratio of mitochondrial phosphorylation. J. Biol. Chem. 254, 2450-2455.

Iwata, S., Ostermeier, C., Ludwig, B., & Michel, H. (1995). Structure at 2.8 Å resolution of cytochrome *c* oxidase from *Paracoccus denitrificans*. Nature (London) 376, 660-669.

Jagendorf, A.J. & Uribe, T. (1966). ATP formation caused by acid-base transition of spinach chloroplasts. Proc. Natl. Acad. Sci. USA 55, 170-177.

Mitchell, P. (1961a). Approaches to the analysis of specific membrane transport. In: Biological Structure and Function (Goodwin, T.W. & Lindberg, O., Eds.), Vol. 2, pp. 581-599. Academic Press, New York.

Mitchell, P. (1961b). Chemiosmotic coupling in oxidative and photosynthetic phosphorylation. Biochem. J. 79, 23-24P.

Mitchell, P. (1961c). Coupling of phosphorylation to electron and hydrogen transfer by a chemiosmotic type of mechanism. Nature (London) 191, 144-148.

Mitchell, P. (1966). Chemiosmotic Coupling in Oxidative and Photosynthetic Phosphorylation. Glyn Research Ltd.

Mitchell, P. (1968). Chemiosmotic Coupling and Energy Transduction. Glyn Research Ltd.

Mitchell, P. (1976). Vectorial chemistry and the molecular mechanisms of chemiosmotic coupling: power transmission by proticity. Trans. Biochem. Soc. 4, 399-430.

Mitchell, P. (1979). Keilin's respiratory chain concept and its chemiosmotic consequences. Science 206, 1148-1159.

Nicholls, D.G. & Ferguson, S.J. (1992). Bioenergetics 2. Academic Press, London and San Diego.

Racker, E. (1977). A new look at mechanisms in bioenergetics. Academic Press, New York.

Racker, E. (1980). From Pasteur to Mitchell: a hundred years of bioenergetics. Fed. Proc. 39, 210-215.

Slater, E.C. (1953). Mechanisms of phosphorylation in the respiratory chain. Nature (London) 172, 975-978.

Slater, E.C. (1994). Peter Dennis Mitchell. Bibliographical Memoirs of the Fellows of the Royal Society 40, 283-305.

Thauer, R.K. & Morris, J.G. (1984). Metabolism of chemotrophic anaerobes: old views and new aspects. Soc. Gen. Microbiol. Symposium 36 Part II (Kelly, D.P. & Carr, N.G., Eds.), pp. 121-168.

Tsukihara, J. et al. (1996). The whole structure of the 13-subunit oxidized cytochrome *c* oxidase at 2.8Å. Science 272, 1136-1144.

Weber, B.H. (1991). Glynn and the conceptual development of the chemiosmotic theory; a retrospective and prospective view. Bioscience Reports 11, 577-617.

West, I.C. (1970). Lactose transport coupled to proton movements in *E.coli.* Biochem. Biophys. Res. Commun. 41, 655-661.

Williams, R.J.P. (1959). In: The Enzymes, 2nd ed. (Boyer, P.D., Lardy, H., & Myrback, K., Eds.), Vol. 1, pp. 391-442. Academic Press, New York.

Williams, R.J.P. (1961). Possible functions of chains of catalysts. J. Theoret. Biol. 1, 1-17.

Williams, R.J.P. (1993). The history of proton-driven ATP formation. Bioscience Reports 13, 191-212.

Yu, C.A. et al. (1996). Crystallization and preliminary structures of beef heart mitochondrial cytochrome bc_1 complex. Biochim. Biophys. Acta 1275, 47-53.

Chapter 3

CHANGING VIEWS OF PHOTOSYNTHESIS

F. R. Whatley

INTRODUCTION

The way in which ideas on photosynthesis have developed forms the subject of this chapter. A modern description of photosynthesis would include reference to the conversion of light into chemical energy in a form used for the fixation of CO_2, principally into carbohydrates, a process which serves as an energy storage mechanism. It would also refer to the redox nature of the reduction of CO_2 at the expense of water, with O_2 being evolved as a waste product. The description would probably also identify a number of electron flow intermediates operating in the initial energy conversion process and state that two photoreactions working in sequence are involved. It was by no means always so!

Although crop plants have been of importance for centuries the experimental search for the essential basis of the way in which plants grow (the "principle of vegetation", as it was called) dates only from the seventeenth century.

EARLY OBSERVATIONS

In 1640 the Flemish chemist van Helmont had looked for this "principle of vegetation". He took a willow shoot weighing 5 pounds and planted it in 200 pounds of dry soil kept covered against the accumulation of dust and kept watered by the addition of rainwater alone. In 5 years the weight of the willow had increased to 164 pounds but the weight of the soil after drying had diminished by only 2 ounces. Logically for van Helmont the source of all plant substance was therefore water. Further progress in unravelling the essential features of photosynthesis could not be made until methods of measuring gases had been developed. van Helmont's concept thus remained the general view until late in the eighteenth century when scientists working in many different fields and in many different European countries provided new information.

It fell to Priestley, the English chemist and philosopher and one of the discoverers of oxygen, to observe in 1772 that "plants, instead of affecting the air in the same manner as with animal respiration, reverse the effect of breathing, and tend to keep the atmosphere sweet and wholesome, when it is become noxious in consequence of animals either living and breathing, or dying and putrifying in it." Priestley used a burning candle or a mouse to test for the presence of oxygen (then called dephlogisticated air). He found that a sample of air in which a mouse would suffocate or which could not maintain a burning candle (i.e. was deficient in oxygen) was restored after a sprig of mint was placed in the foul air and left for a time. This revealed that plants cleanse and purify the atmosphere and the principle of planetary ventilation by plants was established.

This view was enthusiastically received by other scientists but unfortunately the Priestley experiments could not be consistently repeated by his contemporaries, including Scheele, the Swedish chemist and co-discoverer of oxygen, or even by Priestley himself a few years after his original work.

LIGHT: THE UNRECOGNIZED INGREDIENT

The puzzle was cleared up by the Dutchman, Ingenhousz, who was physician to the Empress Maria Theresa of Austria. Ingenhousz greatly admired Priestley's original work and hoped to clarify it. He carried out an amazing series of simple but definitive experiments in England over a period of 3 months in 1779. Unlike most of his contemporaries he set out his experimental data and conclusions in a very concise and logical format. Ingenhousz found that plants could not only "correct bad air in 6 to 10 days, by growing in it" (as shown by Priestley in 1772) but could do so in only a few hours. The operation was carried out by the plants only under the influence of sunlight, was dependent on light intensity, and ceased in the dark. Only the green leaves and green stalks could restore the air to its former state by evolving oxygen, but all parts of the plant decomposed air, converting the oxygen into carbon dioxide ("fixed air"). Plants, like animals, respire. Priestley and others had all apparently failed to note the requirement for light.

CARBON DIOXIDE AND WATER

Senebier, a Swiss pastor, worked at the same time as Ingenhousz but more slowly. Senebier published his results in 1782 establishing that carbonic acid was a necessary prerequisite for oxygen evolution and that the plant retained the carbon in substances needed for its growth. He also confirmed most of the observations that Ingenhousz had made during his great burst of activity. The view that emerged from this work was that the function of light was to split CO_2 and liberate O_2, the carbon being retained to produce plant substance. Although Senebier leaned towards the idea that CO_2 was taken into the roots as carbonic acid it was later confirmed that there is a direct exchange of CO_2 between the air and the leaf.

In 1804 the Swiss botanist and physicist de Saussure, working in the same city as Senebier, observed that the increase in dry weight produced by the assimilation of a certain quantity of CO_2 was consistently greater than the amount of carbon contained in it. Since the O_2 evolved was equivalent to that liberated from CO_2 the only source of the extra weight was water, taken up into a form not removable by drying. Thus water was involved as a substrate in photosynthesis and led to the formation of "hydrated carbon compounds" (i.e. carbohydrates).

WHAT IS LIGHT USED FOR?

The function of light remained unknown prior to the proposal of Julius Robert Mayer in 1845, a German physicist, who clearly stated that plants take in energy in the form of light and use it to create an energy source in the form of a chemical difference. This conversion of light energy into chemical energy is the fundamental basis of photosynthesis. The history of the discovery of photosynthesis has been treated in more detail by Rabinowitch (1955) and Arnon[1] (1961b).

As a result of all these observations we may write a simple equation to represent photosynthesis:

$$CO_2 + \text{water} + \text{light} \xrightarrow{\text{leaf}} \text{organic matter} + \text{chemical energy} + O_2$$

Can we identify which component of the leaf carries out this photosynthetic reaction?

ABSORPTION OF LIGHT IN GREEN PLANTS

Photosynthesis, detected by O_2 evolution, was shown by a Dutch biologist, Engelmann[10], in 1880 to be associated with chloroplasts, the chlorophyll-containing organelles. He made use of the filamentous green alga, *Spirogyra* (which has a large spiral chloroplast), and a suspension of oxygen-requiring motile bacteria (*Bacillus termo*) that always swim towards the highest concentration of oxygen and accumulate there. Engelmann placed a filament of the alga on a microscope slide in a suspension of the motile bacteria, putting on a coverslip and carefully sealing the edges. After a period of illumination the bacteria had used up all the oxygen originally in the liquid and now swam to where the highest concentration of oxygen was being produced. The bacteria could soon be seen under the microscope to be concentrated close to the chloroplast bands in the algal cells (see also Chapter 5).

As a variant of this experiment Engelmann used small spots of white light to illuminate either parts of the chloroplast or parts of the cytoplasm where there was no chloroplast. The oxygen-requiring motile bacteria accumulated in the spot of light falling on the chloroplast but not in the spot illuminating only non-green parts. So light had to be absorbed by the chloroplast to evolve oxygen. By using a prism to generate a spectrum of light of different colors falling on different parts of the chloroplast, Engelmann was also able to show that green light (550 nm) was ineffective but red (650 nm) or blue (450 nm) light caused oxygen evolution as indicated by the pattern of accumulation of the motile bacteria. Since chlorophyll absorbs light well in the red and blue regions of the spectrum but poorly in the green Engelmann concluded that it was the chlorophyll in the chloroplasts that absorbed the light needed for photosynthesis.

CHLOROPLASTS

Light Microscopy

Chloroplasts were first observed in the light microscope by Nehemiah Grew in 1702. He noted that they contained the pigment responsible for the green color of leaves. In the modern light microscope chloroplasts of higher plants are seen to be discoid organelles usually ranging in size from 3 to 10 μm in diameter and 1 to 2 μm in thickness. Their numbers can vary in different species and at different stages of growth between 1 and as many as 500 per cell. In green algae the chloroplasts may have a very different shape, e.g. star-shaped in *Zygnema*, cup-shaped in *Chlorella*, and a ribbon in *Spirogyra.*

Photosynthetically functional chloroplasts of all higher green plants are basically similar in internal structure. Starch was first observed in chloroplasts by Von Mohl in 1873. Soon afterwards Sachs demonstrated that starch accumulation was associated with the assimilation of CO_2 in the light. By 1883 Meyer had identified within the chloroplasts spots or grains which were a darker green; these are the grana. In 1885 Schimper proposed that chloroplasts are semiautonomous organelles and that they reproduce by binary fission. Thus by the end of the nineteenth century the basic structure of chloroplasts was already known. Further details of the internal structure had to await development of the electron microscope (EM) which, by using a beam of electrons instead of light, greatly increased the limits of resolution. Investigations with the EM produced a major step forward in our knowledge of the structure of the chloroplast envelope, the internal membranes including grana, and the aqueous matrix (the stroma) in which these membranes lie.

Electron Microscopy

Ultrastructural studies of chloroplasts or plastids only began about 1950 and the fundamental ultrastructure was well known by 1970. Many laboratories in Germany, the U.S., and elsewhere contributed to our current views. The development of ideas on the ultrastructure of chloroplasts is beautifully summarized in Chapter 9 of the excellent book by Gunning and Steer (1975).

The chloroplasts are delimited by an envelope of two membranes (Figure 1). Internally they contain an aqueous matrix in which the enzymes of the dark reactions of photosynthesis are found. The chloroplast DNA appears as fibrils, which are circular chromosomes. It has a different buoyant density from the

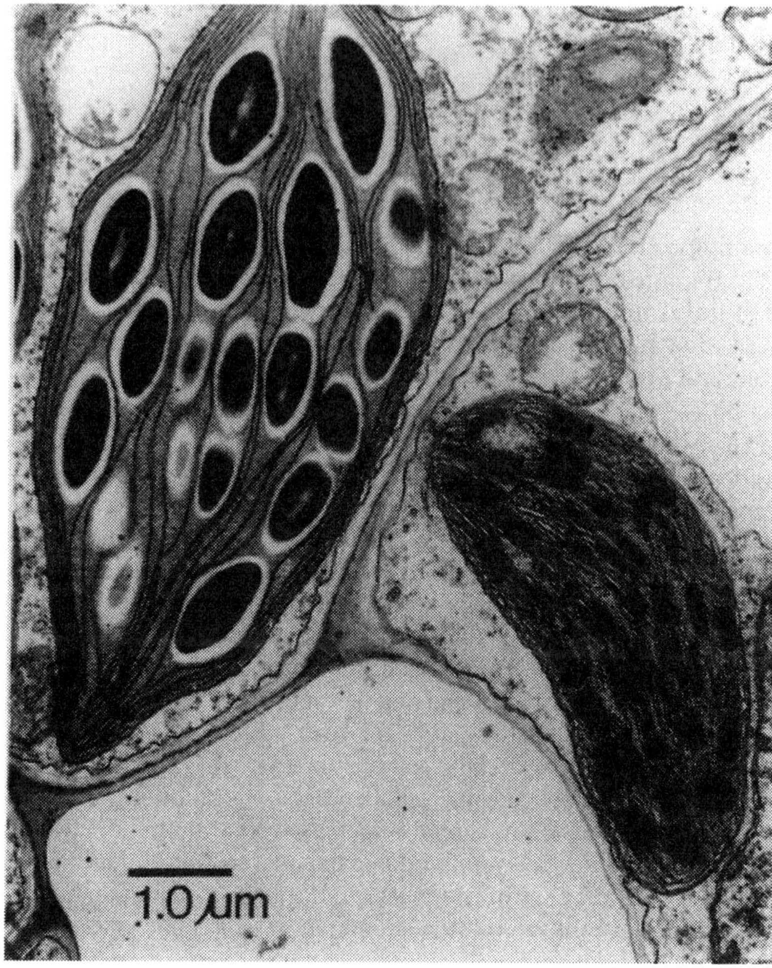

Figure 1. Electron micrograph of chloroplasts in the leaf of maize (*Zea mays*). The smaller chloroplast with grana lies in a mesophyll cell. The larger chloroplast with starch grains and single thylakoids lies in an adjacent bundle sheath cell. (Courtesy of Dr. J.M. Whatley).

nuclear DNA and resembles bacterial DNA in many respects. The chloroplast RNA is located within the stroma in ribosomes that are smaller than those present in the cell cytoplasm. These prokaryotic features support the hypothesis that "green" chloroplasts evolved from prokaryotic endosymbionts belonging to the cyanobacteria (blue-green algae).

Embedded in the stroma is a system of membranes—lamellae or thylakoids—which contains the chlorophyll and the enzymatic apparatus for the primary photochemical reaction sequences and for ATP formation. The thylakoid system comprises an essentially spiral basic framework of interconnected single lamellar ribbons linked at irregular intervals by discrete circular regions of stacked lamellae called grana. In each chloroplast the narrow intramembrane space of the complex thylakoid system forms a continuum which, in effect, is the interior of a closed sac or vesicle whose contents are quite different from those of the surrounding stroma. The thylakoid membranes exhibit strong "sidedness" since the proteins and other components embedded in them face either inwards towards the intra-thylakoid space or outwards towards the stroma, though a few may span the membrane.

In freeze-fractured and etched chloroplasts examined in the EM, the "sidedness" of the membranes can be detected as particles of different sizes on the inner or outer surfaces and at the fracture faces. These particles represent various pigment arrays (quantasomes) and protein complexes involved in photosynthetic electron transport. Particles corresponding to part of the ATP-synthesizing complex (CF_1) can be seen at the external thylakoid surface after freeze etching.

The components of photoreaction I (see below) appear to be concentrated mainly in the stromal lamellae, whereas those of photoreaction II are generally associated with the granal stacks. However, in many C_4 plants, which initially fix CO_2 into malate or aspartate in their mesophyll cells and transfer the C_4 compound to adjacent bundle sheath cells to give an enhanced CO_2 source for the Calvin cycle, the chloroplasts in the bundle sheath cells that carry out the final fixation of CO_2 via PGA have a thylakoid system which lacks grana (Figure 1). By contrast, the chloroplasts in the adjoining mesophyll cells have a normal granal system, although they do not produce the PGA characteristic of normal C_3 plants.

The plastids of non-green algae contain accessory pigments different from those in higher plants and green algae (Table 1). Although the presence of different accessory pigments is often reflected in the distinctive stacking arrangement of the thylakoids, the basic compartmentalization into stroma and lamellae remains the same. In the cyanobacteria which are prokaryotic in organization and hence lack discrete organelles, the photochemical system is nevertheless concentrated in a thylakoid system within the general cytoplasm of the cell. In all these groups oxygen is evolved and water serves as the donor, in contrast to the photosynthetic bacteria, which use other reductants such as hydrogen sulfide.

Plant Pigments

The green matter in leaves that carry out photosynthesis was extracted in 1817 by two French chemists, Pelletier and Caventou, and named chlorophyll.

Table 1. Major Photosynthetic Pigments in Some Plants

	Chlorophylls			Phycobilins		Carotenoids
Organisms	a	b	c	*P-cyan*	*P-eryth*	*Fucoxanthin*
Land plants	+	+				
Green algae	+	+				
Euglenoids	+	+				
Cyanobacteria	+			+	+	
Red algae	+			+	+	
Cryptomonads	+		+	+	+	
Brown algae	+		+			+
Diatoms	+		+			+
Chrysophyte algae	+		+			+
Xanthophyte algae	+					

Although they suspected their chlorophyll to be a mixture, it was not until 1906 that Tswett, a Russian botanist, separated chlorophylls *a* and *b* from leaf extracts by chromatography on powdered sucrose columns. In 1912 the German and Swiss chemists Willstätter and Stoll concluded that chlorophylls *a* and *b* are present universally in green plants and that the ratio *a*:*b* is rather constant at about 3 for most green plants. Alpine plants, typically adapted to high light conditions (sun plants), have a much higher chlorophyll *a*:*b* ratio, while plants adapted to low light conditions (shade plants) have a higher proportion of chlorophyll *b*. Willstätter and colleagues established that the two chlorophylls have the molecular formulae:

chlorophyll *a* $C_{55}H_{72}O_5N_4Mg$
chlorophyll *b* $C_{55}H_{70}O_6N_4Mg$

They are esters of two dibasic acids: chlorophyllin $a = C_{32}H_{30}ON_4Mg(COOH)_2$ and chlorophyllin $b = C_{32}H_{28}O_2N_4Mg(COOH)_2$. The esterifying alcohols are methanol (CH_3OH) and phytol ($C_{20}H_{33}OH$).

By degradation experiments Willstätter found that the chlorophyllins contain four pyrrole nuclei arranged around a central nonionizable magnesium atom. He pointed out that hemin, too, contains four pyrrole nuclei, this time with an iron atom in the center. From the 1920s onwards the German chemist Hans Fischer carried out experiments which included the degradation and partial synthesis of chlorophyll *a*. Fischer (1940) published a structure for chlorophyll *a* (Figure 2) which, while based largely on his own work, obviously took into account the observations of other workers. In the U.S. Woodward (Woodward et al. 1960, 1990) later confirmed this structure by the total synthesis of chlorophyll *a* with the correct stereoconfiguration. Chlorophyll *b* has a very similar structure, differing only in the oxidation of the $-CH_3$ group on ring 2 to $-CHO$.

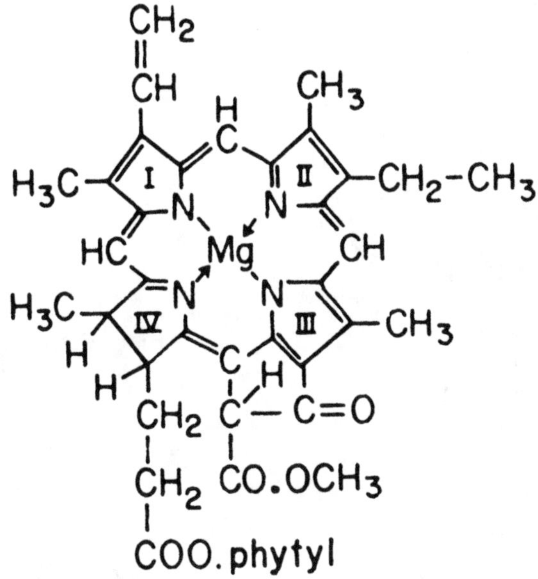

Chlorophyll a

In chlorophyll b the $-CH_3$ of ring II is oxidized to $-CHO$

Figure 2. The chemical structure of chlorophyll a. In chlorophyll *b* the CH_3 group on ring 2 is oxidized to $-CHO$.

Chlorophylls dissolved in ether or 80% acetone show very sharp absorption bands in the blue and red areas of the spectrum and appear green because they absorb little light between blue and red (Figure 3). The spectrum of these pigments within the leaf is less sharp and rather more complicated because the chlorophylls are present as chlorophyll/protein complexes. Chlorophyll *a* is present in all plants. In green plants and green algae it is accompanied by chlorophyll *b*; in brown and some other algae it is accompanied by chlorophylls c_1 and/or c_2 (in which the side chain of ring 4 is oxidized). In blue-green algae or cyanobacteria and in red and cryptophyte algae chlorophyll *a* is accompanied by one or more water-soluble phycobiliproteins (Table 1), which contain the linear metal-free tetrapyrroles phycocyanin and phycoerythrin, related to chlorophylls *a* and *b* respectively. Chlorophyll *a* then is the constant feature of all these photosynthetic organisms and the other tetrapyrroles are

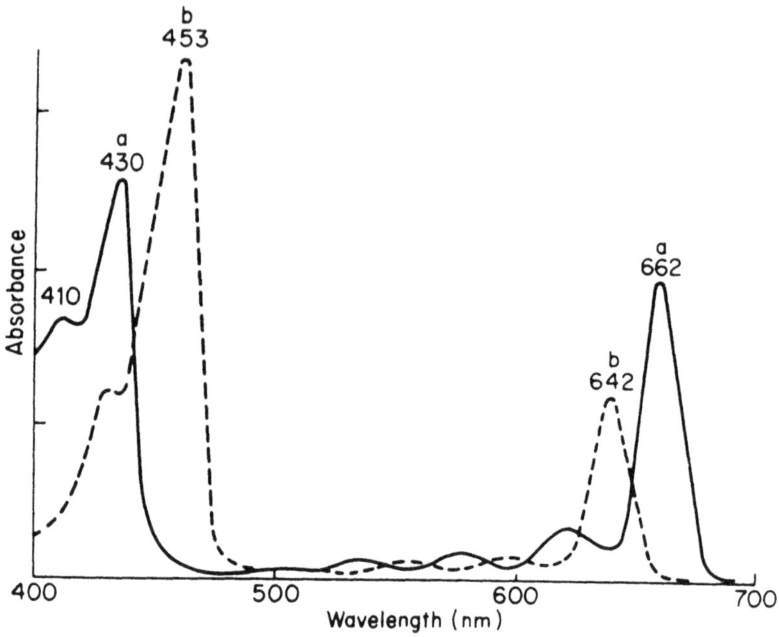

Figure 3. Absorption spectra of chlorophylls in ether.

usually termed accessory pigments (see review by Blinks, 1954). In many chlorophyll *c*-containing algae the role of the accessory pigment tends to be supplemented by the carotenoid fucoxanthin. By absorbing in the green-to-orange region of the spectrum, the accessory pigments partly fill in the gap between the main peaks of chlorophyll *a*. Light energy absorbed by the accessory chlorophyll and phycobilin pigments and by fucoxanthin is transferred to chlorophyll *a* before it is used in photosynthesis.

Absorption of Light into Chlorophyll

One quantum of red light is sufficient to raise a single electron from the resonating system of the chlorophyll to the first excited state (Figure 4). One quantum of blue light, which is more energetic, is sufficient to raise a single electron to the second excited state, but this almost instantly (10^{-12} s) falls back to the level of the first excited state and emits the extra energy as heat. Irrespective of the energy in the quantum of light absorbed the effect is to raise an electron to the first excited state.

If the energy of the excited electron is not otherwise used it is released in 10^{-9}–10^{-6} s as fluorescence quanta when the electron returns to the ground state.

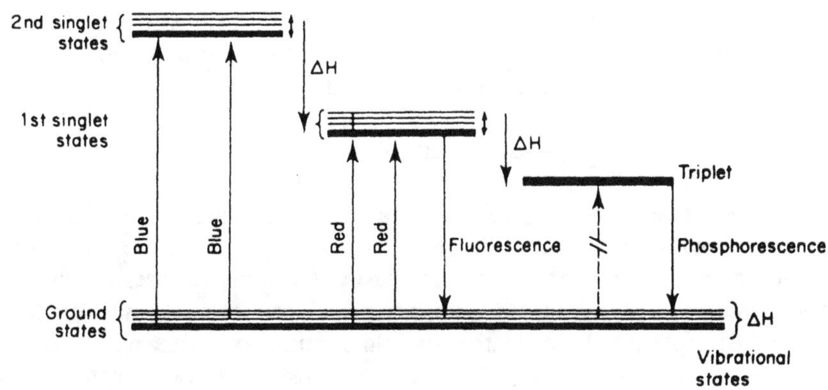

Figure 4. The electronic states of chlorophyll.

The fluorescent light is emitted at a longer wavelength than that of the light absorbed, but its spectrum is independent of the wavelength of the absorbed light. Occasionally the spin on the excited electron can be reversed, so that it is the same as its partner remaining in the ground state. It is then said to be in the triplet state. Decay of the triplet state to the ground state requires a prior spin reversal, which is improbable, and the decay is therefore delayed. When the triplet state decays the delayed light is called phosphorescence, which again is at a longer wavelength than fluorescence. Although triplet states of chlorophyll have been detected in green plants they are uncommon and play no part in photosynthesis.

All these electronic states represent temporary storage and reemission of the energy of light quanta (photons). They belong to the realm of physics. In photosynthesis the energy of the excited electron is captured by an organic molecule and becomes a reducing electron, as described in the section on photophosphorylation. This is the realm of photochemistry and is the essential energy-capturing reaction of photosynthesis—the conversion of light into chemical energy.

Light is not absorbed equally at all wavelengths by chlorophyll, as illustrated in Figure 3. Blue (450 nm) and red light (650 nm) photons correspond to the energy needed to move the electron to the second or first excited state and are efficiently absorbed; however, quanta of green light (550 nm) do not correspond to an appropriate energy transition and are not absorbed—hence the green color of chlorophyll!

ISOLATION OF CHLOROPLASTS

Major advances in the biochemical investigation of photosynthesis were greatly facilitated by the development of methods for the rapid isolation of chloroplasts. Isolated chloroplasts can be studied independently as organelles without the complications of the metabolic interactions in which they may be involved in the intact cell (cf. Arnon, 1967), and individual segments of the overall photosynthetic system can be analyzed.

Requirements of Isolation

Chloroplast isolation requires an initial grinding of leaves for a short time in a mortar with silver sand or in a blender, and for this purpose it is essential to choose soft leaves that disintegrate easily. Grinding releases the contents of the vacuole. Many plants with very acid cell sap are unsuitable for chloroplast isolation because they would require excessive buffering. The majority of plants also accumulate phenolic substances in the vacuole which when released react with proteins and cause enzyme inactivation, or they may be oxidized to yield melanin-like substances in a "browning reaction". This latter reaction may be combatted by adding ascorbate. Leaves of only a few species have been found to be generally suitable as a source of isolated chloroplasts, notably spinach (*Spinacea oleracea*), wheat (*Triticum aestivum*), sunflower (*Helianthus annuus*) and pea (*Pisum sativum*).

The accumulation of starch grains within the chloroplasts is a potential hazard because the grains tend to break out of the chloroplasts during centrifugation. It is therefore best to use leaves harvested at the beginning of the day before starch has accumulated.

A Method of Isolating "Whole" Chloroplasts

A typical preparation of isolated "whole" chloroplasts is described by Whatley and Arnon (1963) and has formed the basis for many later investigations. After initial grinding of spinach leaves in a mortar with sand in buffered 0.35 M NaCl at pH 8.3 the slurry is filtered quickly through muslin to remove large fragments. The homogenate is centrifuged at low speed (200 g) for 30 s to remove nuclei and "cell debris", the chloroplasts are sedimented from the supernatant at 1000 g for 4 min, resuspended quickly in 0.35 M NaCl (unbuffered) and resedimented. Speed is essential. If resuspended in 0.35 M NaCl the chloroplasts appear whole in the light microscope but examination in the EM shows that they have mostly lost their envelope membranes. A consequence of this is that many of the stromal contents are progressively leaked.

Whole chloroplasts can carry out active photosynthetic phosphorylation but can fix CO_2 at only a small percentage of the rate seen in the leaf. If the resedimented whole chloroplasts are resuspended in 0.035 M NaCl they break into chloroplast fragments, which can still phosphorylate well when artificial cofactors are added, but they fix no CO_2. The lost CO_2-fixing ability is restored by adding back an aqueous extract of freshly prepared whole chloroplasts to form a reconstituted system suitable for biochemical analysis.

A Method of Isolating Intact Chloroplasts

Intact chloroplasts retaining their envelope membranes are prepared by substituting 0.33 M sorbitol at pH 6.1 or 7.6 (see Walker, 1971) for the 0.35 M NaCl, adding a "cocktail" of additives to the initial grinding medium and using a very brief homogenization time. After being squeezed through several layers of muslin the intact chloroplasts are collected by centrifugation at 2000 g for 50 s. The sedimented chloroplasts are resuspended in a 0.33 M sorbitol medium suitably buffered and containing the same cocktail. These intact chloroplasts can fix CO_2 at rates comparable on a chlorophyll basis to those in intact leaves. In the EM they can be seen to retain their envelope membranes.

Because these chloroplasts are intact they are impermeable to $NADP^+$, ADP, or ferricyanide. If intact chloroplasts had been used initially in Arnon's laboratory, cyclic and noncyclic photophosphorylation would not have been discovered since the essential substrates could not have reached the thylakoid system. The whole but leaky chloroplasts isolated in 0.35 M NaCl were by good fortune appropriate for the initial investigation of both ATP and CO_2 metabolism and for later analyses of both systems. The greatly enhanced CO_2 metabolism of intact chloroplasts made it possible for other workers to interpret the earlier analytical data and to put them on a physiological basis.

An excellent overview of chloroplast preparations is given by Halliwell (1981).

Catalytic Components of Chloroplasts

In addition to the bulk chlorophyll *a* and *b* molecules which are responsible for light absorption, a number of catalytic components involved in electron transport have been identified in chloroplasts (Table 2).

P_{680} and P_{700} represent the absorption maxima associated with chlorophyll *a* molecules at the reaction centers of photosystems II and I, respectively (see below). Associated with P_{680} are covalently bound Mn^{2+} ions involved in oxygen evolution. The functioning of the Mn ions appears to require the presence of chloride ions, probably as counterions to stabilize Mn complexes as they undergo redox changes, although chloride has not so far been found to be localized within the chloroplast. Associated with P_{700} are pheophytin (chlorophyll *a* minus Mg) and at least two insoluble proteins related to ferredoxin. The P_{680} and P_{700}, the pheophytin, the two iron sulphur centers in the ferredoxin-like proteins, and the cytochrome *b* have all been identified spectroscopically. The other components have been isolated and biochemically characterized. In the case of plastocyanin, plastoquinone, and ferredoxin it has

Table 2. Catalytic Components of Chloroplasts

Catalyst	*Type of Compound*	*Redox (v)*
P_{680}[a]	A chlorophyll *a* dimer in PSII	(+0.54)[g]
P_{700}[a]	A chlorophyll *a* dimer in PSI	(+0.49)[g]
Cytochrome *f*[b]	Cytochrome c-type heme protein	+0.4
Cytochrome b_6[a]	Cytochrome *b*-type heme protein	0.0
Cytochrome b_{559}[a]	Cytochrome *b*-type heme protein	− >0.05
Plastocyanin[c]	Copper protein	+0.39
Plastoquinone[d]	2,3-dimethyl-5-solanesyl-benzoquinone	0.0
Ferredoxin[e]	$2Fe^{3+}$:2S per molecule	−0.42
Ferredoxin/pyridine nucleotide reductase[f]	Flavoprotein	

Notes: [a] Insoluble; recognized spectrophotometrically
[b] Insoluble; isolated after alkaline EtOH treatment
[c] Insoluble; released into solution by sonication
[d] Insoluble; extracted by hexane
[e] Readily soluble in water
[f] Insoluble; released by acetone treatment
[g] Redox value determined only indirectly

been possible to remove them from the chloroplast and restore them later by which procedure photosynthetic activity is first lost and later regained.

CHEMICAL CONSEQUENCES OF LIGHT ABSORPTION

The occurrence of photosynthesis was early characterized by the accumulation of certain products in the green cell and there was considerable interest over the identification of the first product of photosynthesis. As more information was obtained successively earlier "first products" in the chain of photosynthetic activity have been proposed.

Starch

By 1864 Sachs had already shown that starch grains accumulated when certain leaves were illuminated. He removed the chlorophyll with ethanol and used iodine vapor to stain for the starch. Sachs found that those parts of a leaf kept shaded failed to make starch, whereas illuminated parts of the same leaf produced starch granules which accumulated within the chloroplasts. Starch was therefore identified at that time as the first recognizable product of photosynthesis.

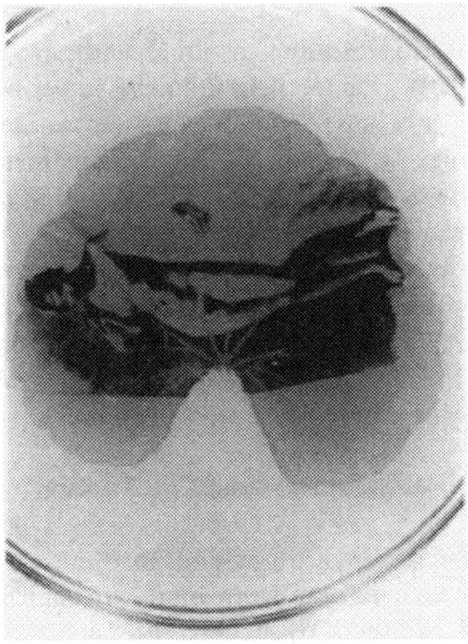

Figure 5. A starch print. (Courtesy of Prof. D.A. Walker). The original is *Study of a Woman* by Paul Pierre Prud'hon (1758-1823).

Starch Prints

The early demonstrations of "starch prints" were with stencils cut to spell out simple words like Stärke (starch). It was clear that the initial product of photosynthesis was accumulated in the illuminated areas in the same position as it was formed and there was no general diffusion of newly synthesized material to other cells in the obscured part of the leaf. The chloroplasts were the site of production and accumulation of the starch in the light.

Later experiments by Molisch and others (1920 et seq.) showed that it was possible to obtain starch prints of much greater complexity. The use of a photographic negative of suitable contrast as a template, and attention to the precise time of exposure, led to very recognizable human figures being produced as a starch print positive by the Sachs method (Figure 5). The graininess of the final picture depended on the size and numbers of the chloroplasts; the choice of a leaf which has many chloroplasts per cell optimized the quality of the picture. The garden geranium *Pelargonium* is very suitable for this purpose and was used in the production of the starch print in Figure 5.

Sugars

Starch remained the first product of photosynthesis to be experimentally observed until the 1920s. There then followed a period in which soluble carbohydrates (especially glucose, fructose, and sucrose) were identified as early products of photosynthesis. In many instances their formation precedes that of starch; they therefore became good candidates as first products, since it was argued that starch was made secondarily from sugars.

In the leaves of many monocots (e.g. tulip, onion) starch is never accumulated under natural conditions and sucrose is the carbohydrate stored. Sucrose was sometimes considered as a possible first product, but was soon generally recognized as a secondary product involved in carbohydrate translocation in many plants and in storage in some.

Radioisotopes and CO_2 Fixation

In the 1950s Calvin and his colleagues (cf. Bassham et al., 1954) used $^{14}CO_2$ and short light exposures in experiments on the green alga, *Scenedesmus*, and separated the labeled products by paper chromatography. They identified phosphoglyceric acid (a C_3 compound) as the earliest product to become labeled. This initial labeling was followed by sequential labeling of many of the sugar phosphates that are now recognized as intermediates in the Calvin cycle. Some authors, especially Arnon, consistently referred to this cycle as the reductive pentose cycle, arguing that many of the enzymes involved had been previously described by Horecker, Racker, and others as participating in an oxidative pentose cycle operating in animals and bacteria and that their contributions ought to be acknowledged, even if only indirectly. Most authors are happy to refer to the Calvin cycle (see also Volume 1 of this series).

Some of the sugar phosphates identified are known to be the precursors of starch, sucrose, and hexoses, earlier proposed as first products of photosynthesis.

Assimilatory Power

Later it was established by biochemical analysis of the photoreactions of isolated chloroplasts that NADPH and ATP produced in light are able to drive all of the dark reactions of photosynthesis leading to CO_2 fixation. By 1960 these two compounds were regarded as the earliest stable products of photosynthesis to be identified and were often referred to jointly as "assimilatory power" by Arnon et al. (1958). However, soon after this it was discovered that ferredoxin, which has a redox potential equivalent to that of hydrogen, is an intermediate in the formation of NADPH and, by the later 1960s it was appropriate to talk of ferredoxin and ATP as the observed first stable products of photosynthesis. Since that time a number of intermediate precursors in the formation of ferredoxin have been identified spectroscopically, but in no case are they sufficiently stable to be regarded as the first products of photosynthesis.

Alternatives to CO_2 Fixation

While there can be no argument that the fixation of CO_2 into carbohydrate via the Calvin cycle represents the main flux of light into chemical energy in photosynthesis, it is important to recognize that this energy conversion may also be seen in a number of other ways, each of which is an example of photosynthesis. Thus the reduction of SO_4^{2-} to S^{2-} needs reduced ferredoxin together with a supply of ATP. The reduction of NO_2^- to NH_3 in chloroplasts again depends directly on reduced ferredoxin, although the reduction of NO_3^- to NO_2^- occurs in the cytoplasm and requires NADPH. The reductive amination of ketoacids to yield amino acids also depends on reduced ferredoxin and their subsequent polymerization to proteins in the chloroplast requires an ample supply of ATP produced by photophosphorylation. The photoreduction of dinitrogen gas to NH_3 in cyanobacteria, though not in chloroplasts, is another photosynthetic reaction.

The external conditions may also affect the nature of the secondary products accumulated by photosynthesizing plants. With high light intensities and high CO_2 concentrations carbohydrates are accumulated preferentially. With high light and low CO_2 concentrations there is a wasteful production of glycollic acid, due to the use of O_2 rather than CO_2 by the enzyme ribulose-bis-phosphate carboxylase/oxygenase under these conditions. The glycollate is not accumulated but metabolized by photorespiration. This reaction sequence involves export of the glycollate from the chloroplast and its subsequent oxidative processing via glycine in microbodies and mitochondria, the loss of CO_2 and the return of glycerate to the chloroplast. At low light intensity and normal CO_2 concentrations the amino acids, glutamate and aspartate, are accumulated. Under other conditions not yet identified the formation of fats is favored.

LIGHT AND DARK REACTIONS

In 1905 F. Blackman studied the effect of light intensity on the rate of photosynthesis by measuring the rate of oxygen evolution from *Chlorella* suspensions. Increase in light intensity resulted in an approximately linear increase in the rate of photosynthesis up to a certain "limiting" intensity, beyond which there was little further increase. Blackman suggested that only part of the photosynthetic process was directly driven by light and that the rest was

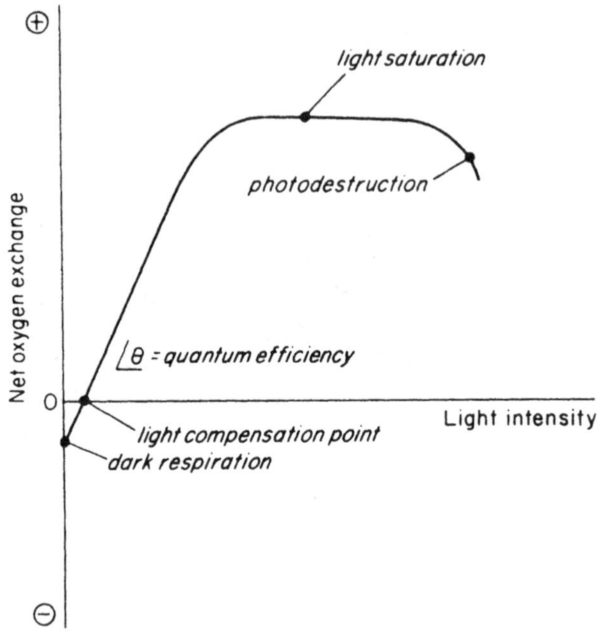

Figure 6. A generalized plot of oxygen exchange against light intensity. Quantum efficiency = quantum requirement^{-1}.

independent of light, i.e. that the overall process included "dark" reactions (Figure 6).

This suggestion was supported by the experiments done at the same time by Brown and Escombe. They used a *Chlorella* suspension illuminated by a beam of intense light variably interrupted by a rotating sector which controlled the amount of light that was received without lowering the intensity at which the light was administered. Up to three-quarters of the light could be removed without affecting the rate of photosynthesis. Clearly light and dark reactions were involved.

Flashing Light Experiments

These observations were taken further by Emerson and Arnold in 1932 (see Emerson, 1958). Photosynthesis was measured by oxygen evolution. They used neon lamps to generate 10^{-5} s flashes of bright light, each sufficient to saturate the photochemical system, and various dark periods between the flashes to allow the product of the light reaction to be expressed as O_2 evolution in the subsequent dark period. The number of O_2 molecules produced per flash increased with increased duration of the dark period. With *Chlorella* cells at 10 °C, the time beyond which there was no further increase in yield (saturation value) was about 0.4 s; it was about 0.04 s at 25 °C. At 10 °C the time to reach half the maximum yield per flash was about 0.04 s, which gives a good measure of the time taken for the dark, temperature-sensitive enzymic reactions of photosynthesis to occur. The effect of temperature is conveniently expressed as Q_{10}, which is defined:

$$Q_{10} = \frac{\text{rate of reaction at } T + 10\ °\text{C}}{\text{rate of reaction at } T\ °\text{C}}$$

The Q_{10} for the dark reaction was 2.9, which suggested an enzymic reaction; it is too high for a photoreaction, for which the Q_{10} would be 1. The existence of an enzymic part in photosynthesis was positively established by these experiments.

Oxygen Evolution by Chloroplasts

The physical separation of the two steps of photosynthesis at the biochemical level followed some time later. In 1937 Hill[2] isolated chloroplasts from leaves and found that they evolve oxygen in light in the presence of suitable electron acceptors, but were unable to fix CO_2 into sugars. This went a long way towards separation of light and dark phases and pinpointed the photoreactions associated with oxygen evolution and the initial reduction steps which in the leaf led to CO_2 uptake. The chloroplast reaction (later named the Hill reaction by C.S. French) formed the basis of many studies of the light reactions of photosynthesis.

Light and Dark Phases in Chloroplasts

The physical separation of the dark reactions was only accomplished in 1958 when Trebst and colleagues demonstrated that in the fixation of CO_2 by chloroplast fragments supplemented with a chloroplast extract, the light and dark steps can be separated in time. Light was first applied to the chloroplast fragments in the absence of CO_2; subsequently the fragments were presented with $^{14}CO_2$ in the absence of light (Trebst et al., 1958; see also Trebst, 1974). The $^{14}CO_2$ was fixed into the normal products of photosynthesis detected in Calvin's laboratory. Trebst then separated the two steps morphologically and showed that the light reactions occurred in the green grana and the dark reactions in the surrounding stroma.

Arnon had suggested at this time (Arnon et al., 1958) that the light step of photosynthesis provides "high energy bonds" in the form of reduced pyridine nucleotide and ATP, and that the dark step is the utilization of this energy for the reduction of CO_2 to the level of sugars and other intermediates. ATP and NADPH are accumulated in the Trebst experiments in the initial light phase before CO_2 is provided and these compounds (assimilatory power) drive the dark reactions when CO_2 is added. As expected the addition of NADPH and ATP to chloroplast extracts in the dark did in fact permit the fixation of radioactive CO_2 into normal intermediates without the need for light. The chemosynthetic nature of the CO_2 fixation reactions of the Calvin cycle was emphasized by these experiments, and was further confirmed by the identification of all the Calvin cycle enzymes in a number of chemosynthetic bacteria (e.g. *Thiobacillus denitrificans*) that fix CO_2 in the dark at the expense of redox reactions.

Photosynthetic Unit

Using *Chlorella* suspensions with different concentrations of chlorophyll, Emerson and Arnold (see Emerson, 1958) measured the yield of O_2 per flash on giving repeated flashes separated by sufficient intervals to allow the dark reactions to go to completion. They found that the maximum number of O_2 molecules evolved per flash was related to the number of chlorophyll molecules present in the suspension. The ratio they observed was 1 O_2 evolved per 2500 chlorophyll molecules. They therefore suggested that for every 2500 chlorophyll molecules there was present one "unit" capable of evolving one molecule of O_2. Each time a photosynthetic unit (PSU) is activated by light all the chlorophyll molecules in the unit work together to cause the evolution of one O_2 molecule via a number of reaction centers.

It was later found that the evolution of each O_2 molecule requires the use of 8 light quanta. It has been calculated by Gaffron that if individual chlorophyll molecules had to accumulate the necessary 8 quanta in sequence, as they would be required to do in the absence of photosynthetic units, there would be a long lag in oxygen evolution after illumination at "saturating" light intensities. Such a lag does not occur and this is a cogent reason in support of the existence of the photosynthetic unit. Moreover the evolution of oxygen in the Hill reaction requires chloroplast fragments of a minimum size, each containing hundreds of chlorophyll molecules.

For oxygen evolution in *Chlorella* the association of about 2500 chlorophyll molecules is required. For each partial reaction involving the transfer of a hydrogen atom from water to $NADP^+$ the cooperation of about 600 chlorophyll molecules would be required and since the transfer of each hydrogen requires the input of 2 quanta each primary photoact must involve a photosynthetic unit size of about 300 chlorophyll molecules. Each light quantum absorbed by a single chlorophyll molecule within the unit must be moved by excitation energy transfer throughout the unit until it reaches the specialized chlorophyll

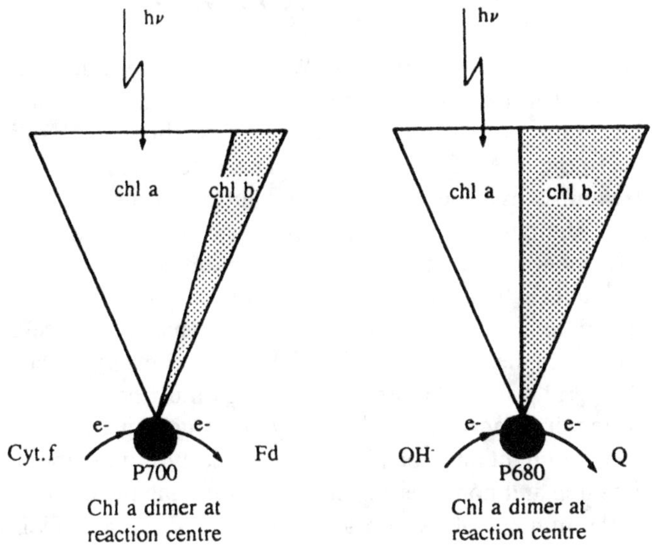

***Figure** 7.* The essential features of photosynthetic units.

a molecules called energy traps or reaction centers. Once the excitation energy has reached the trap the photoactivated chlorophyll *a* molecules bring about the transfer of an electron from a donor to an acceptor molecule, both of which must be held in the immediate vicinity of the trap. Trap molecules undergo light-induced oxidation and subsequent reduction in the dark and this may be detected through changes in the light absorption accompanying redox changes in intermediates. Since the total number of trap molecules is extremely small, the magnitude of the light-induced absorption changes due to their redox reactions is too small to be detected by conventional spectrophotometers. But they can be measured by the use of sensitive difference spectrophotometers.

The current view of a photosynthetic unit then is of a group of about 300 chlorophyll molecules associated with a single trap of a chlorophyll *a* dimer molecule identified spectrophotometrically as P_{700} or P_{680}. Each unit is responsible for a primary photoact and has associated with it the appropriate donor and acceptor molecules to enable the primary electron transfer to be carried out in light (Figure 7). Two types of PSU have been identified in both of which the bulk of the chlorophyll molecules is composed of a mixture of chlorophylls *a* and *b*. The ratio of *a*:*b* is different for the units corresponding to P_{700} and P_{680}. The two units have been separated by detergent treatment followed by differential centrifugation. The bulk of chlorophyll molecules associated with each trap chlorophyll molecule is usually referred to as the antenna pigment.

QUANTUM REQUIREMENT

Following early work in the 1920s in Warburg[1]'s laboratory (reviewed by Emerson, 1958) much effort was expended in the 1940s and 1950s in discovering the minimum number of quanta required to bring about the fixation of one CO_2 molecule to the level of carbohydrate. This reaction requires 4 [H] atoms (as NADPH) and is associated with the evolution of one O_2 molecule as a waste product. The reaction requires an input of energy of at least 120 kcal per C atom. At the wavelength of the red absorption maximum of chlorophyll (the least energetic light capable of driving photosynthesis) the energy per einstein (mole quanta) is only about 40 kcal. This means that several inputs of light energy must be accumulated for the overall chemical change to be possible and several primary photochemical steps must be involved, no single one of which requires more than the energy of a single quantum of red light. If every quantum of light absorbed in the chloroplast brought about a primary photochemical step and none were wasted, then dividing the number of quanta absorbed by the number of carbon atoms converted to carbohydrate would tell us how many photochemical steps are involved per carbon atom. The evolution of oxygen is the most convenient measure of yield since oxygen is easy to measure manometrically or with the oxygen electrode and one O_2 is evolved per CO_2 fixed into carbohydrate.

Warburg reported experimental values of the quantum requirement approximating 4. Emerson consistently found experimentally the quantum requirement to be 8 or more, with 10 being observed most frequently (Emerson, 1958). Since the reduction of each CO_2 requires 4[H], the early quantum requirement reported by Warburg appears to fit a credible mechanism in which the elevation of each [H] from the redox level of water to that of $NADP^+$ requires a single quantum of light. However, it was becoming apparent from a consideration of the likely losses involved in the conversion of light energy into chemical energy and its storage in photosynthetic intermediates (now identified as NADPH and ATP) that 4 quanta are inadequate to drive photosynthesis. But 8–10 quanta would easily provide enough energy and a mechanism incorporating two inputs of light energy per [H] would be possible. If 8 quanta were used the overall efficiency of chemical energy storage from light would be of the order of 30% (see Emerson, 1958).

THE "RED DROP"

Light energy absorbed by the photosynthetic antenna pigments is ultimately transferred to a chlorophyll *a* molecule at the reaction center where it is converted into chemical energy. If all the pigments absorbing light were equally efficient in transferring energy to the reaction centers the quantum requirement of photosynthesis should remain constant throughout the absorbing spectrum. Emerson and Lewis (see Emerson, 1958) showed that a plot of the quantum yield as a function of wavelength of light in green algae is, however, not constant throughout the spectrum. There is a decline at the blue end where the absorption by carotenoids is significant (no energy transfer to the reaction center, so that light absorbed in carotenoids is wasted) and there is a drop at the far red end of the spectrum (beyond 700 nm) even though chlorophyll *a* is still able to absorb at these wavelengths. The decline in the quantum yield at the far red end of the spectrum is called the "red drop".

The red drop could be avoided and the quantum yield in this region increased if supplementary light were added to the far red light absorbed by chlorophyll *a* but inefficiently used. It was suggested that the supplementary light would be absorbed by accessory pigments, chlorophyll *b* in green algae and higher plants, phycobilins in cyanobacteria and red algae, or chlorophyll *c* and fucoxanthin in the brown algae. This phenomenon has been called the enhancement effect and it was suggested that a simultaneous excitation of chlorophyll *a* and the accessory pigments was necessary for high quantum yields. Light absorbed by chlorophyll *a* alone was inefficient.

The "red drop" and enhancement effects are consistent with a two-light reaction and two-pigment system hypothesis, although it was not so interpreted at the time of its discovery. Myers and French (1960) showed that the enhancement effect in the quantum yield could still be observed even when the supplementary light and the far red light were given not simultaneously but a few seconds apart. This implies that the product of one light reaction could exist for a few seconds and still be effective in enhancing the rate of another photoreaction.

Blinks (1957) has reviewed his work on "chromatic transients" in the photosynthesis of the red alga *Porphyra perforata*. When *Porphyra* was exposed to alternating red light (absorbed by chlorophyll *a*) and green light (absorbed by phycoerythrin), which were equally effective in the steady state, it was found that the rates of oxygen evolution were not constant during the transition period. On going from red to green light extra oxygen production was observed, followed by a decrease and recovery to the steady state. By contrast, in the transition from green to red light no extra oxygen production was observed but there was actually a decrease in rate followed by recovery to the steady state. This effect could again be a consequence of a two-pigment system (the two light reactions hypothesis) and is so interpreted today. However, at the time Blinks proposed alternative explanations involving either partial inactivation of chlorophyll molecules in different lights or altered respiration rates on changing the wavelength of the light.

BACTERIAL PHOTOSYNTHESIS

Green plants and algae are not the only organisms to carry out photosynthesis. There exists an interesting group of photosynthetic bacteria which can fix CO_2 anaerobically in the light but which do not evolve oxygen. They turned out to be very relevant to the study of photosynthesis. The photosynthetic bacteria include the Thiorhodaceae, purple sulfur bacteria (e.g. *Chromatium*), the Athiorhodaceae, purple non-sulfur bacteria (e.g. *Rhodospirillum, Rhodopseudomonas*), the Chlorobiaceae, green sulfur bacteria (e.g. *Chlorobium*), and the Chloroflexaceae, green facultative aerobes (e.g. *Chloroflexus*). When grown photosynthetically all these types of bacteria are strict anaerobes and can grow only in the complete absence of oxygen. They cannot use water as a substrate, which is why they do not evolve oxygen. As electron donors the purple sulfur bacteria use hydrogen sulfide, the purple non-sulfur bacteria use simple organic acids and alcohols, the green sulfur bacteria use hydrogen sulfide or thiosulfate and the green non-sulfur bacteria again use organic acids and alcohols.

In the 1930s van Niel (1941) characterized many of these bacteria and made an extensive comparative study of their photosyntheses. His work drew attention to the essential redox nature of the photosynthetic process and he was able to describe the bacterial photosynthesis under the general formulation:

$$2H_2A + CO_2 \xrightarrow{\text{light}} [CH_2O] + 2A + H_2O$$

In the sulfur bacteria H_2A is supplied as H_2S and the waste product is sulfur; in non-sulfur bacteria H_2A may, for example, be isopropanol and the waste product acetone. Van Niel pointed out the similarity between the bacterial photosynthesis and green plant photosynthesis, in which H_2A is water and the waste product oxygen. In an attempt to provide a unified mechanism for photosynthesis in every group he proposed that in all organisms the primary photosynthetic reaction was the splitting of water into [H] and [OH], that the [H] was used for the reduction of CO_2, and that the [OH] either underwent dismutation into O_2 and water, as in green plants, or was reduced by H_2S or an organic substrate so that oxygen evolution did not take place, as in photosynthetic bacteria. This concept of water splitting retained currency for 20 years before its replacement by Arnon's electron donor proposal in the 1950s.

THE HILL REACTION

The possibility of studying partial photosynthetic reactions in isolated components of cells—a biochemical approach—received a great impetus from the work of R. Hill in 1937 (Hill, 1939). He isolated chloroplasts from leaves and showed that in the presence of an artificial electron acceptor (to replace CO_2) they could evolve oxygen in light. In his earliest experiments Hill identified a number of ferric compounds as useful electron acceptors and was able to measure significant and continuous oxygen evolution upon illumination by observing the oxygenation of myoglobin. The partial pressure of oxygen evolved was low in the initial experiments. It was therefore important to use myoglobin to measure the oxygen because myoglobin (muscle hemoglobin) has a high affinity for oxygen and also a hyperbolic dissociation curve, unlike blood hemoglobin, which has a sigmoidal dissociation curve. Using a modified two-stage Dubosq colorimeter with a microspectroscope ocular it was possible for Hill to study many details of the oxygen evolution by the isolated chloroplasts. One important observation was that four electrons were transferred to ferric oxalate (the experimental substrate of choice) for each O_2 molecule evolved.

An alternative electron acceptor was methemoglobin (ferric hemoglobin!). Four molecules of methemoglobin were reduced for each O_2 evolved. Under aerobic conditions each molecule of methemoglobin reduced to hemoglobin appeared as oxyhemoglobin. However, under anaerobic conditions one oxyhemoglobin and three hemoglobin molecules were accumulated for each four methemoglobins reduced. This gave a simultaneous measure of O_2 evolved and methemoglobin reduced, and confirmed experimentally that four electrons are transferred during the Hill reaction per O_2 evolved, i.e. there are no hidden products.

When ferric oxalate was used as the substrate only a low partial pressure of O_2 could be attained because the ferrous oxalate produced was readily reoxidized by the oxygen evolved and the experiment had to be started anaerobically. Hill later added ferricyanide to reoxidize the ferrous oxalate by a rapid chemical reaction so that the oxygen released by the chloroplast could accumulate readily in the air. Under these conditions the manometric measurement of oxygen evolution became possible. Curiously ferricyanide is now known to be a Hill reagent (electron acceptor) in its own right and leaving out the ferric oxalate also results in a perfectly normal Hill reaction. Many other electron acceptors have since been shown to function in the Hill reaction, including a whole variety of quinones and redox dyes.

After methemoglobin was found to act as an electron acceptor it became apparent that chloroplasts contained a protein, the methemoglobin reducing factor, that was essential for the reaction. By the later 1950s this small protein had been purified considerably by Hill and his associates. Although it behaved as the first electron acceptor in photosynthesis its redox potential was not known. It was more than 10 years before its true importance and identity as ferredoxin (see below) was recognized.

Analysis of the Hill reaction, usually by manometric methods, has led to considerable progress in our knowledge of the photochemical reactions of photosynthesis. One important conceptual result was the obvious conclusion that the oxygen evolved must have originated from water. Together with van Niel's observations it became clear that the conversion of CO_2 to carbohydrate was an oxidation–reduction process with the [H] atoms needed to reduce CO_2 to the level of CH_2O coming from water, leaving the oxygen to be discarded as a waste product. ^{18}O labeling experiments with algae and higher plants gave results in agreement with this conclusion. The original, generally accepted proposal by Ingenhousz in the 1780s that the carbon in the CO_2 assimilated was retained by the plant, the O_2 eliminated, and water incorporated into the substance of the plant to produce carbohydrates, could at last be safely rejected.

THE PHOTOCHEMICAL REDUCTION OF PYRIDINE NUCLEOTIDES

The details of the fixation of $^{14}CO_2$ by the Calvin cycle drew attention to the need for ATP and reduced pyridine nucleotide for the reduction of phosphoglyceric acid (PGA) which had been identified as the first product of CO_2 fixation. Ruben (1943) had earlier indicated the importance of ATP in the reduction of CO_2, although he had put forward no mechanism for its production.

In 1951 workers in three laboratories (Arnon, Tolmach, Vishniac and Ochoa[2]) obtained clear indications that isolated chloroplasts can photoreduce pyridine nucleotides. However, none ever accumulated and their reduction could only be detected by allowing the accumulation of another reaction product (e.g. malate or lactate) in a coupled reaction. In keeping with the then current view, Vishniac and Ochoa coupled the photoreduction of NAD by chloroplasts to its aerobic reoxidation by adding mitochondria from mung beans and were able to detect ATP formation as a consequence. This experiment was taken to indicate that chloroplasts had a function which was restricted to producing a reductant in light that could subsequently be used to generate ATP in an oxidative reaction sequence in mitochondria (reviewed by Vishniac, 1955). Such a conclusion was also supported by the fact that the Hill reaction itself represented only a partial reaction of photosynthesis.

CHLOROPLASTS AS THE UNIQUE SITE OF PHOTOSYNTHESIS

In 1954 Arnon and colleagues had difficulty in accepting the above view because mitochondria are scarce in the palisade parenchyma cells in the leaf, yet these cells are photosynthetically the most active. It was therefore essential to show that ATP was formed by chloroplast suspensions when illuminated. The initial experiments of Arnon et al. (1954) using $^{32}P_i$ showed that ATP was indeed formed on illumination of isolated chloroplasts at pH 7.2. Since care had been taken to eliminate the mitochondria from the preparation it was concluded that light-driven ATP synthesis was confined to the chloroplasts. Unlike the ATP synthesis by mitochondria, the light-induced ATP synthesis by chloroplasts occurred without net consumption of oxygen and without the need to supply any substrate to serve as an external energy source. The only "substrate" consumed in photosynthetic phosphorylation was light.

Isolated chloroplasts able to make ATP were also found to incorporate ^{14}C into intermediates of the carbon cycle when illuminated at pH 7.2 in the presence of $^{14}CO_2$. The ^{14}C uptake was small but continued for an hour and was accompanied by an equivalent evolution of oxygen, which could be determined manometrically. The preliminary conclusions were (1) that chloroplasts can both make ATP and fix CO_2, (2) that chloroplasts are the unique site of photosynthesis in green plants, and (3) that they do not require the participation of other organelles to carry out photosynthesis.

Although the initial rates of ATP synthesis and CO_2 fixation were very low they were consistently observed. The rate of ATP synthesis was soon increased to very high levels on a chlorophyll basis by adding artificial "cofactors" such as FMN or vitamin K_3, substituting ADP for AMP as the substrate to be phosphorylated, and adjusting the pH to 8.3. Other investigators later showed that phenazine methosulfate could catalyze phenomenal rates of ATP synthesis.

Although Whatley and his colleagues in Arnon's laboratory were able to increase the rates of CO_2 fixation on a chlorophyll basis, particularly in a reconstituted system, rates comparable with those in the leaf were never achieved (Whatley et al., 1956). This was due mainly to the fact that the chloroplasts initially used, although appearing intact under the light microscope, had actually lost their outer membranes and had leaked essential components from the stroma, as could later be seen in the electron microscope.

In the mid-1960s CO_2 fixation rates similar to those in the leaf were eventually achieved by Jensen and Bassham (1966) and Walker (1971) using intact chloroplasts isolated in a way that still retained their outer membranes.

In the reconstituted system for CO_2 fixation, an aqueous extract of freshly isolated chloroplasts (CE) was combined with a preparation of broken chloroplasts. The products of CO_2 fixation in this reconstituted system were similar to those in the initial system and in the leaf. In Arnon's laboratory all the enzyme systems needed to catalyze the Calvin cycle were shown to be present in CE and it was this observation that made possible the experimental separation of light and dark reactions referred to earlier (Trebst et al., 1958).

There was general agreement that the evidence of Engelmann and of Sachs in the nineteenth century showed that the complete process of photosynthesis was confined to chloroplasts. The subsequent experiments of Hill and of Vishniac and Ochoa on isolated chloroplasts had allowed some doubts to be entertained about the competence of chloroplasts alone to carry out photosynthesis without the cooperation of other organelles. However, the later experiments of Arnon and others clearly confirmed the original concept of chloroplasts as the unique site of photosynthesis.

PHOTOPHOSPHORYLATION

Cyclic Photophosphorylation

Although photophosphorylation by isolated chloroplasts when first discovered (Arnon et al., 1954; Arnon, 1961a, 1984) depended on the presence of air, it was apparent that oxygen was needed only as a catalyst. In Arnon's laboratory it was soon found that the rate of ATP formation was speeded up by the addition of flavin mononucleotide or vitamin K_3 and the need for oxygen disappeared, the synthesis of ATP proceeding in the light in an atmosphere of nitrogen or argon.

These latter observations coincided with the demonstration by Frenkel (1954) that chromatophores (the bacterial equivalent of thylakoids) of the photosynthetic bacterium *Rhodospirillum rubrum* can carry out an active light-dependent synthesis of ATP under strictly anaerobic conditions. For bacterial chromatophores to make ATP in the light the addition of a small amount of a reducing substance such as succinate, ascorbate or NADPH was necessary under certain conditions, especially when the particles had been allowed to age. These additives are not used up during photophosphorylation and appeared to affect only the redox balance of the system. Their importance is to bring all the electron transport intermediates into a suitable redox state to allow a cyclic electron flow to occur, thus "poising" the system. A common anaerobic mechanism for photophosphorylation thus appeared to be shared by photosynthetic bacteria and green plants.

The earliest formulation in Arnon's laboratory of a mechanism (a working hypothesis) for light-dependent ATP synthesis in chloroplasts involved the photolysis of water, yielding reduced ([H]) and oxidized ([OH]) moieties, followed by their recombination through a series of electron carriers known to be present in chloroplasts and arranged in order of their redox potentials to release the energy for ATP synthesis (Figure 8). The energy source was light and there was no net consumption of an external electron donor or oxidant. However the work on chromatophores initiated by Frenkel suggested that there was no need to postulate a photolysis of water, either for chromatophores or chloroplasts.

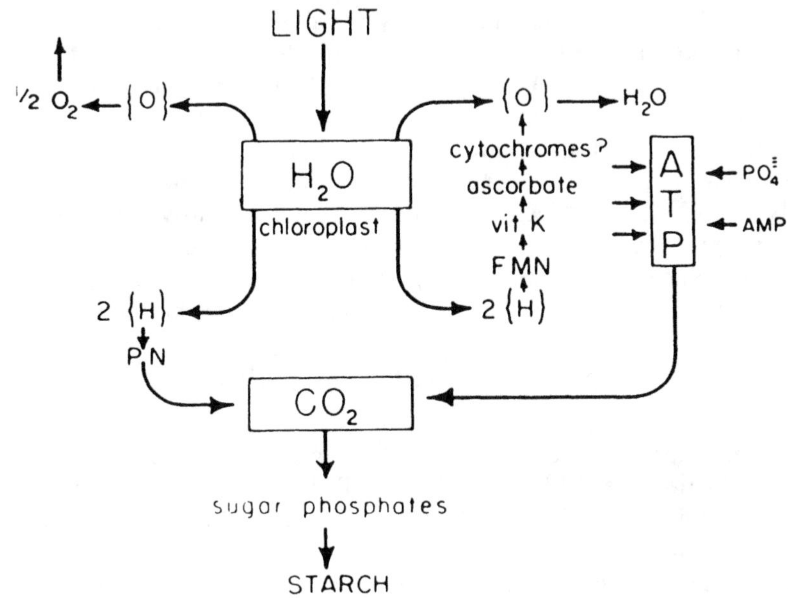

Figure 8. An early representation of photosynthesis in chloroplasts based on "water splitting".

It was proposed (Arnon, 1955) that on illumination a chlorophyll *a* molecule in the ground state becomes excited (energized) on receiving a photon (quantum of light) either directly or more probably indirectly by resonance transfer from bulk chlorophyll *a* and *b* antenna pigments in the PSU. The electron raised to the first excited state is donated to an acceptor molecule of low redox potential physically located adjacent to the chlorophyll *a* trap, temporarily leaving an oxidized (chl^+) molecule. The chl^+ immediately receives an electron from a donor molecule, such as reduced cytochrome *f*, also located next to the chlorophyll *a* molecule, which then reverts to the ground state. The transient photooxidation of cytochrome *f* has indeed been detected spectroscopically. This primary photooxidation is essentially temperature-independent (Q_{10}=1) and no collision chemistry is involved. In the early formulation the first electron acceptor proposed was FMN, vitamin K_3 or a related physiological equivalent. The initial acceptor molecule has subsequently been identified in green plants as a pheophytin molecule, an intermediate in the reduction of ferredoxin (see below).

The reducing electron captured by the initial acceptor is returned to the oxidized cytochrome *f* (which thereby becomes reduced) through a series of intermediate carriers and the free energy released is coupled to ATP synthesis.

This formulation is analogous to the sequence of electron carriers of oxidative phosphorylation in mitochondria. However all the reactions in mitochondria occur in the dark and reductant (substrate donor) and oxidant (electron acceptor molecules) must be supplied externally. In mitochondria the electrons participate in a linear flow. In photophosphorylation both reductant and oxidant molecules originate in the initial photoreaction and their recombination provides the energy for ATP formation. Because of the cyclic pathway followed by the electrons in this type of light-driven ATP synthesis, Arnon suggested the term "cyclic photophosphorylation" for this reaction sequence. Such a sequence appears to operate in green plants, algae, and cyanobacteria as well as in the anaerobic photosynthetic bacteria.

The Generation of Assimilatory Power

In an important advance over previous investigators interested in photoreductions (see Vishniac, 1955), San Pietro and Lang (1956) showed that the direct photoreduction of pyridine nucleotide (NAD^+ or $NADP^+$) and its accumulation in the reduced form could be accomplished in the presence of large amounts of isolated chloroplasts. A factor required for this activity was identified and named photosynthetic phosphopyridine nucleotide reductase (PPNR). At the same time Arnon and colleagues had found that under certain conditions $NADP^+$, but not NAD^+, could double the rate of cyclic phosphorylation catalyzed by suboptimal amounts of FMN or vitamin K_3. The photoreduction of $NADP^+$, but not of NAD^+, by small amounts of chloroplasts was shown to require the presence of a specific "TPN($NADP^+$)-reducing factor" present in a chloroplast extract (CE) of the sort used in the $^{14}CO_2$ fixation experiments (reconstituted system) described earlier. The factor (PPNR) was purified by San Pietro; it then became specific for $NADP^+$. The ability of the unpurified protein to catalyze the reduction of NAD^+ and $NADP^+$ is attributed to the presence of a transhydrogenase.

Arnon et al. (1958) showed subsequently that the reduction of $NADP^+$ was accompanied by oxygen evolution and the formation of ATP. The evolution of oxygen was measured manometrically, $NADP^+$ reduction spectrophotometrically at 340 nm and $AT^{32}P$ formation from ADP and ^{32}P, radioactively. The observed stoichiometry is represented by the equation:

$$2ADP+2P_i+2NADP^++4H_2O \xrightarrow[\text{chloroplasts + CE}]{\text{light}} 2ATP+O_2+2NADPH+2H^++2H_2O$$

A single light-driven reaction sequence thus produces the two components of the assimilatory power—ATP and NADPH—needed for CO_2 fixation by the

Calvin cycle. Because of the linear flow of electrons this variation is called noncyclic photophosphorylation.

The noncyclic photophosphorylation just described must be regarded as the physiological equivalent of the Hill reaction, viz. the evolution of oxygen in the presence of an added electron acceptor. A nonphysiological variant uses ferricyanide in place of $NADP^+$. The photoreduction of ferricyanide, the stoichiometric evolution of oxygen, and the synthesis of ATP were again observed although no "TPN-reducing factor" was required. In the absence of a phosphate acceptor system (ADP + P_i) the rate of ferricyanide reduction was diminished, i.e. the simultaneous conservation of energy as pyrophosphate bonds caused a stimulation in the rate of electron transport. This indicated that the chloroplasts can show a degree of phosphate control analagous to the metabolic control seen in mitochondria, a conclusion later confirmed more elegantly by West and Wiskich (1968).

The fixation of each CO_2 through the Calvin cycle requires 2 NADPH + 3 ATP molecules. This is not fully supplied by noncyclic photophosphorylation. What then is the source of the additional ATP required? Arnon believed that it was supplied by cyclic photophosphorylation. Experimentally the addition of small amounts of FMN or other cofactors of cyclic phosphorylation to the system photoreducing $NADP^+$ (i.e. noncyclic photophosphorylation) partly suppressed the accumulation of NADPH and the rate of ATP synthesis was increased. This increase occurred by way of cyclic photophosphorylation. Other workers have done experiments which show that under some conditions there appears to be a higher yield of ATP in noncyclic photophosphorylation and that the stoichiometry should be represented as the formation of 1.5 ATP per NADPH accumulated. This author tends to the Arnon view and supposes that the increased ATP formation observed by some experimenters results from a leakage of electrons from a noncyclic to a cyclic pathway.

Mechanism of ATP Synthesis

A coupling factor (CF_1) is required for the synthesis of ATP by isolated chloroplasts. Isolated CF_1 from chloroplasts exhibits ATPase activity and will hydrolyze ATP to ADP + P_i. The physical and catalytic properties of CF_1 from chloroplasts are very similar to those of F_1 (ATPase) in mitochondria (reviewed by McCarty, 1979).

Because of the similarity of the membrane-bound ATPases in chloroplasts and mitochondria it has been assumed that the mechanisms of ATP formation in both systems are similar. According to Mitchell's[1] (1966) chemiosmotic theory mitochondria pump protons across the semipermeable mitochondrial membrane at the expense of the electron transfer resulting from substrate oxidation. The protons are returned to the other side of the membrane via F_0 and F_1, the proton channel in the membrane, and the ATP synthesizing complex, respectively. In chloroplasts it seemed likely that the synthesis of ATP again depends on the pumping of protons (at the expense of light energy) across the semipermeable membrane of the thylakoid into the lumen of the intramembrane space, their return to the stromal side via CF_o and CF_1 in the membranes being coupled to ATP synthesis. The "sidedness" of membrane required for this to happen has already been referred to above (Chloroplast Ultrastructure).

An important experiment that appears to support the chemiosmotic theory directly was performed by Jagendorf and Uribe in 1966 (reviewed by Avron and Neumann, 1968). They found that ATP formation could take place in chloroplasts solely at the expense of an imposed pH gradient. Chloroplasts were first placed in the dark under acid conditions (pH 4) and, after equilibrating, the medium was then suddenly changed to basic conditions (pH 8). The change of pH in the dark from pH 4 to pH 8 in the presence of ADP and P_i caused the synthesis of a small amount of ATP without any need for light. The pH gradient alone drove the synthesis of ATP. An interpretation of this result in terms of chemiosmotic theory was also supported by observations on the transport of organic acid anions during illumination and the subsequent formation of ATP in the dark as the acid within the thylakoids dissociated, thereby acting as a proton source.

Perhaps the most surprising aspect of ATP synthesis in isolated chloroplasts is the apparently irreversible nature of the reaction, which leads to the complete conversion of added ADP to ATP in the light. Experimentally, isolated chloroplast fragments also completely reduce added ferredoxin and subsequently $NADP^+$ in the light. A description of the formation of ATP and NADPH in terms of energetics requires the use of nonequilibrium thermodynamics. A consequence of the complete photochemical conversion of ADP and $NADP^+$ to ATP and NADPH is the ability of these substances to "drive" the Calvin cycle firmly in the direction of CO_2 fixation and reduction to the level of carbohydrate in a series of dark equilibrium reactions.

FERREDOXIN

The requirement for a photosynthetic phosphopyridine nucleotide reducing factor (PPNR) or a "TPN-reducing factor" in the light-dependent reduction of $NADP^+$ and accumulation of NADPH drew attention to the work on the methemoglobin (heme protein) reducing factor (see above), a small protein that seemed to function as if it were the initial acceptor in photosynthesis by isolated chloroplasts. When this protein was highly purified it was shown to be identical to PPNR (Davenport, 1963). At this time Mortensen (Mortensen et al., 1962) had isolated from the anaerobic bacterium *Clostridium* a small protein which he named ferredoxin because it contained Fe and acted as a redox carrier between hydrogen gas and various electron donors and acceptors.

The bacterial ferredoxin turned out to be the first of a new class of electron-carrying proteins of low redox potential. PPNR of spinach chloroplasts was found to resemble the bacterial ferredoxin and was termed "spinach ferredoxin" (Tagawa and Arnon, 1962; reviewed by Arnon, 1967). It contained $2Fe^{3+}$ and $2S^{2-}$ atoms per molecule. When reduced by hydrogen gas in the presence of an added hydrogenase it was shown to have a redox potential very close to that of hydrogen (−0.42v at pH 7).

Reduced spinach ferredoxin is rapidly reoxidized by oxygen alone or by $NADP^+$ in the presence of a flavin enzyme readily extractable from chloroplasts after acetone treatment. The enzyme, now called ferredoxin-$NADP^+$ reductase, was highly purified and crystallized. It forms a complex with ferredoxin and $NADP^+$ which can be seen spectroscopically. The photoreduction of $NADP^+$ is a two-step process in which ferredoxin is first reduced by illuminated chloroplast fragments and subsequently reoxidized in a dark step by $NADP^+$ (Whatley et al., 1963). The reduction of each ferredoxin molecule requires the transfer of one electron. The reduction of each $NADP^+$ molecule is a two-electron step. It is the function of the flavin enzyme to act as a gating mechanism for this reaction by accepting in sequence two single electrons from ferredoxin (when it becomes fully reduced), and subsequently passing on a pair of electrons to $NADP^+$ when the enzyme becomes fully oxidized.

Alternative electron acceptors for the reoxidation of reduced ferredoxin include H^+ ions leading to the evolution of H_2 gas in light when hydrogenase is added, and O_2 when H_2O_2 is the initial product. In all cases the electron transport is accompanied by ATP formation. Cyclic photophosphorylation is also brought about under conditions when reduced ferredoxin donates its electrons to components of the cyclic electron flow pathway (Tagawa et al., 1963). In the absence of $NADP^+$ an alternative pseudocyclic photophosphorylation can proceed in which the reduced ferredoxin is reoxidized by O_2 and the H_2O_2 is broken down to $\frac{1}{2}O_2$ and water, so that no net oxygen evolution results. This is a variant of noncyclic electron flow leading to $NADP^+$ reduction. It is likely that the ATP synthesis observed in Arnon's earliest experiments (Arnon, 1988) were due to pseudocyclic phosphorylation.

TWO LIGHT REACTIONS

Hill and Bendall Formulation

Hill and Bendall (1960) put forward a scheme for the involvement of two light reactions in photosynthesis. It was based on the conclusion that cytochromes (cytochrome b_6 = 0.0 v; cytochrome f = +0.4 v) having redox potentials between H_2/H^+ (−0.42 at pH 7) and O_2/H_2O (+0.81) are necessary intermediates in the photosynthetic reaction. Their scheme implies two inputs

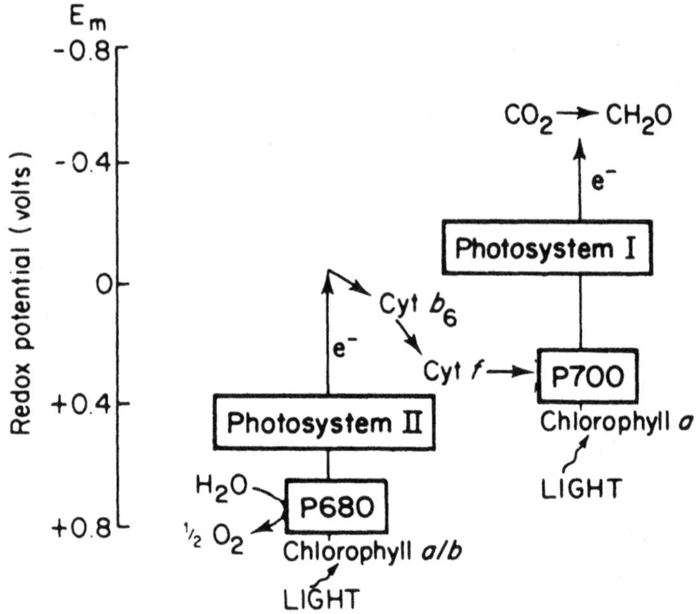

Figure 9. The proposal of Hill and Bendall (1960): the Z-scheme.

of light quanta for each electron raised from +0.81 to −0.42 v rather than the single quantum previously envisaged. Hill and Bendall proposed that the two light reactions should be in series, not in parallel. A diagrammatic representation is given in Figure 9. It is commonly known as the "Z-scheme". The way in which the components of the Z-scheme might be arranged within a thylakoid membrane is suggested in Figure 10. There is abundant evidence, summarized by Trebst (1974), for these components within the membrane. Their arrangement is important for the light-driven transfer of protons from the chloroplast stroma into the lumen of the thylakoid.

The Hill and Bendall formulation represented a big conceptual step forward and, although they did not specifically point this out, was consistent with a mass of physiological evidence already on hand. The scheme has many advantages over the previous formulations which supposed that a repeated single quantum event was sufficient to produce the reductant for CO_2 fixation.

On the basis of a single quantum per [H] the minimum quantum requirement would be 4 per CO_2. This was the quantum requirement initially published by Warburg in 1936. However, as discussed above, Warburg's former student, Emerson (1958), had been unable to confirm these results and his measurements indicated the quantum requirement to be 8–10 per CO_2 fixed. This fitted the

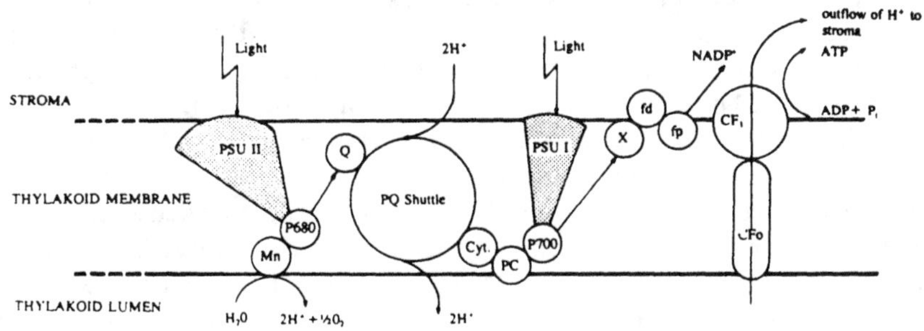

Figure 10. Photosynthetic electron flow from water to $NADP^+$ in a chloroplast thylakoid membrane. Based on ideas in Trebst (1974) and McCarty (1979). CF_o: proton channel. CF_1: coupling factor (ATPase). cyt *f*: cytochrome *f*. PC: plastocyanin. PQ: plastoquinone. Q: primary acceptor of PS II. X: primary acceptor system of PS I. fd: spinach ferredoxin. fp: ferredoxin/$NADP^+$ reductase. P_{700} and P_{680}: reaction center chlorophyll *a* dimers of PS I and PS II, respectively.

Hill and Bendall proposal, whereas Warburg's result could not be accommodated in the Z-scheme.

The observations of the red drop and of enhancement, which were difficult to explain on the basis of one light reaction, were now clearly seen as supportive of the Z-scheme with its two sequential light reactions. Incidentally the Hill reaction itself also shows enhancement. It is interesting that although Emerson earlier discussed the possibility that two light reactions might be involved, he was not then ready to accept this view in print and maintained a complicated explanation based on repetitive single light reactions. The Z-scheme was also seen to be the easiest and most satisfactory explanation of the experiments of Myers and French and of Blinks chromatic transients.

Testing the Z-Scheme

The Hill and Bendall formulation fell upon a biochemical fraternity that was already flirting with the idea of two cooperative light reactions but had not been prepared to abandon the apparently simpler one light reaction accepted up to then. Consequently the new formulation encouraged a large burst of new experiments designed to test and support the Z-scheme. A few of these experiments are noted below.

A Push-Pull Reaction

Duysens (Duysens and Amesz, 1962) carried out a fascinating experiment with *Porphyridium cruentum*, a red alga with phycobilins absorbing in the green at 562 nm as well as chlorophyll *a* absorbing in the red at 680 nm. He observed the effect of illumination at these two wavelengths on the redox state of the cytochrome *f*, which he followed spectroscopically from changes in the absorption at 420 nm. On illuminating with red light the cytochrome *f* became oxidized; if green light was now added a partial reduction of cytochrome *f* was observed and on switching off the green light the redox state was shifted back towards oxidized cytochrome *f*. This was excellent evidence in favor of the Hill and Bendall formulation and was interpreted to mean that red light was absorbed into photoreaction I (P700) and that green light was absorbed via the accessory pigments into photosystem II (P680). Light absorbed by photosystem I then caused the oxidation of the cytochrome *f* situated in the electron transport chain between the two reactions, whereas green light brought about its reduction at the expense of electrons originating in water. Since photoreactions I and II are both operating in this case, an equilibrium position was set up in which the cytochrome *f* is only partially reduced. Cytochrome *f* becomes fully reduced only in the dark when it receives electrons from a reductant previously accumulated in the light.

Electron Paramagnetic Resonance Spectroscopy

On illuminating chloroplasts or algae a light-induced electron paramagnetic resonance (EPR) is observed (reviewed by Weaver, 1968). Analysis of its decay shows two components. Signal I has a g-value of 2.002, which indicates an unpaired electron associated with an organic molecule. It appears very quickly on illumination with white light; its decay on turning off the light is somewhat slower but still fast. This signal is due to oxidized P_{700} (P_{700}^{+}) at the reaction center of photosystem I. Signal II has a g-value of 2.0046, again an unpaired electron associated with an organic molecule. On illumination it appears quite quickly but in the dark it decays very slowly. It is probably due to the production of a semiquinone of plastoquinone by the operation of photosystem II.

The EPR signal also appears on giving far red light (713 nm) but it is smaller on simultaneous illumination at 635 nm. Photosystem I oxidizes P_{700} and photosystem II tends to keep it more reduced by supplying electrons from water. This is good evidence for the operation of an electron transport system between two light reactions.

Functional Separation of Partial Reactions

In Arnon's laboratory (reviewed by Arnon, 1961b) the use of indophenol dyes to react with intermediates of the electron transport chain between the two photoreactions led to the operational separation of photosystems I and II. In the presence of the herbicide DCMU (dichloromethylurea), which effectively inhibits oxygen evolution, the flow of electrons from reduced dichlorophenolindophenol ($DCPIPH_2$) to $NADP^+$, accompanied by ATP formation, can still be observed and cyclic photophosphorylation can also operate. The $DCPIPH_2$ was maintained in the reduced state by ascorbate. Photosystem I drives both these systems and photosystem II is not involved. Only the second part of the electron transport chain is operating, without the intervention of electrons from water via photosystem II. If oxidized DCPIP is supplied (maintained with ferricyanide) in the absence of DCMU, the electron flow is short circuited and only photosystem II operates. Oxygen evolution occurs but ATP formation is not observed.

In other experiments, Mn^{2+} and Cl^- were shown to be required for oxygen evolution (photosystem II) but not to be involved in photosystem I. The removal of plastoquinone from isolated chloroplasts led to the loss of their oxygen-evolving ability, which was restored when plastoquinone was added back. The photoreduction of $NADP^+$ by water was thus prevented in the absence of plastoquinone, but the substitution of DCPIP as electron donor instead of water restored $NADP^+$ photoreduction.

Experiments with Monochromatic Light

A more elegant demonstration of the operation of photosystem I alone or photosystems I plus II together without the use of external donors or inhibitors was first made in Arnon's laboratory with monochromatic light. Using chloroplast fragments illuminated with far red light (704 nm), in which only photosystem I was activated, only cyclic photophosphorylation could occur. With light at shorter wavelengths (e.g. 650 nm), in which photosystems I and II are both activated, noncyclic photophosphorylation leading to $NADP^+$ reduction was readily observed.

Chloroplast Fractionation Experiments

Digitonin treatment of isolated spinach chloroplasts followed by differential centrifugation has in several laboratories led to the separation of two types of chloroplast fragments, one relatively enriched in chlorophyll *b* and the other enriched in chlorophyll *a*. The chlorophyll *a* enriched system is more effective in catalyzing photosystem I reactions. The chlorophyll *b* enriched system is more effective in catalyzing photosystem II reactions. The two types of chloroplast fragments represent the partial physical separation of reaction center complexes (photosynthetic units) responsible for the initial light capture in the two photosystems. The distribution of these two chloroplast fractions within the thylakoid system indicates that photosystem I is mostly located on its own in the stroma lamellae, whereas photosystems I and II are present together in the granal stacks of spinach chloroplasts.

Spillover

Red light absorbed only by photosystem I cannot spill over into photosystem II because the transfer of energy from system I to system II is energetically forbidden (hence the "red drop"). Light of the shorter, more energetic wavelengths absorbed by photosystem II can be spilled over into photosystem I which operates with longer, less energetic wavelengths. The spillover of light from photosystem II to photosystem I appears to be controlled by the state of phosphorylation of a key protein whose degree of phosphorylation is a measure of the ATP/ADP ratio in the chloroplasts. If more ATP is required for the photosynthetic reactions of the Calvin cycle the ratio falls, protein phosphorylation decreases, and spillover into photosystem I is increased.

Thirty Years On

The flood of new experiments, of which those given above are just a few examples, all support the Z-scheme so well that Hill was heard to say "with all this support there must be something wrong with the scheme!" More than 30 years after the publication of the Z-scheme there is little to support his pessimism. But Hill would undoubtedly have been delighted to have seen any definitive experiment that opposed the two light reaction formulation.

THE PRINCIPLE OF PLANETARY VENTILATION REVISITED

With the benefit of the mass of physiological and biochemical work that followed the initial discovery of photosynthesis, perceived by Priestley in 1772 as the "principle of planetary ventilation", it is instructive to examine what are the roles of carbohydrate and the atmosphere in this system.

It is easy to construct a diagram (Figure 11) that shows the throughput of light energy via photosynthesis in the chloroplast into carbohydrates (CH_2O), which can subsequently be used as substrates for respiration in mitochondria either in the same cell, or in another plant tissue after translocation, or in an animal cell after the plant has been eaten. The waste product of photosynthesis, oxygen, and the waste product of respiration, carbon dioxide, are recycled through the atmosphere to make these compounds again available as substrates for respiration and photosynthesis, respectively. What Priestley emphasized was the importance of plants in planetary ventilation (the atmospheric exchange) but he left out the carbohydrates and appeared to have no idea of the energetics involved. He was also unaware of the need for light and so could never have named the process "photosynthesis", a term first used only in 1893 but in common usage in English by 1900.

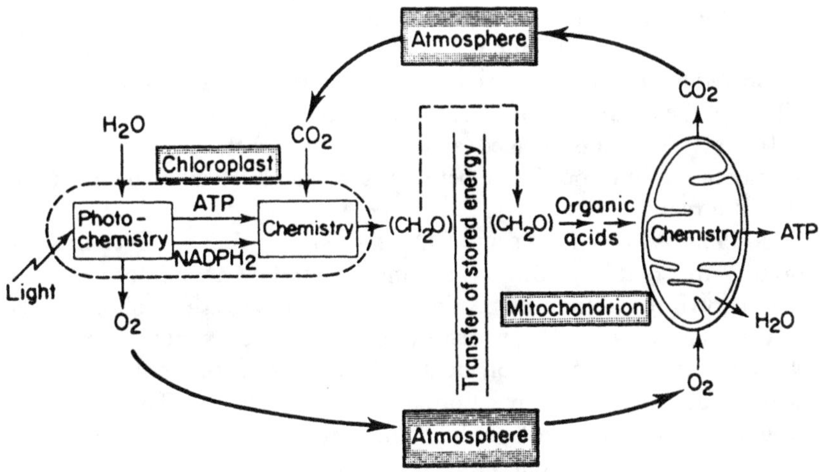

Figure 11. Energy flow and the carbon cycle.

QUANTUM LEAPS IN PHOTOSYNTHESIS

It is perhaps worth drawing attention to the existence of a number of distinct phases in the progress of photosynthesis research.

In the initial exploratory phase (1772–1880) individual workers recognized, in turn, O_2 evolution, the requirement for light and for the green parts of plants, the use of CO_2 and water as substrates, and fundamental conversion of light into chemical energy. This resulted in the accumulation of carbohydrates, and identified the chloroplast as the site of absorption of light and of photosynthesis. These discoveries were largely based on simple physiological experiments on higher plants and involved gas analysis and weighing.

This was followed by a phase (1905–1950) in which the effects of light on whole organisms, usually algal cultures, were analyzed. From these experiments developed the concept of light and dark reactions, the photosynthetic unit, and quantum requirement. These observations depended on elaboration of manometric techniques to measure O_2 and CO_2 exchanges, the accurate measurement of light absorption, the employment of white and monochromatic light, and the techniques of giving repetitive flashes of high intensity followed by variable dark intervals.

The next phase (1945–1954) was the application of $^{14}CO_2$ for shorter and shorter times to algal suspensions and the analysis, by two-dimensional paper chromatography, of the products. This led to the formulation of the Calvin cycle and drew attention to phosphoglycerate as the first product of $^{14}CO_2$ fixation in whole organisms.

The most recent phase was the biochemical approach. It began in the 1930s with van Niel's comparative studies on photosynthetic bacteria and the idea of water splitting as a common reaction in all photosynthetic organisms. This was followed by the isolation of chloroplasts by Hill and a clear demonstration that H_2O is the source of O_2. The thread was temporarily broken by World War II, but the results formed a basis for the great surge of biochemical experimentation that took place between 1950 and 1975 and which continues, though at a reduced rate today.

In the 1950s ATP synthesis and CO_2 fixation were first observed in Arnon's laboratory in isolated chloroplasts using $^{32}P_i$ and $^{14}CO_2$. Chloroplasts were thereby shown to be the unique site of photosynthesis. The photoreduction of $NADP^+$ at the expense of H_2O and the coupled synthesis of ATP in photosynthetic phosphorylation identified the source of the assimilatory power needed to drive the dark reaction of the Calvin cycle. The discovery of photophosphorylation must rank in importance with the identification of oxidative phosphorylation in mitochondria.

Numerous investigators have shown that chloroplasts contain catalytic amounts of proteins, including cytochromes, plastocyanin, ferredoxin–$NADP^+$ reductase, ferredoxin and plastoquinone, all with characteristic redox potentials. These have been arranged in order of their midpoint potentials to form working models of a photosynthetic electron transport chain that could be coupled with ATP synthesis. Kinetic analysis of changes in the redox state on illumination support this in general, but the published schemes are still tentative working models liable to modification.

The two-light reaction proposal by Hill and Bendall was eagerly embraced since it made it easier to interpret much earlier data that had been difficult to accommodate in the supposedly simpler single-light reaction formulation. Many ingenious experiments that supported the two-light reaction proposal were subsequently carried out in several laboratories. There can be no doubt that the concept of two-light reactions acting in sequence was a major step forward.

The mechanism of ATP synthesis in chloroplasts is thought to be similar to that in mitochondria. On the present evidence the Mitchell hypothesis involving proton pumping and return of protons through a $CF_o + CF_1$ ATPase complex seems to be favored. But the evidence does not categorically rule out the Boyer hypothesis (involving conformational changes driven by protons)—perhaps we are talking of different manifestations of the same phenomenon.

The analysis of chloroplasts has depended on many conventional techniques, not least the removal of components from the system and adding them back in purified form to restore lost activity.

Many electron transport intermediates were isolated and purified prior to extensive biochemical characterization; others have only been seen spectroscopically following the development of sophisticated difference

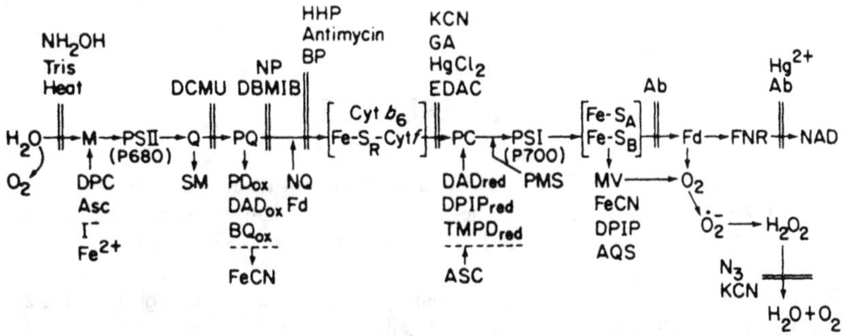

Figure 12. Inhibitors of electron transport and artificial electron donors and electron acceptors. Vertical double lines indicate probable site of inhibitor action; arrows towards the chain indicate site of electron donation; and arrows out of the chin indicate site of electron acceptance. **Ab**, antibody; **AQS**, anthraquinone sulfonate; **Asc**, ascorbate; **BP**, bathophenanthroline (iron-chelator); **BQ**, benzoquinone; **DAD**, diaminodurene (2,3,5,6-tetramethyl-*p*-phenylene diamine); **DBMIB**, 2,5-dibromo-3-methyl-6-isopropyl-*p*-benzoquinone; **DCMU**, 3-(3,4-dichlorophenyl)-1,1-dimethylurea; **DPC**, diphenylcarbazide; **DPIP**, dichlorophenol indophenol; **EDAC**, 1-ethyl-3-(3-dimethylaminopropyl)-carbodiimide; **FeCN**, ferricyanide; **GA**, glutaraldehyde; **HHP**, halogenated hydroxypyridine (cyclic electron transport inhibitor); **MV**, methylviologen; **NP**, nitrophenol; **NQ**, naphthaquinone; **O_2^-**, superoxide; **PMS**, phenazine methosulfate; **SM**, silicomolybdate; **TMPD**, *N*-tetramethyl-*p*-phenylene diamine.

spectrophotometers. Analysis of the separated photosynthetic units for photosystems I and II showed the composition of their bulk antenna pigments by conventional spectrophotometry, but the reaction center chlorophyll *a* dimers, P_{700} and P_{680}—can only be seen by difference methods. Studies using inhibitors and mutants have contributed greatly to the formulation of electron transport chains, as well as studies on alternative electron donors and acceptors, summarized in Figure 12. New techniques now being applied to the outstanding problem, the mechanism of oxygen evolution, include the use of antibodies which *inter alia* can detect sidedness in the photosynthetic membranes, and genetic engineering, whereby individual proteins of particular interest may be modified or rendered inactive. The era of molecular biology in photosynthesis has already begun. Nevertheless biochemical techniques will continue to play a major role, but will become increasingly dependent on the development of highly sophisticated spectrometers and other physical instruments.

ACKNOWLEDGMENTS

I wish to thank my wife, Dr J.M. Whatley, for her significant contributions to the manuscript. Without her sustained help this chapter might never have been written.

The editors have made a number of helpful suggestions which I have been pleased to incorporate.

REFERENCES

Arnon, D.I. (1955). The chloroplast as a complete photosynthetic unit. Science 122, 9-16.

Arnon, D.I. (1961a). Cell-free photosynthesis and the energy conversion process. In: Light and Life (McElroy, W.D. & Glass, B. Eds.), pp. 489-566. The Johns Hopkins Press, Baltimore, MD.

Arnon, D.I. (1961b). Changing concepts of photosynthesis. Bull. Torrey Bot. Club 88, 215-259.

Arnon, D.I. (1967). Photosynthetic activity of isolated chloroplasts. Physiol. Rev. 47, 317-358.

Arnon, D.I. (1984). The discovery of photophosphorylation. Trends in Biochem. Sci. 9, 258-262.

Arnon, D.I. (1988). The discovery of ferredoxin; the photosynthetic path. Trends in Biochem. Sci. 13, 30-33.

Arnon, D.I., Allen, M.B., & Whatley, F.R. (1954). Photosynthesis by isolated chloroplasts. Nature (London) 174, 394-396.

Arnon, D.I., Whatley, F.R., & Allen, M.B. (1958). Assimilatory power in photosynthesis. Science 127, 1026-1034.

Avron, M. & Neumann, J. (1968). Photophosphorylation in chloroplasts. Ann. Rev. Plant Physiol. 19, 137-166.

Bassham, J.A., Benson, A.A., Kay, L.D., Harris, A.Z., Wilson, A.T., & Calvin, M. (1954). The path of carbon in photosynthesis. XXI. The cyclic regeneration of carbon dioxide acceptor. J. Am. Chem. Soc. 76, 1760-1770.

Blinks, L.R. (1954). The photosynthetic function of pigments other than chlorophyll. Ann. Rev. Plant Physiol. 5, 93-114.

Blinks, L.R. (1957). Chromatic transients in photosynthesis of red algae. In: Research in Photosynthesis. Proc. Gatlinburg Conf. 1955 (Gaffron, H. et al., Eds.), pp. 444-449. Wiley (Interscience), New York.

Davenport, H.E. (1963). Pathway of reduction of metmyoglobin and nicotinamide adenine dinucleotide phosphate by illuminated chloroplasts. Proc. R. Soc. B. 157, 332-345.

Duysens, L.N.M. & Amesz, J. (1962). Function and identification of two photochemical systems in photosynthesis. Biochim. Biophys. Acta 64, 243-260.

Emerson, R. (1958). The quantum yield of photosynthesis. Ann. Rev. Plant Physiol. 9, 1-24.

Fischer, H. (1940). Fortschritte der Chlorophyllenchemie. Naturwissenschaften 28, 401-405.

Frenkel, A.W. (1954). Light induced phosphorylation by cell-free preparations of photosynthetic bacteria. J. Am. Chem. Soc. 76, 5568-5569.

Gunning, B.E.S. & Steer, M.W. (1975). In: The Ultrastructure of Plant Cells. Edward Arnold, London.

Halliwell, B. (1981). Chloroplast Metabolism, Chapter 1, especially pp. 13-21. Clarendon Press, Oxford.

Hill, R. (1939). Oxygen produced by isolated chloroplasts. Proc. R. Soc. B. 127, 192-210.

Hill, R. & Bendall, F. (1960). Function of the two cytochrome components in chloroplasts: a working hypothesis. Nature (London) 186, 136-137.

Jensen, R.G. & Bassham, J.A. (1966). Photosynthesis by isolated chloroplasts. Proc. Natl. Acad. Sci. USA 56, 1095-1101.

McCarty, R.E. (1979). Roles of a coupling factor for photophosphorylation in chloroplasts. Ann. Rev. Plant Physiol. 30, 79-104.

Mitchell, P. (1966). Chemiosmotic coupling in oxidative and photosynthetic phosphorylation. Biol. Rev. Cambridge Phil. Soc. 41, 445-502.

Mortensen, L.E., Valentine, R.C., & Carnahan, J.E. (1962). An electron transport factor from *Clostridium pasteurianum*. Biochem. Biophys. Res. Commun. 7, 448-452.

Myers, J. & French, C.S. (1960). Relationships between time course, chromatic transient and enhancement phenomena of photosynthesis. Plant Physiol. 35, 963-969.

Rabinowitch, E.I. (1955). Photosynthesis and Related Processes, Vol. I, pp. 12-28. Interscience, New York.

Ruben, J. (1943). Photosynthesis and phosphorylation. J. Am. Chem. Soc. 65, 279-282.

San Pietro, A. & Lang, H.M. (1956). Accumulation of reduced pyridine nucleotides by illuminated grana. Science 124, 118-119.

Tagawa, K. & Arnon, D.I. (1962). Ferredoxins as electron carriers in photosynthesis and in the biological production and consumption of hydrogen gas. Nature (London) 195, 537-543.

Tagawa, K., Tsujimoto, H.Y., & Arnon, D.I. (1963). Role of chloroplast ferredoxin in the energy conversion process of photosynthesis. Proc. Natl. Acad. Sci. USA 49, 567-572.

Trebst, A.V. (1974). Energy conservation in photosynthetic electron transport of chloroplasts. Ann. Rev. Plant Physiol. 25, 423-458.

Trebst, A.V., Tsujimoto, H.Y., & Arnon, D.I. (1958). Separation of light and dark phases in the photosynthesis of isolated chloroplasts. Nature (London) 182, 351-355.

Van Niel, C.B. (1941). The bacterial photosyntheses and their importance for the general problem of photosynthesis. Advan. Enzymol. 1, 263-328.

Vishniac, W. (1955). Biochemical aspects of photosynthesis. Ann. Rev. Plant Physiol. 6, 115-134.

Walker, D.A. (1971). Photosynthesis Part A. [18]. Chloroplasts (and grana); aqueous (including high carbon dioxide ability). Methods Enzymol. 23A, 211-220.

Weaver, E.C. (1968). EPR studies of free radicals in photosynthetic systems. Ann. Rev. Plant Physiol. 19, 283-294.

West, K.R. & Wiskich, J.T. (1968). Photosynthetic control by isolated pea chloroplasts. Biochem. J. 109, 527-532.

Whatley, F.R., Allen, M.B., Rosenberg, L.L., Capindale, J.B., & Arnon, D.I. (1956). Photosynthesis by isolated chloroplasts. V. Phosphorylation and carbon dioxide fixation by broken chloroplasts. Biochim. Biophys. Acta 20, 462-468.

Whatley, F.R. & Arnon, D.I. (1963). Photosynthetic phosphorylation in isolated chloroplasts. Methods Enzymol. 6, 308-313.

Whatley, F.R., Tagawa, K., & Arnon, D.I. (1963). Separation of light and dark reactions in electron transfer during photosynthesis. Proc. Natl. Acad. Sci. USA 49, 266-270.

Woodward, R.B. et al. (1960). The total synthesis of chlorophyll. J. Am. Chem. Soc. 82, 3800-3802.

Woodward, R.B. et al. (1990). The total synthesis of chlorophyll *a*. Tetrahedron 46 (22), 7599-7659.

Chapter 4

MUSCLE CONTRACTION AND RELAXATION

S. V. Perry

EARLY WORK LEADING TO THE DEFINITION OF THE CONTRACTILE SYSTEM IN PROTEIN TERMS

The early biochemical studies of striated muscle were principally directed to understanding the reactions involved in the production of lactic acid which Fletcher and Hopkins[1] had shown in 1907 to accompany contractile activity. Indeed the two processes were thought to be so closely associated that there were early theories which gave lactic acid a role in the contractile process itself. Up to the 1920s it was widely believed that the primary reaction in muscle contraction was the release of lactic acid from a hypothetical large precursor molecule referred to as "lactacidogen". It was postulated that the hydrogen ions produced neutralized negative charges on the contractile protein filaments, allowing them to fold and shorten. In 1930 the independence of contraction from lactic acid production was clearly demonstrated by Lundsgaard who made the discovery that contraction could take place in iodoacetate-poisoned muscle in which glycolysis is inhibited. Indeed in these conditions a slight increase in alkalinity occurred due to the breakdown of phosphocreatine that accompanied contraction. It is of interest that this important discovery, which dramatically changed the dogma of the time, was the accidental by-product of an investigation designed to study the specific dynamic action of proteins (see Huxley[7], 1986).

The structure of the muscle cell had been well-defined by the nineteenth century histologists. They recognized that the intracellular components occupying most of the cytoplasm, the myofibrils, were unique to muscle and were probably involved in the contractile process. During contraction the band pattern of the myofibrils changed but there was much controversy as to the nature of the changes. There was evidence that the glycolytic activity and other energy-yielding systems were located in different regions of the cell from those containing the contractile apparatus. For example the enzymes of glycolysis were readily extracted by low ionic strength buffers, indicating their localization in the soluble cytoplasm, whereas extraction at higher ionic strengths was required to solubilize the myofibrils.

A number of early workers investigated the structural proteins of muscle. The name myosin was originally given by Kuhne (1859) to the substance in muscle press juice, consisting largely of the soluble cytoplasm, which on standing at room temperature set to a gel. Because it was confused with a protein having the properties of a true globulin, the name was later reserved for the protein which we now know as myosin. This protein, which is the major component of striated muscle, is more abundant in extracts made at a higher ionic strength than exists in the simple press juice. In 1881 Schipiloff and Danilevsky reported that the extraction of myosin resulted in a loss of birefringence of the myofibrils. The properties of isolated myosin were defined in the detailed studies of Weber[4] (1925), Edsall (1930), and von Muralt and Edsall (1930a,b). As a result of their studies of the birefringence of myosin solutions the latter two investigators concluded that this protein was responsible for the anisotropy of the A-band (Figure 1)—a view confirmed when convincing evidence that this protein was the major component of the A-band of the myofibril was provided by Weber (1934).

By 1939 myosin had been accepted as the contractile substance and the discovery in that year by Engelhardt[1] and Ljubimova that the ATPase activity of minced muscle was associated with the myosin isolated from it, was a major advance. Its importance was clear from the fact that ATP, considered to be the source of energy for muscle activity, was hydrolyzed by a protein forming part of the contractile apparatus. Also, it was the first report of a very asymmetric structural protein of high molecular weight that possessed enzymic activity. Up to that time all the enzymes that had been purified and characterized had been shown to be protein molecules of relatively small size

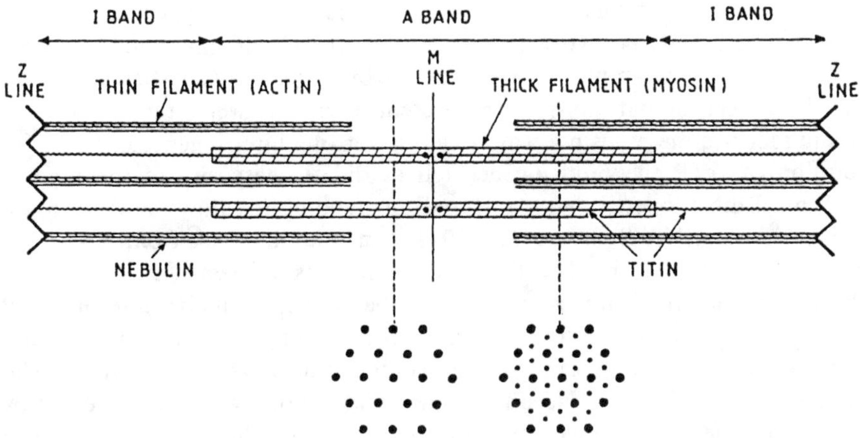

Figure 1. Diagramatic representation of the myofibril of vertebrate skeletal muscle indicating the location of the proteins and illustating the arrangement of the filaments when seen in cross section (from Perry, 1996).

and globular in shape. Conditioned by this background many attempts were made to show that the enzymic activity of myosin was due to a contamination of the structural protein with an ATPase enzyme of smaller size. This led to the publication of at least one claim of separation of the enzyme from myosin, a claim that later was shown to be false and by the late 1940s it was clear that myosin was indeed an enzyme.

Definition of the contractile system took a second major step forward as a result of observations made by Szent-Gyorgyi[1] and collaborators in 1941–1942 working in occupied Hungary (Szent-Gyorgyi, 1941, 1942, 1945). Up to this time the standard method of myosin preparation was to extract minced muscle with solutions at ionic strengths of about 0.6 and purify the protein by taking advantage of its globulin properties, i.e. by repeated precipitation at low ionic strength and redissolving at higher ionic strengths. This procedure produced what was thought to be a relatively pure preparation of myosin. Szent-Gyorgyi noticed that if the extraction of the muscle mince was prolonged, overnight for example, the extract became extremely viscous. The viscosity of this extract could be dramatically reduced by low concentrations of ATP but subsequently returned to the original value with time as the ATP was hydrolyzed by the myosin.

Szent-Gyorgyi's group were able show that on prolonged extraction a new, protein, actin, passed into solution with myosin to form a viscous complex which they called actomyosin. The implication of this finding was that all preparations of myosin that had been studied previously were probably actomyosin preparations containing varying amounts of actin, the presence of which would modify their enzymic and physicochemical properties.

While the Hungarian work was in progress the Needhams group, working in Cambridge, U.K., reported that the flow birefringent properties of myosin varied between preparations of the protein. If rabbits were starved for 24 h before sacrifice the myosin always exhibited strong birefringence of flow. On addition of ATP the birefringence fell as did the viscosity of the myosin solution. Both properties were restored on standing to allow the ATP to be broken down (Needham et al., 1941). In the light of Szent-Gyorgyi's independent results the Needhams' myosin was in fact an actomyosin. Presumably actomyosin was extracted in their experiments because the ATP level in the muscle extract was low due to the lack of glycogen in muscle from starved animals. Actomyosin is more readily extracted from the myofibrils in the absence of ATP. The Needhams and collaborators were so impressed by the physicochemical changes resulting from the addition of ATP to myosin that they put forward the possibility that contraction was "essentially an enzyme substrate combination" (Needham et al., 1942). An idea well ahead of its time.

By the application of a novel preparative procedure involving selective denaturation of the myosin by organic solvents, Straub (1942) developed an excellent method for preparing pure actin. This method has stood the test of time and is the basis of the methods used today to prepare actin from muscle. Actomyosin prepared from the individually isolated component proteins forms a viscous solution at high ionic strength which is dissociated by low concentrations of ATP. The significance of this finding was that for the first time ATP, an important metabolite for the production of which much of the cell chemistry is directed, had been shown to produce a marked change of state in a complex of two structural proteins which were the major components of the muscle cell. This finding also hinted that the nucleotide might have a dual role. Thus, in addition to being a substrate for the contractile protein system, ATP also had the capacity to dissociate the protein complex responsible for its hydrolysis. Clearly these properties were of significance for the contractile event, but precisely how was not clear at that time.

ATP AND CONTRACTION

Much more pertinent was the striking demonstration that if the actomyosin was precipated by lowering the ionic strength and formed into a thread, the addition of ATP induced contraction. If the thread was formed with actomyosin filaments oriented longitudinally, it became shorter and thicker on contraction as is the case with muscle itself, thus providing a true model (Figure 2). An even more satisfactory model was obtained by immersing whole

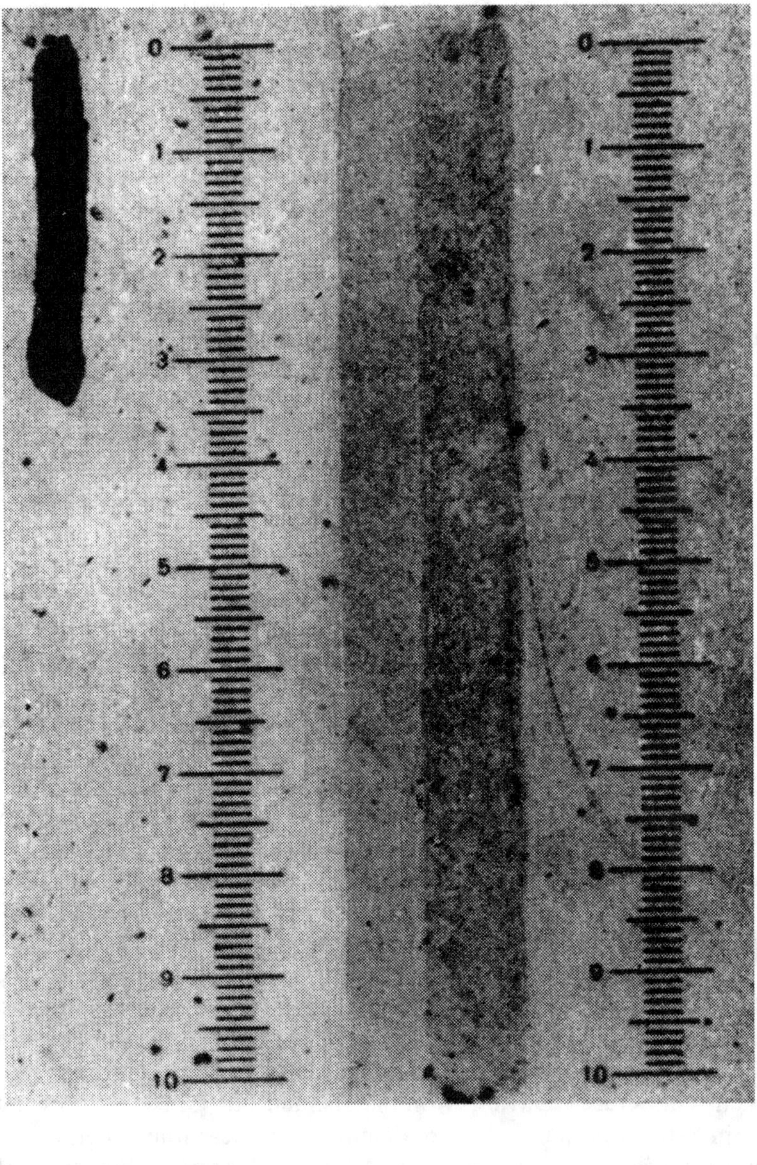

Figure 2. The first illustration in the literature of the contraction of an actomyosin thread. A thread of precipated actomyosin, then known as "myosin B" (shown on the right), was immersed in a fresh aqueous extract of muscle. Within seconds the transparent thread synereses and turns white and opaque as it shortens due to the action of ATP in the extract (shown on the left) (from Szent-Gyorgyi, 1941-1942).

muscle strips in 50% glycerol at subzero temperatures for a few days to produce so-called glycerated fibers (Szent-Gyorgyi, 1949). This treatment resulted in destruction of the membrane and extraction of much of the soluble protein from the cells, leaving the myofibrillar skeleton of the muscle cell intact. Thin filaments of this preparation can be readily teased out and mounted for physiological studies since they respond to ATP by developing tension comparable to that obtained with intact muscle

It became clear that actin and myosin were the only proteins required to produce a system which in the correct ionic conditions, exhibited a model contractile response on addition of low concentrations of ATP. If the ATPase activity was inhibited, contraction of the systems *in vitro* did not occur implying that contraction in the model systems required the hydrolysis of ATP. Lundsgaard (1930, 1934) had shown that in iodoacetate-treated muscles the breakdown of phosphocreatine accompanied contraction and appeared to be the source of energy for the process. He demonstrated later, however, that ADP was the cofactor for phosphocreatine breakdown (by the so-called Lohmann reaction). It was therefore concluded that the ADP produced was rapidly converted to ATP which was the direct source of the energy for contraction under these conditions:

$$\text{Phosphocreatine} + \text{ADP} = \text{Creatine} + \text{ATP}$$

Nevertheless, despite Lundsgaard's findings and the overwhelming evidence for ATP as the substrate for contraction with the isolated protein systems, A.V. Hill (1949) challenged biochemists to demonstrate that changes in the ATP concentration actually occurred in living muscle during a single isolated contraction. Such a demonstration with intact muscle is difficult for a number of reasons. The ATP hydrolyzed during a single twitch is a small fraction of the total ATP concentration and the associated biochemical changes are very fast, occurring in milliseconds. Also, skeletal muscle contains an active creatine phosphokinase with creatine phosphate present in much higher concentrations than the ADP produced in a single contraction. This ensures that it is rapidly converted back to ATP. Satisfactory uncontracted control muscles are difficult to obtain because the rapid chilling of muscles to fix the metabolite levels before contraction can itself produce a spontaneous contraction, unless special precautions are taken.

The challenge was taken up in the 1950s by a number of groups including Fleckenstein, Krebs[1], Davies, Mommaerts, and others. Nevertheless none of these investigators were able to demonstrate measurable net changes of energetically equivalent quantities of ATP, ADP, or phosphocreatine during single contractions of frog rectus abdominus and turtle leg muscles. Turtle muscles were used for some of these investigations in the hope that their slower response times would facilitate analysis of the biochemical changes. Inorganic phosphate was shown to be liberated during a single muscle twitch, but despite testing a wide range of phosphorus compounds none were found to undergo a change that would account for the amounts produced in these experiments.

The situation was finally resolved by Cain et al. (1962) who used the fact reported by Kuby and Mahowald (1959) that crystalline creatine phosphokinase was completely inhibited by 1-fluoro-2,4-dinitrobenzene (FDNB, the Sanger[2] reagent). When frog rectus abdominus was treated with this reagent it still contained the normal amounts of phosphocreatine but would only carry out about three contractions rather than the normal of 30 or more obtained with the normal unpoisoned muscle. With these preparations the only source of ATP was the endogenous content . It was shown that during a single twitch the nucleotide was broken down with a linear relationship between the work done and the amount hydrolyzed. The earlier failures to demonstrate breakdown of the nucleotide had been due to its extremely rapid resynthesis from the ADP produced in a single twitch by the combined action of creatine phosphokinase and adenylate kinase. These enzymes are present in relatively large amounts in normal skeletal muscle. Also the high levels of phosphocreatine in skeletal muscle make the small changes that occur after a single twitch difficult to determine with any precision.

Once the contractile system had been defined in terms of the proteins involved and the substrate used for the mechanochemical process, two main lines of approach to the biochemical study of muscle emerged. The first of these was directed to the understanding of the molecular mechanism of contraction, whereas the other was concerned with the the mode of its regulation. The model systems made from the purified actin and myosin contracted in the presence of ATP but an active relaxation process, which is an essential feature of muscle function, could not be demonstrated with them. With some model contractile systems which corresponded more closely to the situation in the cell, such as the glycerated fiber, and which had not been exhaustively washed, it was possible to demonstrate both contractile and relaxation responses. This suggested that there were components in addition to actin and myosin that had a role in regulating the contractile response.

THE CONTRACTILE SYSTEM AND THE MECHANOCHEMICAL PROCESS

Much of the effort of the 1950s was directed towards determining the structures of myosin and actin and relating them to the myofibril. The histological evidence which had suggested that the myofibrils were the site of the contractile process was confirmed and extended when it was shown that myofibrils could be isolated free of other intracellular components in quantities that could be used for biochemical study (Schick and Hass, 1949; Perry, 1951). These preparations contained the actin and myosin of the muscle, and in the presence of ATP contracted exhibiting band changes identical with those obtained in intact muscle. It was apparent that there were proteins present in the myofibril in addition to actin and myosin, but at this period research into mechanism of the contractile process concentrated on the latter two proteins.

Actin

In his original preparation Straub (1942) extracted acetone-dried muscle with distilled water to give a preparation which he called “inactive actin”. It was given this name because on its addition to myosin the viscosity of the complex did not change significantly. Further, it was not effected by the addition of ATP as was the case with actomyosin obtained by long-term extraction of fresh muscle. By the addition of salts, “inactive actin” was converted to “active actin” which formed the viscous complex of actomyosin with the normal properties (Straub, 1943). It was concluded that the so-called “inactive actin” was the monomeric form of the protein now known as G-(globular) actin which in the presence of divalent cations polymerizes to F-(fibrous) actin which has the property of forming the viscous complex with myosin.

The report by Straub and Feuer (1950) that G-actin preparations contained ATP, known as bound nucleotide, which was converted to ADP on polymerization, stimulated suggestions that a cycle of actin polymerization–depolymerization might be involved in the contractile process. It is now known that this is not the case, but the original observation that actin polymerization was accompanied by ATP hydrolysis was at the time a unique observation and even today its function has not been satisfactorily explained. By the late 1940s when electron microscopy was beginning to be used to examine isolated proteins it was pointed out that the filaments of F-actin were remarkably like the I-filaments observed in longitudinal sections of muscle. It became clear that isolated F-actin filaments seen in the electron microcope (EM) were indistinguishable from the thin filaments of both striated and smooth muscle (Hanson[6] and Lowy, 1963). The thin filaments of muscle were established as a double-helical array of actin monomers with dimensions and a periodicity which are very similar to those of actin filaments prepared from the isolated protein.

Because of its importance in the contractile process and in view of its size and globular shape, G-actin seemed to be a promising candidate for crystallization and determination of its three-dimensional study by X-ray analysis. Its tendency to polymerize into filaments which form paracrystals prevents normal crystallization which requires that it is held in the G form. Success in producing satisfactory crystals arose from the entirely unconnected observation that actin is a naturally occuring inhibitor of deoxyribonuclease 1 (Lazarides and Lindberg, 1974). G-actin forms a 1:1 complex with the the enzyme that can be crystallized. The atomic structure of deoxyribonuclease 1 was determined in 1984 and that of the complex with G-actin by Kabsch et al. (1990). A similar structure has also been obtained from its complex with profilin.

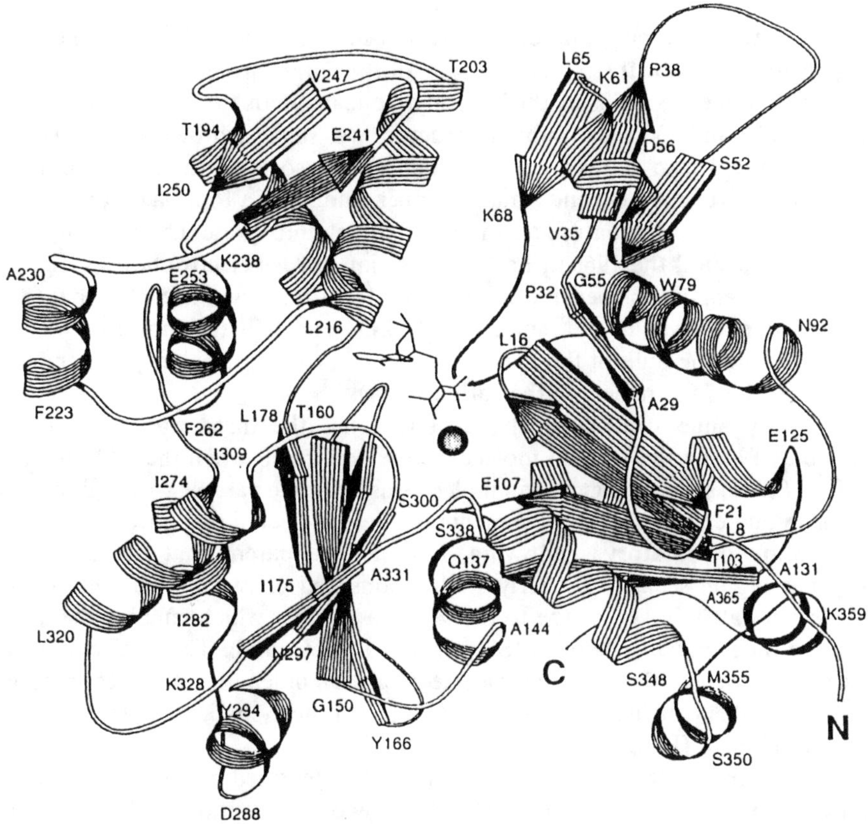

Figure 3. The structure of G-actin as determined by X-ray crystallography of the actin-DNAse complex (Kabsch et al., 1990). The ADP molecule and the calcium ion (hatched circle) are shown at the base of the cleft in the molecule.

The G-actin molecule consists of two major domains each of which can be divided into two subdomains (Figure 3). The bound nucleotide, ATP in G-actin, and the divalent cation associated with the molecule are both located in the cleft between the two major domains. An unexpected result, which was not suspected from the amino acid sequence, is a remarkable structural similarity between the four subdomains of actin and those of the N-terminal ATPase fragment of the 70-kDa bovine heat shock protein (HSC 70).

Myosin

Due to the asymmetric nature of the myosin molecule and its tendency to aggregate, values for the molecular mass of the native protein obtained in the late 1940s and early 1950s by the hydrodynamic methods then available varied between 800 and 400 kDa. Subunit structure was suggested by the early work of Tsao (1953) who showed concentrated urea dissociated the molecule into large subunits of 165 kDa and a much smaller amount of a 16-kDa polypeptide. Controversy as to the nature and number of subunits present in myosin was not resolved until the 1960s when the technique of gel electrophoresis under dissociating conditions became avalable. By this time it was accepted that the molecular mass was in the region of 500 kDa and that the molecule was a hexamer with the bulk of the mass made up of two major polypeptide chains (see Perry 1967 for review of the earlier work).

Hydrodynamic studies which suggested that the mass was not evenly distributed along the myosin molecule were confirmed when the EM studies of Rice (1961) showed the myosin molecule as a rodlike particle with a globular mass at one end. With improvements in the EM technique, particularly by application of rotatory shadowing, it became apparent that the terminal globular region seen in the early studies consisted of two separate globular heads separately linked to the to the tail in a manner which enabled them to take up independent positions (Slayter and Lowey 1967). Thus has emerged the current view of the myosin molecule with a tail of length 150 nm consisting of two parallel α-helices in a coiled coil and two pear-shaped heads of about 17 nm in length (Figure 4).

The low molecular weight fraction originally identified by Tsao (1953) was shown by electrophoresis to consist of two types of polypeptides with molecular masses in the region of 15 to 25 kDa, known as the light chains, to distinguish them from the large polypeptides that make up the bulk of the molecule,the heavy chains (Gershman et al., 1969; Weeds and Lowey, 1971). The light chain more readily extracted from native myosin was originally called the DTNB (dithiodinitrobenzoic acid) light chain because of its extraction by treatment with the sulphydryl reagent, DTNB. This polypeptide of about 18 kDa molecular mass is now usually known as the regulatory phosphorylatable, or P light chain. The latter name applies in vertebrate muscle where it contains an N-terminal site that is phosphorylated by a specific enzyme, myosin light-chain kinase. The other type of light chain with a variable molecular weight

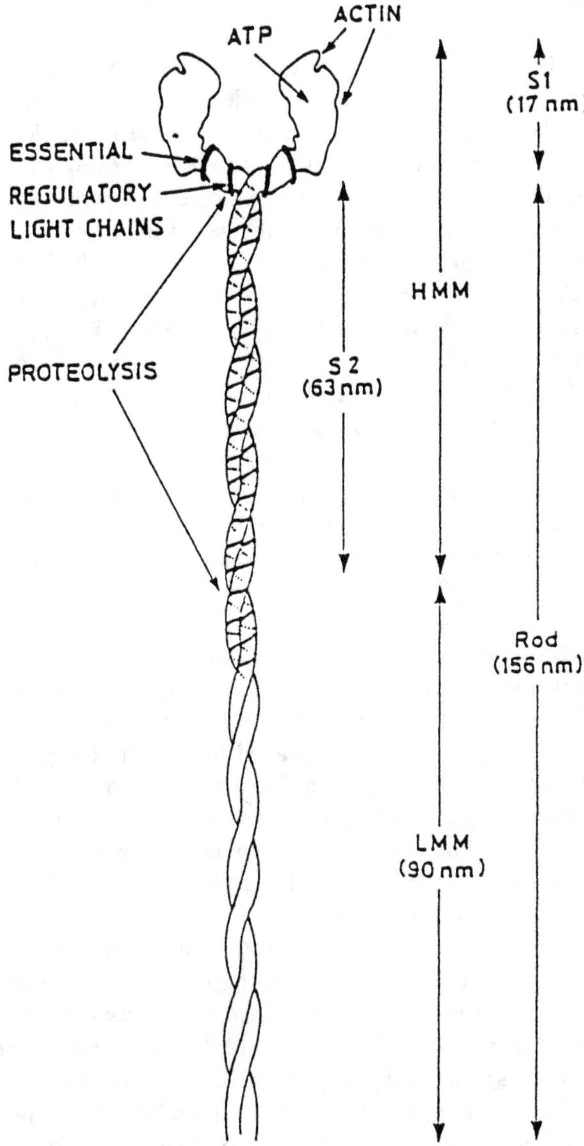

Figure 4. Schematic representation of the myosin molecule indicating the helical coiled tail and the fragments obtained by controlled proteolysis. In vertebrate muscle the regulatory light chain has a site near the N-terminus that is phosphorylated. The regions on the myosin near where ATP and actin are presumed to bind during the contractile cycle are approximately indicated. S1, subfragment 1; S2, subfragment 2; LMM, light meromyosin; HMM, heavy meromyosin.

is known as the alkali or essential light chain and requires more extreme conditions for removal from the heavy chains.

As details of the size and shape of the myosin molecule have emerged, a variety of techniques have been used to determine the structure–function relationships. Selective proteolytic cleavage of the native molecule has been particularly valuable in this respect and provided early examples of biologically active fragments derived from native protein molecules (Figure 4). Such studies have identified a region in the myosin tail susceptible to cleavage by trypsin and other enzymes to produce light meromyosin, an α-helical fragment of about 120 kDa and heavy meromyosin of 350 kDa representing the remainder of the tail and the head of the molecule (Szent-Gyorgyi, 1953). The former fragment possesses the globulin-like solubility properties, whereas the latter was soluble at low ionic strength, interacts with actin, and retains the ATPase activity. The region susceptible to proteolysis in the coiled coil region of the myosin tail is often referred to as the "hinge" to indicate a flexible region which has been postulated by some investigators to have a role in the contractile process. Heavy meromyosin can be further cleaved by proteolytic enzymes to split off the head, subfragment 1, and the remainder of the tail, subfragment 2 (Mueller and Perry, 1962).

Subfragment I with molecular mass about 130 kDa contains one light chain of each type and corresponds to one of the head regions of the myosin molecule. It forms a complex with actin and possesses ATPase activity. Unlike the intact protein it is soluble and has been much used for enzymic studies on the actomyosin system. It can be further cleaved by trypsin into fragments of 50, 23, and 20 kDa. By the use of cross-linking reagents and other techniques the relation of these regions of the myosin head to the enzymic and actin interaction sites can be defined. These smaller fragments do not exhibit the major functional properties of the myosin head, but their study has provided information about location of the functional sites on the head.

Subfragment 1 is clearly that region of the myosin molecule where the protein motor is located and to understand the mechanochemical process responsible for contraction its structure needs to be defined at high resolution. Unlike intact myosin which does not form crystals suitable for X-ray crystallography, subfragment 1 with its more compact globular shape has much greater potential in this respect. Many attempts were made to crystallize the native protein but success was only achieved after reductive methylation, a procedure which modifies the lysine residues. The availability of these crystals has enabled for the first time the structure of a motor protein to be determined at a resolution of 2.8 Å resolution (Rayment et al., 1993a). With this structure available, and the mass of biochemical information about the involvement of specified residues in the biological activity of the myosin head, it is possible to identify in three dimensions the sites where ATP and actin are bound.

Despite the enormous step forward provided by the description of the motor protein in atomic terms it is still not possible to describe in precise molecular terms the mechanochemical process leading to contraction. Nevertheless the detailed structural information we now have about the myosin head and the actin monomer provides a sound base from which theories can be developed to explain how relative movement between filaments composed of the two proteins can occur (see Rayment et al., 1993b).

There is little doubt that the contractile process involves conformational changes in the myosin molecule. The nature of these has yet to be defined with precision despite the application of a wide range of techniques both *in vitro* and *in vivo*. The structure obtained from the crystals is that adopted by the reductively methylated subfragment 1. How this relates to a stage in the cycle of events at the myosin head during contraction, if at all, has yet to be ascertained. Nevertheless, with a structure of actin and the myosin head now available the prospects of investigating structure–function relationships by site-directed mutagenesis are good. It surely will not be long before the mode of action of the myosin motor will be understood in atomic terms.

Myofibrillar Structure and Function

In parallel with the studies on the isolated proteins of the contractile system attempts were made to define the structural changes that occurred in the myofibril during contraction. The banded appearance of the myofibril at rest had been long recognized but until the 1950s there was little agreement as to the changes in striations that occurred during contraction of living muscle. It was difficult to decide whether the changes in striation were a consequence of activation, of tension development or, of shortening. Also the ordinary light microscope cannot provide a satisfactory image of unstained striations of high and low refractive index with a repeat distance of 2–3 μm in a single fiber, diameter 50–100μ. By using the interference microscope, A.F. Huxley and Niedergerke (1954) avoided this problem and were able to show that over a wide range of sarcomere lengths the A-band width was unchanged and that the shortening was accounted for by the decrease in length of the I-band. Similar results were obtained *in vitro* by using ATP to induce contraction in isolated myofibrils (Hanson and H.E. Huxley[5], 1954). By this time it had become clear from EM studies (Draper and Hodge, 1949; Huxley, 1953) that the myofibril was composed of two main types of filaments: the thick, or A-filaments, and the thin, or I-filaments. The A-filaments were confined to the A-band and the results of selective extraction procedures implied that they consisted of myosin. The I-filaments are located in the I-band and extend into the A-band interdigitating between the A-filaments up to the H-space. For reasons similar to those used to locate the myosin it was clear that actin was the major component of the thin filaments.

It was against this background that both groups of workers proposed independently that the band changes seen in the contracting and relaxing myofibril could be best explained by the sliding-filament hypothesis. This hypothesis postulates that the length of the filaments remains constant but that the overlapping arrays slide past each other as the sarcomere length changes. Shortening occurs by the actin filaments being drawn further into the myosin array and withdrawn when the muscle relaxes or is stretched.

Development of the Concept of the Cross Bridge

To explain how movement between the actin and myosin filaments arrays occurs and leads to force development it is necessary to postulate some type of active interaction between the two protein systems. The clue to the structural nature of this interaction came from early EM observations (Huxley, 1957) of longitudinal sections of muscle. These showed that the two filaments, which are separated by a gap of about 13 nm, are linked at regular intervals by projections which were called cross bridges. These projections are a feature of the A-filament but are not present in the H-zone of resting muscle. They are present in regions of the A-band occupied by actin filaments in resting muscle.

The origin of the cross bridges in the myosin molecule was confirmed when it was shown that synthetic filaments made from purified myosin closely resembled the thick filaments of the myofibril with projections similarly distributed (Huxley, 1963). It soon became clear that the cross bridges corresponded to the head of the myosin molecule where its enzymic and actin binding activities were located. In the early studies little detail could be distinguished in the cross bridge. With improvement in techniques and using muscles such as insect flight muscle, which gave particularly good images in the EM, both heads of the myosin could be recognized, particularly in cross section, giving a "flaired" appearance to the cross bridges (Figure 5). It was noted that often under rigor conditions the cross bridges were aligned at an angle of about 45° to the filament axis, whereas in other conditions they appeared in the EM studies to be much less well orientated. From these observations the concept arose that when attached to the actin the cross bridge rotated through 45° from the 90° position and thus caused relative movement of the two filaments (Figure 6).

Kinetic Studies on the Actomyosin ATPase

The earlier studies *in vitro* on the interaction of actin and myosin provided the background from which ideas about the nature of events during the cross-bridge cycle were developed. In the first place, as has been pointed out earlier, the proteins can interact and form a strong link that is broken by ATP, the

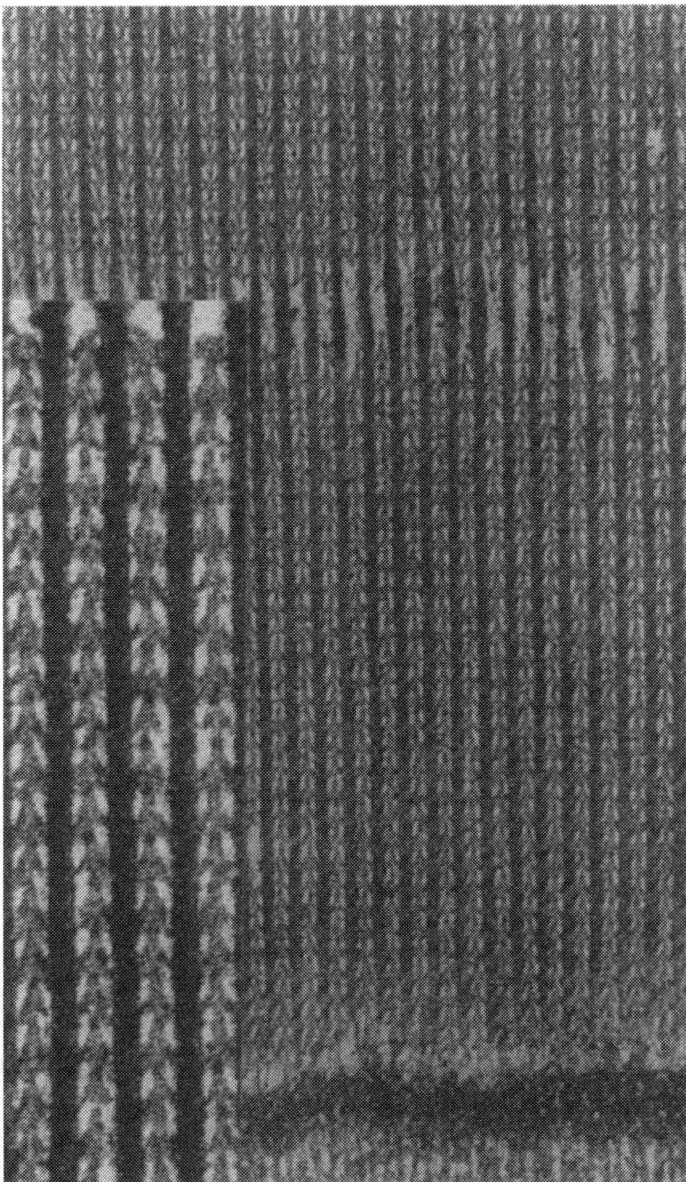

Figure 5. Electron micrograph of insect flight muscle in rigor showing the cross bridges. Section through sarcomere shows Z-line at the bottom with the I-band (much smaller in insect flight than in vertebrate skeletal muscle) and the H-space (magnification x ~75,000). Inset shows detail in the A-band (magnification x ~150,000). Kindly provided by Mary and Michael Reedy.

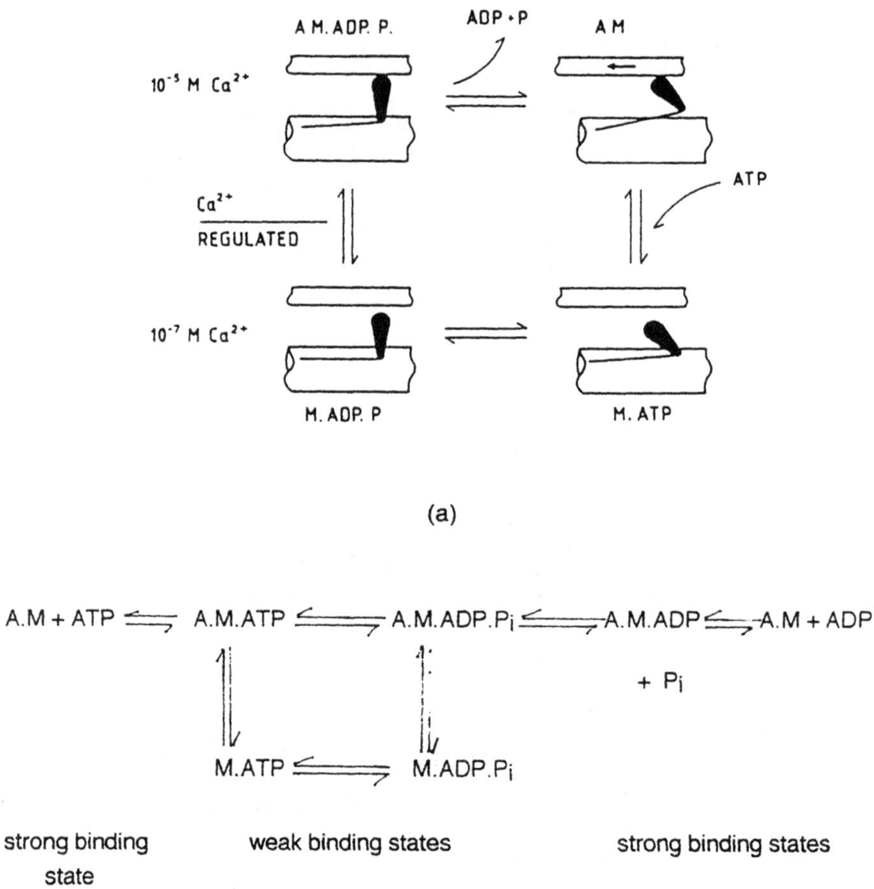

Figure 6. (**a**) The cross-bridge cycle as proposed by Lymn and Taylor (1971). (**b**) Simplified scheme of kinetic events occurring during the hydrolysis of ATP.

substrate used to provide the energy for contraction. Virtually all the ATP in muscle is present as the magnesium complex, which is a very poor substrate for the myosin MgATPase in the absence of actin with a rate constant of less than 0.1 s^{-1}. The latter rate corresponds to the situation in resting muscle. In the presence of actin the rate of MgATP hydrolysis by myosin increases by at least 2 orders of magnitude to produce a rate similar to that found in contracting muscle.

To be able to describe the cross-bridge cycle in detail it was necessary to understand the kinetic properties of the actin-activated ATPase of myosin. One of the problems in investigating this enzyme system under ionic conditions corresponding to those existing in intact muscle (0.1–0.15 M KCl, 1 mM $MgCl_2$) is that the ATPase activity of reconstituted actomyosin is very low. Under these conditions actomyosin exists as a precipitate and addition of ATP causes first a clearing phase, presumably due to dissociation of the actomyosin, followed by precipation, and then by syneresis of the hydrated gel when the filamentous structure contracts, so-called "superprecipitation". Although superprecipation provides a valuable model system *in vitro*, because of its heterogenous nature it does not lend itself to kinetic analysis. For this reason much of the detailed kinetic analysis of the hydrolysis of ATP has been carried out on the enzymically active fragments of myosin, heavy meromyosin, and subfragment 1, usually at ionic strengths much lower that that occuring in the muscle cell.

Early studies (Eisenberg and Moos, 1968; Eisenberg et al., 1968; see Eisenberg and Greene, 1980) were interpreted as indicating that the increase in MgATPase rate obtained with the acto-heavy meromyosin complex was due to activation of the hydrolytic step by actin. A major step forward was the finding that the rate of hydrolysis of the first molecule of ATP was much faster than the maximum steady-state rate obtained with the system (Tokiwa and Tonomura, 1965; Lymn and Taylor, 1971). From the transient state kinetics it was concluded that the effect of actin was not on the hydrolytic step but on the release of products from the myosin active site. This latter reaction is the rate-limiting step in the hydrolysis of ATP by myosin.

In light of their kinetic studies, Lymn and Taylor proposed a model which for the first time related the biochemical events to the cross-bridge cycle in a satisfactory manner (see Figure 6). The feature of this scheme is that, in resting muscle, actin is detached from the myosin head to which are bound the products of ATP hydrolysis. On stimulation actin interacts with myosin, the hydrolysis products are released, and the head undergoes a conformational change that results in movement between the two filament systems. After release of the products, ATP is immediately bound to to the enzymic site once, again causing actin to dissociate and the cross bridge to return to the original state. If stimulation continues the cycle will be repeated. In a simple manner this scheme incorporates the long-known dual function of ATP, namely to act as a substrate for myosin and to dissociate the actomyosin complex.

Although the basic tenets of the Lymn–Taylor scheme are probably correct it is becoming clear that it is an oversimplication of the cross-bridge cycle.The kinetics of the hydrolysis of ATP by subfragment I have now been analysed in great detail. The hydrolysis of ATP in solution has been shown to take place in seven stages, involving three conformational forms of the enzyme, and the rate constants for each step determined (see Trentham et al., 1976 for a review). Introduction of actin complicates the kinetics: although it is well established that actin speeds up the rate-limiting step in the hydrolysis, the precise nature of the actomyosin complex during the earlier stages of the reaction has been a matter for much discussion. From detailed kinetic analysis of the effect of actin on the transient state of ATP hydrolysis by myosin subfragment 1, it has been shown (Stein et al., 1981) that it is not mandatory for actin to dissociate from myosin during each cycle of the cross bridge. Indeed their results showed that the rate of the ATP hydrolysis step was even greater when actin was bound to subfragment 1.

The Eisenberg group also showed that myosin and myosin–nucleotide complexes could form two classes of complexes with actin: those that bound to actin strongly and those that bound weakly (see Geeves 1991, for review). Current views suggest that there may not be a clear detachment of the myosin head from the actin during the cross-bridge cycle, but that contraction occurs when the weak interaction, characteristic of the relaxed state, is converted into a strong force-developing interaction which leads to relative movement between the two filaments. With the atomic structures of subfragment 1 and actin available it is now becoming possible to define with some precision the regions of the two molecules involved with their interaction.

It is clear that there is much to be learned about the detailed molecular changes that occur during contraction. The scheme presented in Figure 6 is clearly an over simplication of the process since it does not incorporate the different molecular conformations of myosin that have been demonstrated to occur from the transient kinetic studies of ATP hydrolysis by myosin (Trentham et al., 1976). It should be pointed out that the kinetic studies, detailed as they are, have been carried out in solution. *In vivo*, the hydrolytic events occur in a heterogenous phase in which some of the conclusions from solution kinetics may not be valid.

The relative movement occurring per cross-bridge cycle, the power stroke, has long been considered to be about 10–15 nm. Movement must occur as a consequence of the transition between at least two structural states and would be expected to involve considerable conformational change in the myosin head. Many models have incorporated a change in the angle that the myosin head makes with the actin filament from 90°–45° to account for this movement. Despite much research, no evidence of gross structural change in the myosin head, which might be expected to account for such a change in angle of attachment, has been obtained. Indeed neutron scattering data indicates that deformation of the myosin cannot involve more than 20% of the mass of the head (Curmi et al., 1988).

Despite the lack of experimental proof for gross conformational changes in the myosin head that could account for the power stroke, there is good evidence for smaller localized conformational changes involving movements of a few angstroms associated with specific domains of the head. For example the pocket which contains the active site closes when ATP is bound to myosin (Yount et al., 1992). Also, fluorescence polarization studies suggest that the regulatory light chain region of myosin tilts during imposed filament sliding and the subsequent quick force recovery (Irving et al., 1995). The latter demonstration is of particular interest since this light chain has been shown by the structural studies (Rayment et al.,1993a) to be wrapped around an α-helical region of subfragment 1 distal from the ATP and actin binding sites. This α-helical tail-like region which extends for 8.5 nm has been suggested to act as a lever which tilts during the power stroke resulting from movement of the actin binding region of myosin (Rayment et al., 1993a). The tilt observed in the fluorescence polarization studies, however, is only about 3%, which is much less than would be required to produce a cross-bridge movement of 10 nm for example.

MODEL CONTRACTILE SYSTEMS

Using the model contractile systems developed after the discovery of actin—i.e. actomyosin suspensions, oriented actomyosin fibers, and glycerated muscle fibers—much has been learned about the characteristics of the contractile process. To get detailed insight into the cross-bridge cycle, however, it is necessary to investigate the process at the molecular level. The possibility that it might be feasible to observe the relative movement of actin filaments and myosin molecules was first hinted at by the report of Oplatka and Tirosh (1973) of active streaming in actomyosin solutions. Later they demonstrated similar effects with heavy meromyosin and subfragment 1. They speculated that interaction between an actin filament and free myosin molecules caused movement of the molecules relative to each other and the surrounding solution and was responsible for the streaming that they observed.

It was not until some 10 years later that a motility assay *in vitro* that enabled the movement of myosin along an actin filament to be observed in the microscope and measured (Scheetz and Spudich, 1983). In this system, heavy meromyosin was covalently linked to fluorescent plastic microspheres and their movement observed along polar cables of actin in Nitella. Motility assays have been developed so that the movement of actin filaments over a surface on which the myosin or its active subfragments are deposited can also be measured. Further by attaching an actin filament to a glass needle the force developed

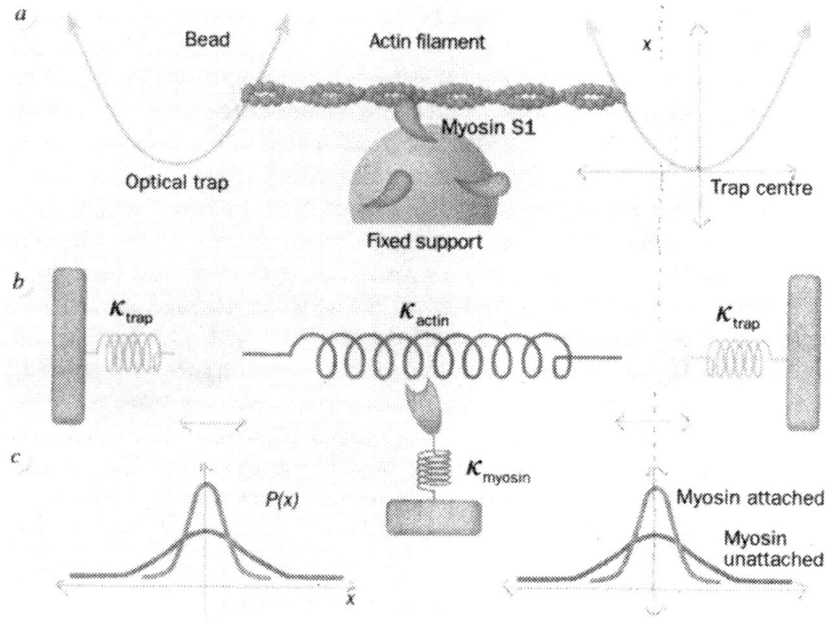

Figure 7. Experimental geometry used to record movement and force developed when single motor protein molecules interact with an actin filament. (**a**) An actin filament is stretched taut by attached polystyrene beads of about 1μ diameter. Each bead is held in one or two optical traps depicted as harmonic potential wells. A photodiode sensor records the x and y coordinates of one of the beads. The filament interacts with the myosin subfragment 1 molecule bound to a glass sphere immobilized on a coverslip which serves as a fixed support. The equilibrium position of the beads (*dotted lines*) are displaced from the trap centers, reflecting the pre-tension in the filament. (**b**) Mechanical equivalent of system in (**a**). The myosin cross-bridge acts as a spring of intermediate stiffness and the actin filament as a stiff spring (all stiffnesses are designated by κ). (**c**) Variance in bead displacement. In the absence of attachment the position is distributed as a broad gaussian due to Brownian movement. Attachment of the myosion cross bridge stiffens the system and reduces the width of this distribution (from Block, 1995).

by subfragment I interacting with it in the presence of ATP can be estimated from the bending produced (Kushino and Yanagida, 1988). From motility assys a number of important properties of the muscle actomyosin system have been determined. In these systems myosin always moves in the same direct direction along an actin filament—towards the "barbed" end, which is the direction it moves in the intact myofibril.

On the other hand an actin filament can move in any direction along a surface of heavy meromyosin molecules, the direction being determined by the polarity of the actin filament. The speed of movement of actin filaments is similar on a surface coated with subfragment 1 to that coated with myosin. This implies that the molecular motor is located in the head of the myosin molecule and that the tail is not of major importance for contractile function. In similar systems the sliding velocity is reduced by 90% in the absence of light chains and only restored fully when both are present (Lowey et al., 1993). Since it is known that the light chains are not essential for the ATPase activity of myosin, these experiments suggest that they may have a structural role in the neck region of the molecule. Both are wrapped around the 8.5-nm α-helical C-terminus of subfragment 1 (see Figure 4) making it more rigid, which would be an important property if this region acts as a lever (Rayment et al., 1993a).

The experiments described above have usually involved the response of a number of protein molecules. With the increased sophistication of the optical trap techique (Figure 7) it has become possible to measure nanometer movements and piconewton forces at millisecond rates using single myosin molecules and single actin filaments (Finer et al., 1994; Molloy et al., 1995). This has enabled the mechanics of contraction to be defined on a molecular basis. From such studies values for the step size ranging from 4 to 11 nm have been obtained. This wide range of values is surprising and may be explained as arising from shortcomings in the techniques employed. If this is not the case, it is possible that the step size depends on the conditions under which it is measured. The single force transients average 3–4 piconewtons. This is significantly greater than the value estimated per myosin head from experiments on intact muscle contracting under isometric conditions (1.3 pN). The latter result can be explained if only some of the heads are exerting tension at any one time in whole muscle.

INVESTIGATIONS INTO THE MECHANISM OF RELAXATION

Striated Muscle

The fiber models of contraction developed by Szent-Gyorgyi exhibited contractile properties very similar to those of muscle, but differed in one essential characteristic. Whereas the contractile response in living muscle was invariably followed by relaxation this was usually not the case with the models. Relaxation could be produced in the model systems by artificial means, e.g. by washing out the ATP and replacing it with pyrophosphate, which although not a substrate for the ATPase, dissociated actomyosin. Also if an ATPase inhibitor was added to a fiber that had been contracted and the ATP washed out, further addition of ATP caused relaxation (see Perry, 1956 for a review of these early investigations). Although the methods used to simulate relaxation were not those occurring in the cell, the results were important because they highlighted the dual function of ATP in the muscle cell (Weber and Portzehl, 1954); namely (1) when acting as a substrate for the actomyosin ATPase contraction occurred, and (2) when the ATPase was inhibited ATP acted as a plasticiser by bringing about the dissociation of actomyosin so that the system readily extended under the imposed tension.

Bozler (1951) was probably the first to demonstrate a complete cycle of contraction followed by relaxation in glycerated fibers induced by the application of ATP. To get the effect, the fibers were freshly prepared and lightly washed to remove the glycerol. On storage, the fibers lost the ability to relax while maintaining their contractile activity unimpaired, observations that hinted at the possible presence of a labile relaxing factor.

At this time Bailey[8] and his student Marsh were studying the postmortem changes in muscle. They were particularly interested in the fact that it was impossible to extract sarcoplasm by squeezing freshly minced muscle. This was readily achieved if the mince was allowed to stand at room temperature for half an hour or so. Marsh (1952) noted that the sedimentation volume of muscle fiber fragments obtained by homogenization of fresh rabbit psoas muscle in isotonic KCl underwent characteristic changes on standing. After a few minutes the sedimented volume of the fragments dropped but could be restored by the addition of ATP. As the ATP disappeared the volume fell again and under favorable circumstances could be restored by further additions of nucleotide. Since the increase in sedimention volume was accompanied by an increase in the length of the muscle fibers the observation was interpreted as a relaxation effect. If the homogenate supernatant was replaced with fresh isotonic KCl, addition of ATP no longer produced the increase in volume of the fiber fragments, but an irreversible syneresis similar to that produced with actomyosin suspensions occurred. It was concluded that muscle sarcoplasm contained a factor, designated the Marsh and later the Marsh–Bendall factor, which in the presence of ATP could bring about the relaxation of the muscle fiber. This investigation clearly established the dual role of ATP in contraction and relaxation that had been hinted at by the study of purified actomyosin systems.

Since it was now clear that relaxation depended upon the regulation of the myofibrillar ATPase, research advanced along two lines. One was concerned with the definition of the ionic conditions required for the activation of the MgATPase of the myofibril washed free from sarcoplasm, and the other with the nature of the Marsh factor in the sarcoplasm that was responsible for relaxation in the intact fiber.

Calcium and the Myofibrillar ATPase

The MgATPase of isolated myofibrils washed free of sarcoplasmic relaxing factor was shown to be strongly inhibited by concentrations of EDTA which were only a few percent of the concentration of the magnesium in the enzymic system (Perry and Grey, 1956a). It was therefore suggested that the MgATPase required very low concentrations of a cation other than magnesium that had a high affinity for EDTA. Earlier work (Kamada and Kinosita, 1943; Heilbrun and Weircinski, 1947) had indicated that microinjection of calcium into intact muscle cells was particularly effective in inducing contraction. In the early 1950s Schwarzenbach and collaborators had synthesized an analogue of EDTA which was much more specific for calcium. This compound [ethylene glycol bis(β-aminoethyl ether) *N, N'*-tetraacetate] was then called "glycolcomplexon". It is now known as EGTA and is widely used to demonstrate the requirement of enzymic and other systems for calcium.

At the time EGTA was not commercially available but one report had appeared in the literature of its use to demonstrate the ability of EDTA and "glycolcomplexon" to inhibit the swelling of liver mitochondria. From the known properties of the latter chelator it was concluded that calcium was responsible for the swelling of mitochondria (Raaflaub, 1955). A sample of EGTA was obtained from Schwarzenbach which at concentrations much lower than those of magnesium in the system inhibited the MgATPase of myofibrils and crude preparations of actomyosin (Perry and Grey, 1956b). As the first reported use of EGTA in a muscle system this implied that the myofibrillar ATPase required traces of calcium for full activation. This observation was confirmed by careful studies using calcium EGTA buffers by Weber (1959) who showed that calcium concentrations of 10^{-6} to 10^{-5} M were required to activate the Mg ATPase and bring about superprecipitation of actomyosin.

Another significant observation made at this time was that the MgATPase of a crude actomyosin preparation (so-called "natural actomyosin", i.e. extracted as the complex from muscle) was inhibited by low concentrations of EGTA and EDTA. This implied that this preparation required a trace of calcium for activation of the ATPase. If actomyosin was prepared from the separately isolated actin and myosin ("synthetic actomyosin") the MgATPase activity was not inhibited in the presence of the chelators (Perry and Grey, 1956b). This observation suggested that there was some factor or factors present in the crude actomyosin preparation in addition to actin and myosin which made the ATPase sensitive to traces of calcium.

Early efforts to determine the nature of the sarcoplasmic factor responsible for relaxation produced a confusing story with enzymes such as myokinase and creatine phosphokinase being claimed at various times as the factor. The common feature was their ability to maintain the ATP concentration, known to be a requirement for the relaxing system and probably to raise it high enough reduce the ATPase activity by substrate inhibition (see Weber and Portzehl, 1954; Perry, 1956; Hasselbach, 1964, for reviews). The breakthrough came when it was reported by Kumagi et al. (1955) that the relaxing activity of muscle sarcoplasm was associated with the ATPase of muscle known as the Kielley

Meyerhof enzyme (Kielley and Meyerhof, 1948) which could be precipitated by 20% ammonium sulphate. This enzyme had earlier been shown to be associated with the microsome fraction obtained by high speed centrifugation of muscle (Perry, 1952). Confirmation of the association of the relaxing factor with this cellular fraction was provided by Portzehl (1957) who found that all activity could be removed from muscle sarcoplasm by high speed centrifugation.

In view of the sensitivity of the myofibrillar MgATPase to calcium, it seemed likely that the role of the microsomal fraction might be to regulate the concentration of this ion in the muscle cytoplasm. In support of this mechanism several groups reported that the granular fraction of muscle had the capacity to remove calcium from solutions containing magnesium and ATP (Ebashi, 1960; Gergely and Parker, 1960; Hasselbach and Makinose, 1960). A more detailed analysis established that the endoplasmic reticulum of muscle, now known as the sarcoplasmic reticulum (SR), possesses an active pump which in the presence of ATP and magnesium is able to transport free calcium from the cytoplasm and concentrate it in the lumen of the reticulum (Hasselbach and Makinose, 1961; Ebashi and Lipmann, 1962). The active component of the pump is a membrane-bound Ca^{2+}ATPase whose structure and function has been studied in detail (see Martonosi, 1994 for review). It has the capacity to reduce the free cytoplasmic calcium to 10^{-8} M and maintain a gradient of at least 1000-fold across the membrane.

Thus by 1960 it was clear that in resting muscle the free sarcoplasmic calcium was very low, maintained at 10^{-7} M or less by the action of the pump in the SR where the calcium is stored. By a mechanism which even today is not fully understood, on stimulation the SR membrane momentarily becomes permeable to calcium which floods out from the stores and raises the cytoplasmic level to 10^{-5} M. This leads to the activation of the myofibrillar ATPase and the myofibril contracts.

The Troponin System

The next important development in the investigation of the mechanism of relaxation was the elucidation of the system associated with the myofibril that was responsible for the calcium sensitivity of its MgATPase. In 1963 Ebashi reported that the superprecipitation of actomyosin reconstituted from purified actin and myosin [the ATPase of which unlike the "natural actomyosin" was not sensitive to calcium (Perry and Grey 1956b)] could be made calcium-sensitive by the addition of a third protein component which he called "natural tropomyosin". Although this component resembled tropomyosin in many properties, it clearly was not identical with the protein prepared by the method of Bailey (1946), which does not confer calcium sensitivity on reconstitituted actomyosin. Subsequently a protein fraction was isolated which when added to normal tropomyosin restored its viscous properties to those of "natural tropomyosin" (Ebashi and Kodama, 1965). This fraction, which also had the property of confering calcium sensitivity on reconstituted actomyosin, was given the name troponin (Ebashi and Endo, 1968).

It soon became clear that troponin was not a homogenous protein and several groups worked on its fractionation. Clearly troponin had a number of different properties all of which might be ascribed to a single molecule or alternatively could be associated with different components. In the absence of calcium the MgATPase of reconstituted actomyosin was very low in the presence of troponin. This implied that it had a inhibitory action on the ATPase. The first clue that troponin might contain a component with this property was the report of the isolation of a fraction from a crude troponin preparation which had inhibitory activity on the MgATPase of reconstituted actomyosin (Perry et al., 1966; Hartshorne et al., 1967). Significantly this inhibitory protein, as it was called, had no action on the CaATPase of actomyosin. This indicated that it was in some way involved in the interaction of actin with myosin upon which the activation of the MgATPase depended.

The fact that the action of troponin on the actomyosin ATPase was markedly affected by 10^{-6} to 10^{-5} M calcium implied that troponin possessed high affinity site(s) for calcium which when occupied relieved its inhibitory action on the actomyosin MgATPase. Evidence was obtained for two proteins in the troponin complex with distinct properties, troponin A and B (Hartshorne and Mueller, 1968) which corresponded to the calcium sensitizing factor and the inhibitory protein, repectively, of Schaub and Perry (1969). Subsequently it was shown that the inhibitory protein preparations contained another protein, also strongly basic like the inhibitory protein, but without inhibitory activity. It had a higher apparent molecular mass than the other troponin components, and initially was known as the 37-kDa protein.

By the early 1970s it had become clear that troponin was a equimolar complex of three distinct proteins, each of which when isolated exhibited unique properties. These properties were remarkably stable for the standard procedure of isolation which at that time was by chromatography in 8 M urea. Nevertheless to restore calcium sensivity to the MgATPase of actomyosin all three components were required. Nomenclature of the components of the troponin complex became very confusing for each of the groups working on the problem used their own (see Table 1). Order was restored in 1972 when the names troponin C (the calcium binding protein), troponin I (the inhibitory protein), and troponin T (the 37-kDa protein that binds tropomyosin and is responsible for the viscous properties of "natural tropomyosin") were adopted.

The troponin complex is well defined in terms of the sequences of the proteins involved and their sites of interaction (see Perry,1996). Sequence studies on troponin C soon established that it belonged to the family of calcium binding proteins, all of which contain the helix-loop-helix domain first identified by

Table 1[a]

Troponin C	*Troponin I*	*Troponin T*
Troponin A (Hartshorne and Mueller, 1968)	Inhibitory factor (Hartshorne et al., 1967; Schaub and Perry, 1969, 1971)	37000 component (Wilkinson et al., 1971, 1972)
Calcium sensitizing factor (Schaub and Perry, 1969)	Inhibitory component of troponin B (Hartshorne and Mueller, 1968)	
Troponin 4 (Greaser and Gergely, 1971)	Troponin 2 (Greaser and Gergely, 1971)	Troponin 3 (Greaser and Gergely, 1971)
Troponin III, fraction III (Ebashi et al., 1971)	Troponin II (Ebashi et al., 1971)	Troponin I (Ebashi et al., 1971)
Component III (Murray and Kay, 1971)	Component II (Murray and Kay, 1971)	Component I (Murray and Kay, 1971)
Troponin C (Greaser and Gergely, 1973)	Troponin I (Greaser and Gergely, 1973)	Troponin T (Greaser and Gergely, 1973)

Note: [a] Under the appropriate headings are listed the names used by investigators for fractions of similar biological activity and amino acid composition until the generally accepted terminology of troponins C, I and T was adopted in 1972.

Kretsinger and Nockolds (1973) in the cytoplasmic muscle protein, parvabumin. As yet, the three-dimensional structure of the complex at high resolution is not available. An important step towards the goal of understanding how the regulatory system interacts with the motor protein system has been the crystallization of troponin C and the determination of its structure (Figure 8; Herzberg and James, 1988; Satyshur et al., 1988). This has enabled the the structural changes that occur when calcium binds to a high-affinity protein to be defined (Strynadka and James, 1989). Quite apart from the significance of these effects for the function of troponin they are of more general importance in understanding the role of calcium in the many other biological systems in which high affinity calcium-binding proteins have a regulatory role.

The Role of Tropomyosin

The elucidation of the nature of the troponin complex and the function of its components provided for the first time evidence of a role for tropomyosin in muscle. Tropomyosin was isolated by Bailey in 1946 and shown to be the third most abundant protein in the myofibril. Its property of interacting with actin and its location as a linear polymer lying in each of the two grooves of the actin filament strongly suggested a structural role in the myofibril. It is

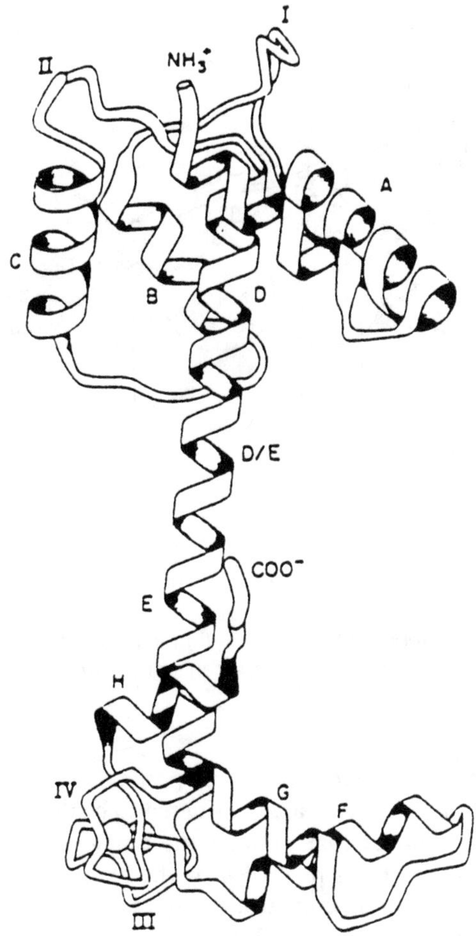

Figure 8. Ribbon representation of the polypeptide chain of turkey skeletal muscle troponin C. The upper domain is the calcium regulatory domain (N-domain); the lower domain is the high affinity Ca^{2+}/Mg^{2+} binding domain (C-domain). The eight helices involved in the four supersecondary helix–loop–helix motifs are labeled sequentially. There is an additional helix at the beginning of the N-domain sequence. The segments of the polypeptide chain forming the calcium-binding loops are labeled I to IV (from Herzberg and James, 1988).

now clear that it has an important role in the regulatory process in striated muscle. One remarkable feature of the regulatory system of striated muscle is the fact that there is only one molecule of the troponin complex for every seven monomer units of actin in the I-filament. When muscle is stimulated,

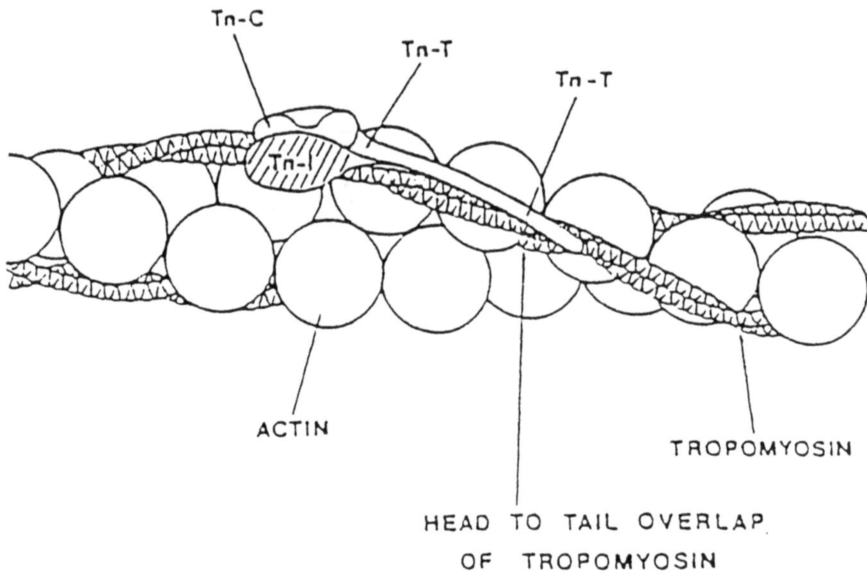

Figure 9. Schematic representation of the thin filaments of vertebrate striated muscle indicating possible locations of the components of the troponin complex (after Heeley et al., 1987).

calcium is bound at the molecule of troponin C that is close to one actin monomer. All actin monomers are considered to be potentially able to interact with myosin in a manner that initiates the cross-bridge cycle. The question arises as to how the changes produced by the binding of calcium at troponin C close to one actin monomer are transmitted to the other six more distal actin monomers (Figure 9). A possible clue is given by the fact that studies *in vitro* indicate that the inhibitory activity of troponin I is much increased in the presence of tropomyosin. In the absence of tropomyosin equimolar amounts of actin and troponin I are required to bring about maximum inhibition of the actomyosin MgATPase. When tropomyosin is present, a ratio of one molecule of troponin I to up to seven actin monomers is adequate for inhibition. It follows that *in vivo* tropomyosin has the ability to extend the action of the troponin complex over seven actin monomers. How it does this is one of the important problems of striated muscle regulation.

X-ray diffraction and EM image reconstruction studies indicate that an early event on stimulation of muscle is a movement of the tropomyosin molecule in the I-filament groove. This has led to the suggestion that tropomyosin in effect controls the interaction of actin and myosin. In resting muscle the tropomyosin molecule, which lies alongside seven actin monomers, is considered to block the sites of interaction with myosin on each actin with the result that the MgATPase is low. When the troponin complex binds calcium, conformational changes occur which result in movement of the tropomyosin molecule so that the interaction sites on the actin are exposed. Activation of the MgATPase and the cross-bridge cycle then takes place. This mechanism, the so-called "steric hypothesis" (Huxley, 1972; Haselgrove, 1972), has attractive features but it should be emphasized that it is only a hypothesis. We still have still much to learn about the significance of the conformational changes that occur in actin, myosin, and troponin during muscle activity and their relation to the mechanochemical cycle.

Smooth Muscle and Myosin Phosphorylation

Until the 1960s most of the biochemical studies of muscle were carried out on the striated tissue of which large samples of homogenous tissue are readily obtained. Investigation of smooth muscle was more difficult: although it is widely distributed, it is present in relatively small amounts in different organs and usually associated with large amounts of connective tissue. It was clear that smooth muscle contained actin and myosin and although the isoforms of these proteins in this tissue differ slightly in sequence, the molecular forms of the proteins are very similar to their skeletal counterparts. Likewise, they occur in separate filaments in the smooth muscle cell and all the evidence is that the contractile process is very similar to that of striated muscle. Regulation did, however, appear to be different for despite the fact a rise in cytoplasmic calcium concentration initiated contraction, none of the components of the troponin complex could be found in smooth muscle.

An early insight into the special importance of the regulatory light chain in the regulation of smooth muscle actomyosin came from the work of Kendrick-Jones et al. (1970). In molluscan adductor myosin this light chain is essential for conferring calcium sensitivity on the actomyosin. It has recently been shown that a novel calcium binding site on the essential light chain of molluscan myosin is stabilized by links with both the heavy and regulatory light chain of myosin (Xie et al., 1994). Thus the binding of calcium to this site on stimulation leads to activation of the actomyosin ATPase.

The report of the phosphorylation of the regulatory (phosphorylatable) light chain of myosin (Perrie et al., 1973) and the presence of a specific kinase (Pires et al., 1974) in skeletal muscle hinted at a possible mechanism for regulating contraction by action directly on the myosin might also exist in striated muscle. Myosin light chain kinase is particularly active in fast skeletal muscle; phosphorylation accompanies repeated contraction and is followed by dephosphorylation when contractile activity ceases. Phosphorylation of the regulatory light chain of striated muscle myosin, however, is not essential for myosin ATPase or contraction, although kinetic studies indicate that it increases the affinity for actin (Pemrick, 1980). There is also a report (Persechini and Stull, 1985) that with skinned fiber preparations calcium sensitivity is increased at suboptimal concentration of the cation. Thus it would appear that in striated muscle phosphorylation of the light chain has a modulating rather than a triggering effect on the cross bridge cycle.

It was a major step forward that changed the emphasis of smooth muscle research when it was reported, in contrast to the situation in striated muscle, that the low MgATPase activity of vertebrate smooth muscle actomyosin preparations was markedly increased when the regulatory light chain was phosphorylated (Gorecka et al., 1976; Sobieszek and Small, 1976; Chacko et al., 1977). From these and subsequent studies has emerged the current dogma that the major event in initiating contraction in vertebrate smooth muscle is the activation of myosin light chain kinase by calcium and the subsequent phosphorylation of the light chain.

Although the mechanism of regulation of smooth muscle is not as well defined as that of striated muscle, it does exhibit marked differences. Other proteins not present in significant amounts in the striated tissue, namely caldesmon (Sobue et al., 1981) and calponin (Takahayashi et al., 1986) may have a regulatory role in smooth muscle. Both caldesmon and calponin bind to actin and calmodulin and, at least *in vitro*, are able to regulate the actomyosin MgATPase in a calcium-dependent manner. These results suggest that regulation in smooth muscle involves unique events at the actin as well as the myosin filaments.

PLASTICITY OF THE CONTRACTILE SYSTEM OF MUSCLE

It has long been clear that contractility is a widespread feature of cells and that muscle exhibits a highly specialized development of this property. The contractile properties of the different muscle types in any given species vary widely. To accommodate this range of properties myosin exists in a number isoforms resulting from sequence differences in the heavy and light chains. In all cases the overall molecular shape is very similar and they cannot be distinguished in the electron microscope. In an important paper Barany (1967) pointed out that there was a good correlation between the contractile speed of the muscle and the specific ATPase of the myosin isolated from it. Later Buller et al. (1969) showed that when a muscle was cross-innervated the myosin isoform changed to the isoform with enzymic activity appropriate for the changed contractile speed. This indicated that the expression of myosin genes was regulated by the activity pattern of the muscle, which itself was under control of the nerve. Similar relationships have been reported for the I-filament proteins, troponin, and tropomyosin (Amphlett et al., 1975; Dhoot et al., 1981;

Heeley et al., 1983). It has now become clear that muscle is a plastic tissue that can change the pattern of gene expression so that the isoforms of the proteins of the contractile and regulatory proteins are those most appropriate for the activity pattern of the muscle.

UNCONVENTIONAL MYOSINS

From the study of nonmuscle cells it became apparent that actomyosin, although widespread, was not the only system responsible for motility. Protein motors other than myosin, such as dynein and kynesin have been identified. Nevertheless, the general pattern of all the motile systems were similar (see also Chapter 5). In eukaryotes this invariably appears to involve a motor protein moving along a filamentous protein track formed by the polymerization of a globular subunit, actin or tubulin. Conditioned by studies on the muscle system it was considered until the early 1970s that the actomyosin contractile system normally functioned with a double-headed myosin molecule. Nevertheless at that time there was no evidence of a specific role for the two heads in the mechanochemical process.

The discovery of Pollard and Korn (1973) that Acanthamoeba contained, in addition to the normal two-headed myosin, another myosin of lower molecular weight with only one head broadened views on myosin. It indicated that two heads were not required for myosin to function as a motor, a conclusion later confirmed with the model motile systems. The single-headed myosin responsible for a variety of motile functions in amoeba is now referred to as myosin I, whereas the conventional two-headed myosin is known as myosin II. Application of molecular biological techniques to muscle has been particularly striking in its effect in defining the role of myosin in general cell contractility. The cloning and sequencing of the gene for the myosin heavy chain, which was first achieved with *unc*54 myosin II from C. elegans (Karn et al., 1983), has now been carried out on many vertebrate and invertebrate myosin IIs (Emerson and Bernstein, 1987).

From this information, consensus sequences characteristic of the myosin head, and presumed to be of significance for its motor function, have been established. Using these consensus sequences, genes corresponding to a superfamily of myosins have been identified in eukaryotic cells. Most of these are single-headed myosins and are known collectively as the unconventional myosins to distinguish them from the conventional double-headed form. To date the genes corresponding to 11 classes of myosin have been identified (Figure 10). The amounts of the unconventional myosin present are small compared with the myosin II of specialized contractile tissues, but they are widely distributed in plant and animal cells. The unconventional myosins usually posses a single short tail adapted for attachment to other cell structures

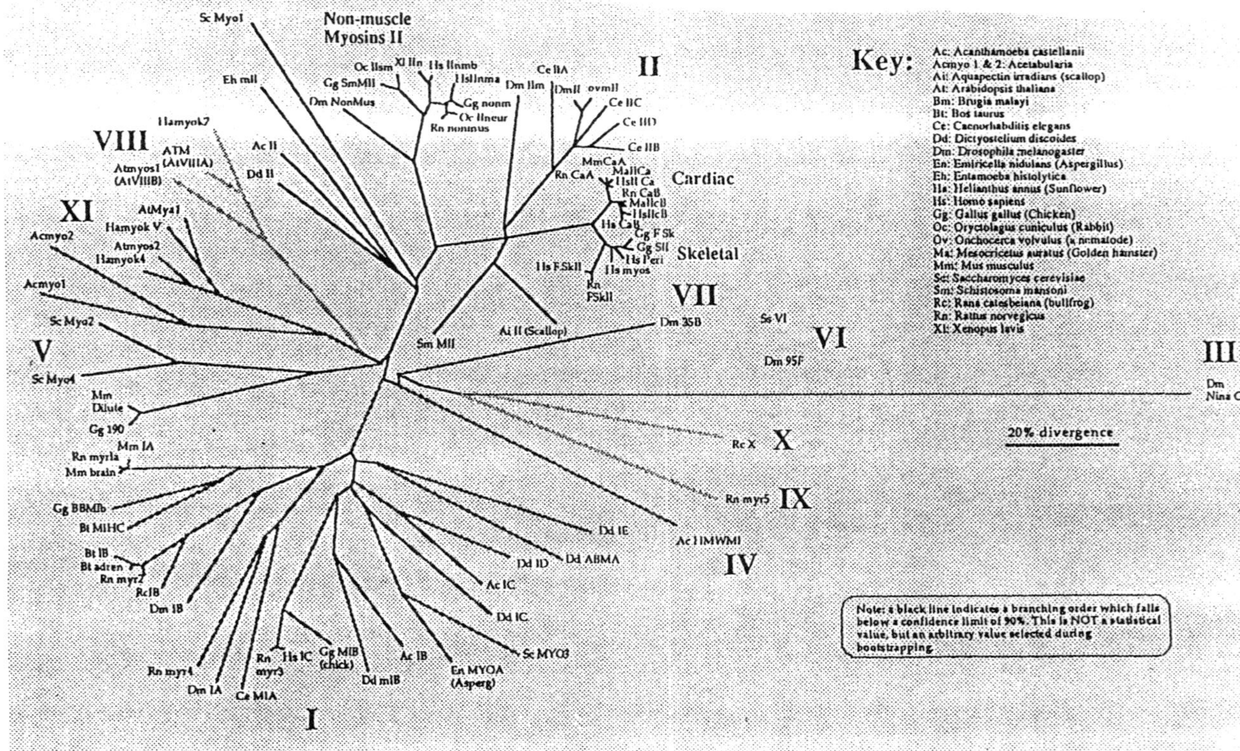

Figure 10. An unrooted phylogenetic tree of the 81 myosin motor domain sequences available as from May 1995, showing the divison into 11 classes (Cope, Knight, and Kendrick-Jones, personal communication).

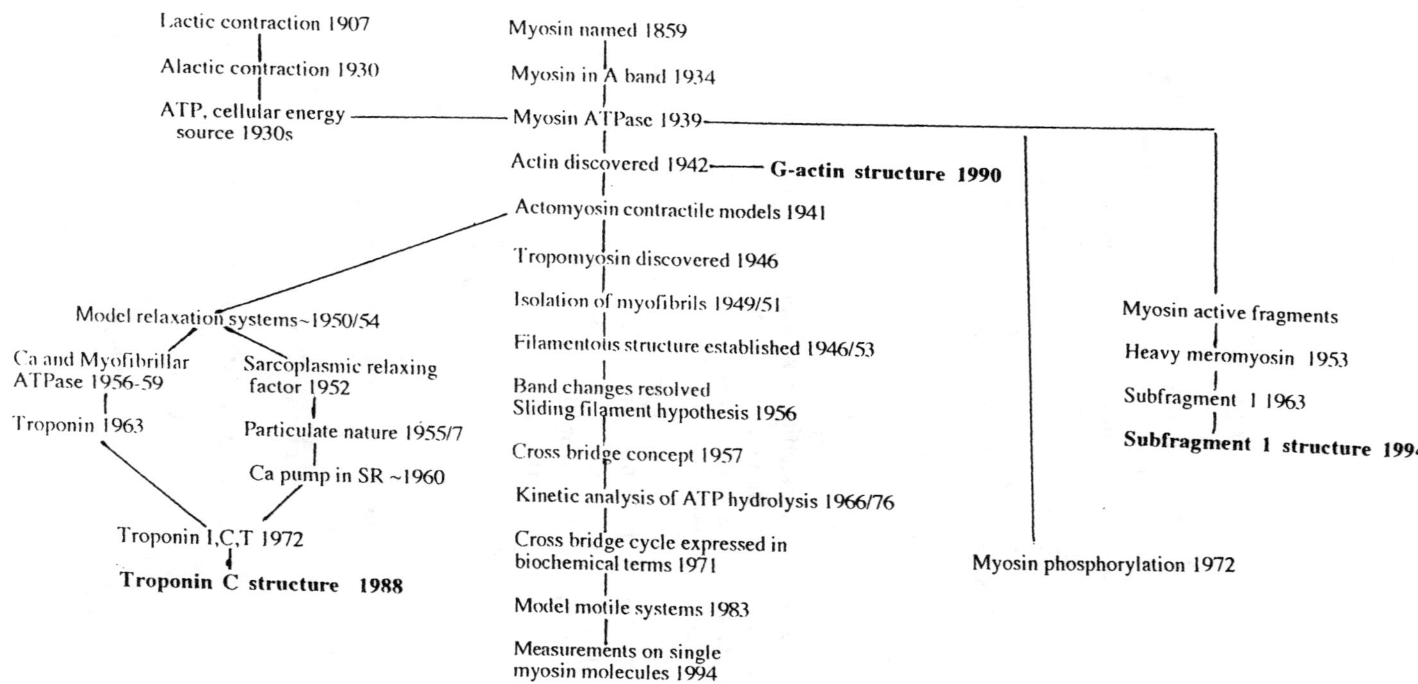

Figure 11. Chart illustrating in chronological order the important stages in the development of the biochemistry of contraction and relaxation in striated muscle.

rather than filament formation, as is the case with myosin II. Phyllogenetic evidence suggests that that the myosin superfamily evolved from an ancestral form of motor protein.

CONCLUDING COMMENTS

Although the histology and physiology of muscle has been studied for well over a century, it was not until the 1930s that biochemists began to direct their attention to the specialized components of the muscle cell involved in the contractile process. Since that time, progress has been impressive and marked by the discovery of new proteins and enzymes involved in the contracto-regulatory system of striated muscle (Figure 11). Unlike many biochemical systems cellular ultrastructural organization and the relation of the protein components to it has been a dominant theme of research into the mechanism of contraction. This has meant that electron microscopy and the application of sophisticated methods of image analysis has advanced hand in hand with biochemistry. The advances in knowledge of the biochemical mechanisms present in skeletal muscle have stimulated parallel investigations into cardiac and smooth muscle. These studies have indicated that the actomyosin contractile system is very similar in all muscle types, whereas the the biochemical mechanisms for its regulation are much more diverse, specific, and specialized for the muscle type.

So far as skeletal muscle is concerned it is probable that all the protein components involved in the contracto-regulatory system of the myofibril are now defined. The structure of three of the six proteins involved are known to a resolution to c. 2 Å. The problem now is to determine the structure of the troponin complex and the thin filament at a similar resolution. Remarkably, techniques have so advanced that the characteristics of the movement at the molecular level can be measured and related to the properties of the whole muscle. The nature of the molecular and atomic events that underly contraction, however, still defeat us since we do not yet have the methods that enable us to define them with precision.

REFERENCES

Amphlett, G.W., Perry, S.V., Syska, H., Brown, M.D., & Vrbova, G. (1975). Cross innervation and the regulatory protein system of rabbit soleus muscle. Nature (London) 257, 602-604.

Bailey, K. (1946). Tropomyosin: A new asymmetric protein component of muscle. Nature (London) 157, 368-369.

Barany, M. (1967). ATPase activity of myosin correlated with speed of muscle shortening. J. Gen. Physiol. 50, 197-218.

Bozler, E. (1951). Mechanism of relaxation in extracted muscle fibres. Amer. J. Physiol. 167, 276-283.

Block, S.M. (1995). One small step for myosin. Nature (London) 378, 132-133.

Buller, A.J., Mommaerts, F.J.M., & Seraydarian, K. (1969). Enzymic properties of myosin in fast and slow twitch muscles of the cat following cross innervation. J. Physiol. 205, 581-579.

Cain, D.F., Infante, A.A., & Davies, R.E. (1962). Chemistry of muscular contraction. Adenosine triphosphate and phosphoryl creatine as energy supplies for single contractions of working muscle. Nature (London) 196, 214-217.

Chako, S., Conti, M.A., & Adelstein, R.S. (1977). Effect of phosphorylated smooth muscle myosin on actin activation and Ca^{2+} regulation. Proc. Natl. Acad. Sci. USA 74, 129-133.

Curmi, P.M.G., Stone, D.B., Schneider, D.K., Spudich, J.A., & Mendelson, R.A. (1988). Comparison of myosin subfragment 1 bound to actin and free in solution. A neutron scattering study using actin made "invisible" by deuteration. J. Mol. Biol. 203, 781-798.

Dhoot, G.K., Perry, S.V., & Vrbova, G. (1981). Changes in the distribution of the components of the troponin complex in muscle fibres after cross innervation. Exper. Neurol. 72, 513-530.

Draper, M.H. & Hodge, A.J. (1949). Studies on muscle with the electron microscope I. Ultrastructure of toad striated muscle. Austr. J. Exp. Biol. Med. Sc. XXVII, 465-503.

Ebashi, S. (1960). Calcium binding and relaxation in the actomyosin system. J. Biochem. (Tokyo) 48, 150-151.

Ebashi, S. (1963). Third component participating in the superprecipitation of "natural actomyosin". Nature (London) 200, 1010.

Ebashi, S. & Endo, M. (1968). Calcium ions and muscular contraction. Prog. Biophys. Mol. Biol. 18, 123-183.

Ebashi, S. & Kodama, A. (1965). A new protein factor promoting aggregation of tropomyosin. J. Biochem. (Tokyo) 58, 107-108.

Ebashi, S. & Lipmann, F. (1962). Adenosine triphosphate-linked concentration of clacium ions in a particulate fraction of rabbit muscle. J. Cell Biol. 3, 389-400.

Ebashi, S., Wakabayashi, T., & Ebashi, F. (1971). Troponin and its components. J. Biochem. (Tokyo) 69, 441-445.

Edsall, J.T. (1930). Studies on the physical chemistry of muscle globulin II. On some physicochemical properties of muscle globulin (myosin). J. Biol. Chem. 89, 289-313.

Eisenberg, E. & Greene, L.E. (1980). The relation of muscle biochemistry to muscle physiology. Ann. Rev. Physiol. 42, 293-309.

Eisenberg, E. & Moos, C. (1968). The adenosinetriphosphatase activity of actoheavymeromyosin. A kinetic analysis of actin activation. Biochemistry, 7, 1486-1489.

Eisenberg, E., zobel, C.R., & Moos, C. (1968). Subfragment 1 of myosin:adenosine triphosphatase. Activation by actin. Biochemistry, 7, 3186-3194.

Emerson, C.P. & Bernstein, S.I. (1987). Molecular genetics of myosin. Annu. Rev. Biochem. 56, 695-726.

Engelhardt, V.A. & Ljubimova, M.N. (1939). Myosin and adenosine triphosphatase. Nature (London) 144, 668-669.

Finer, J.T., Simmons, R.M., & Spudich, J.A. (1994). Single myosin molecule mechanics, piconewton forces and nanometre steps. Nature (London) 368, 113-119.

Fletcher, W.M. & Hopkins, F.G. (1907). Lactic acid in muscle. J. Physiol. 35, 247-309.

Geeves, M.A. (1991). The dynamics of actin and myosin association and the cross bridge model of muscle contraction. Biochem. J. 274, 1-14.

Gergely, J. & Parker, C.J. (1960). Studies on the relaxing factor of muscle. Conference of Contractility, Pittsburg, PA, pp. 6-7.

Gershman, L.C., Stracher, A., & Dreisen, P. (1969). Subunit structure of Myosin III. A proposed model for rabbit skeletal myosin. J. Biol. Chem. 244, 2726-2736.

Gorecka, A., Askoy, M.O., & Hartshorne, D.J. (1976). The effect of phosphorylation of gizzard myosin on actin activation. Biochem. Biophys. Res. Commun. 71, 325-332.

Greaser, M. & Gergely, J. (1971). Reconstitution of troponin activity from three protein components. J. Biol. Chem. 246, 4426-4233.

Greaser, M. & Gergely, J. (1973). Purification and properties of the components of the troponin complex. J. Biol. Chem. 248, 2125-2133.

Hanson, J. & Huxley, H.E. (1954). Changes in the cross striation of muscle during contraction and stretch and their structural interpretation. Nature (London) 173, 973-976.

Hanson, J. & Lowy, J. (1963). The structure of F-actin and of actin filaments isolated from muscle. J. Mol. Biol. 6, 46-60.

Hartshorne, D.J. & Mueller, H. (1968). Fractionation of troponin into two distinct proteins. Biochem. Biophys. Res. Commun. 31, 647-653.

Hartshorne, D.J., Perry, S.V., & Schaub, M.C. (1967). A protein factor inhibiting the magnesium activated adenosine triphosphatase of desensitized actomyosin. Biochem. J. 104, 907-913.

Haselgrove, J.C. (1972). X-ray evidence for a conformational change in the actin-containing filaments of vertebrate striated muscle. Cold Spring Harbor Symp. Quant. Biol. 37, 341-352.

Hasselbach, W. (1964). Relaxing factor and the relaxation of muscle. Prog. Biophys. Mol. Biol. 14, 167-222.

Hasselbach, W. & Makinose, M. (1960). Die calcium bildung durch die muskelgrana und ihre erschaffungswirkung. Pflugers Arch. 272, 45-46.

Hasselbach, W. & Makinose, M. (1961). Die calciumpump der "erschaffungsgrana" des muskels und ihre abhangigkeit von der ATP-spaltung. Biochem. Zeit. 333, 518-528.

Heeley, D.H., Dhoot, G.K., Frearson, N., Perry, S.V., & Vrbova, G. (1987). The effect of cross innervation on the tropomyosin composition of rabbit skeletal muscle. FEBS Lett., 152, 282-286.

Heilbrun, L.V. & Wiercinski, F.J. (1947). The action of various cations on muscle protoplasm. J. Cell. Comp. Physiol. 29, 15-32.

Herzberg, O. & James, M.N.G. (1988). Refined crystal structure of troponin C from turkey skeletal muscle at 2.0 Å resolution. J. Mol. Biol. 203, 761-779.

Hill, A.V. (1949). Adenosine triphosphate and muscular contraction. Nature (London) 163, 320.

Huxley, A.F. (1986). Discoveries in muscle: observation, theory and experiment. Brit. Med. J. 293, 115-117.

Huxley, A.F. & Niedergerke, R. (1954). Interference microscopy of living muscle fibres. Nature (London) 173, 971-973.

Huxley, H.E. (1953). Electron microscope studies on the organisation of the filaments of striated muscle. Biochim. Biophys. Acta 12, 387-394.

Huxley, H.E. (1957). The double array of filaments in cross striated muscle. J. Biophys. Biochem. Cytol. 3, 631-648.

Huxley, H.E. (1963). Electron microscope studies on the structure of natural and synthetic protein filaments from striated muscle. J. Mol. Biol. 7, 281-287.

Huxley, H.E. (1972). Structural changes in actin and myosin-containing filaments during contraction. Cold Spring Harbor Symp. Quant. Biol. 37, 361-378.

Irving, M., Allen, T.S-C., Sabido-David, C., Crack, J.S., Brandmeier, B., Kendrick-Jones, J., Corrie, J.E.T., Trentham, D.R., & Goldman, Y.E. (1995). Tilting of the light chain region of myosin during step length changes and active force generation in skeletal muscle. Nature (London) 375, 688-691.

Kabsch, W., Mannherz, H.G., Suck, D., Pai, E.F., & Holmes, K.C. (1990). Atomic structure of actin: DNAase I complex. Nature (London) 347, 37-44.

Kamada, T. & Kinosita, H. (1943). Disturbances initiated from naked surface of muscle protoplasm. Japan J. of Zoology 10, 469-493.

Karn, J., Brenner, S., & Barnetts, S. (1983). Protein structural domains in the Caenorhabditis elegans unc54 myosin heavy chain gene are not separated by introns. Proc. Natl. Acad. Sci. USA 80, 4253-4257.

Kendrick-Jones, J., Lehman, W., & Szent-Gyorgyi, A.G. (1970). Regulation in molluscan muscles. J. Mol. Biol. 54, 313-326.

Kielley, W.W. & Meyerhof, O. (1948). Studies on the adenosine triphosphatase of muscle II. A new magnesium-activated adenosinetriphosphatase. J. Biol. Chem. 176, 591-601.

Kretsinger, R.H. & Nockolds, C.E. (1973). Carp muscle calcium binding protein. J. Biol. Chem. 248, 3313-3326.

Kuby, S.A. & Mahowald, T.A. (1959). Studies on ATP-transphosphorylases. Fed. Proc. 18, 267.

Kumagi, H., Ebashi, S., & Takeda, F. (1955). Essential relaxing factor in muscle other than myokinase and creatine phosphokinase. Nature (London) 176, 166.

Kuhne, W. (1859). Uber Muskelzuckungen ohne Betheiligung der Nerven. Arch. Anat. Physiol. 78, 314-333.

Kushino, A. & Yanagida, T. (1988). Force measurements by micromanipulation of single actin filaments by glass needles. Nature (London) 334, 74-76.

Lazarides, E. & Lindberg, U. (1974). Actin is the naturally occurring inhibitor of deoxribonuclease I. Proc. Natl. Acad. Sci. USA 74, 4742-4746.

Lowey, S., Waller, G.S., & Trybus, K.M. (1993). Skeletal muscle light chains are essential for physiological speeds of shortening. Nature (London) 334, 74-76.

Lundsgaard, E. (1930). Untersuchungen uber Muskel kontraktionen ohne Milchsåurebildung. Biochem. Zeit. 217, 162-177.

Lundsgaard, E. (1934). Phosphagen und Pyrophosphatumsatz in joessigsåureverggigteten Muskeln. Biochem. Zeit. 269, 308-328.

Lymn, R.W. & Taylor, E.W. (1971). Mechanism of adenosine triphosphate hydrolysis by actomyosin. Biochemistry 10, 4617-4624.

Marsh, B.B. (1952). The effects of adenosine triphosphate on the fibre volume of a muscle homogenate. Biochim. Biophys. Acta 9, 247-260.

Martonosi, A.N. (1994). Regulation of calcium by the sarcoplasmic reticulum. In: Myology (Engel, A.G., & Francini-Armstrong, C., Eds.), 2nd ed., Vol. 1, pp. 553-584. McGraw-Hill, New York.

Molloy, J.E., Burns, J.E., Kendrick-Jones, J., Tregear, R.T., & White, D.C.S. (1995). Movement and force produced by a single myosin head. Nature (London) 378, 209-212.

Mueller, H. & Perry, S.V. (1962). The degradation of heavy meromyosin by trypsin. Biochem. J. 85, 431-439.

Murray, A.C. & Kay, C.M. (1971). Separation and characterisation of the inhibitory factor of the troponin system. Biochem. Biophys. Res. Commun. 44, 237-244.

Needham, J., Shen, S-C., Needham, D.M., & Lawrence, A.S.C. (1941). Myosin birefringence and adenyl pyrophosphate. Nature (London) 147, 766-768.

Needham, J., Kleinzeller, A., Miall, M., Dainty, M,. Needham, D.M., & Lawrence, A.S.C. (1942). Is muscle contraction essentially an enzyme substrate combination? Nature (London) 150, 46-49.

Oplatka, A. & Tirosh, R. (1973). Active streaming in actomyosin solutions. Biochem. Biophys. Acta 305, 684-688.

Pembrick, S.M. (1980). Phosphorylated L2 light chain of skeletal myosin is a modifier of actomyosin ATPase. J. Biol. Chem. 255, 8836-8841.

Perrie, W.T., Smillie, L.B., & Perry, S.V. (1973). A phosphorylated light chain component of myosin from skeletal muscle. Biochem. J. 135, 151-164.

Perry, S.V. (1951). The adenosinetriphosphatase activity of myofibrils isolated from skeletal muscle. Biochem. J. 48, 257-265.

Perry, S.V. (1952). The ATPase of lipoprotein granules isolated from muscle. Biochem. Biophys. Acta 8, 499-509.

Perry, S.V. (1956). Relation between chemical and contractile function and structure in skeletal muscle. Physiol. Revs 36, 1-76.

Perry, S.V. (1967). Structure and interactions of myosin. Prog. Biophys. Mol. Biol. 17, 327-381.

Perry, S.V. (1996). Molecular Mechanisms in Striated Muscle. Cambridge University Press, Cambridge.

Perry, S.V. & Grey, T.C. (1956a). A study of the effect of substrate concentration and certain relaxing factors on the magnesium activated myofibrillar ATPase. Biochem. J. 64, 184-192.

Perry, S.V. & Grey, T.C. (1956b). Ethylenediaminetetraacetate and the ATPase activity of actomyosin systems. Biochem. J. 64, P5.

Perry, S.V., Davies, V., & Hayter, D. (1966)."Natural tropomyosin" and the factors sensitising actomyosin ATPase to ethylenedioxybis(ethyleneamino) tetraacetic acid. Biochem. J. 99, 1c-2c.

Persechini, A., Stull, J.T., & Cooke, R. (1985). The effect of myosin phosphorylation on the contractile properties of skinned rabbit skeletal muscle fibres. J. Biol. Chem. 260, 7951-7954.

Pires, E., Perry, S.V., & Thomas, M.A.W. (1974). Myosin light chain kinase, a new enzyme from skeletal muscle. FEBS Lett. 41, 292-296.

Pollard, T.D. & Korn, E.D. (1973). Acanthamoeba myosin 1: Isolation from Acanthamoeba castellani of an enzyme similar to myosin. J. Biol. Chem. 248, 4682-4690.

Portzehl, H. (1957). Die bindung des eschlaffungs-faktors an die muskelgrana. Biochim. Biophys. Acta 26, 373-377.

Raaflaub, J. (1955). Komplex bildner als cofaktoren isolieren zellgranula. Helv. Chim. Acta 38, 27-37.

Rayment, I., Rypniewski, W.R., Schmidt-Base, K., Smith, R., Tomchick, D.R., Benning, M.M., Winkelmann, D.A., Wesenberg, G., & Holden, H.M. (1993a). Three dimensional structure of myosin subfragment-1: a molecular motor. Science 261, 50-58.

Rayment, I., Holden, H.M., Whittaker, M., Yohn, C.B., Lorenz, M., Holmes, K.C., & Milligan, R.A. (1993b). Structure of the actin myosin complex and its implications for muscle contraction. Science 261, 58-65.

Rice, R.V. (1961). Conformation of individual macromolecular particles from myosin solutions. Biochim. Biophys. Acta 52, 602-604.

Satyshur, K.A., Roa, S.T., Pyzalski, D., Creaser, M., & Sundaralingam, M. (1988). Refined structure of chicken skeletal muscle troponin C in two calcium state at 2 Å resolution. J. Biol. Chem. 263, 1628-1647.

Schaub, M.C. & Perry, S.V. (1969). The relaxing protein system of striated muscle. Resolution of the troponin complex into inhibitory and calcium sensitising factors and their relationship to tropomyosin. Biochem. J. 115, 993-1004.

Schaub, M.C. & Perry, S.V. (1971). The regulatory proteins of the myofibril. Characterisation and properties of the inhibitory factor (Troponin B). Biochem. J. 123, 367-377.

Schick, A.F. & Hass, G.M. (1949). A new method for isolation and purification of mammalian striated myofibrils. Science 109, 986-987.

Schipiloff, C. & Danilevsky, A. (1881). Uber die natur der anisotropen substanzen des quergestreiften muskels und ihre raümliche Vertheilung im muskelbunden. Zeit. Physiol. Chem. 5, 349-365.

Scheetz, M.P. & Spudich, J.A. (1983). Movement of myosin coated fluorescent coated beads on actin cables *in vitro*. Nature (London) 305, 31-35.

Slayter, H.S. & Lowey, S. (1967). Structure of the myosin molecule as visualized by electron microscopy. Proc. Natl. Acad. Sci. USA 58, 1611-1618.

Sobieszek, A. & Small, J.V. (1976). Myosin-linked calcium regulation in vertebrate smooth muscle. J. Mol. Biol. 102, 75-92.

Sobue, K., Muramoto, Y., Fujita, M., & Kakuichi, S. (1981). Purification of a calmodulin-binding protein from chicken gizzard that interacts with actin. Proc. Natl. Acad. Sci. USA 78, 5652-5655.

Stein, L.A., Chock, P.B., & Eisenberg, E. (19981). Mechanism of the actomyosin ATPase: effect of actin on the ATP hydrolysis step. Proc. Natl. Acad. Sci. USA 78, 1346-1350.

Straub, F.B. (1942). Actin. Studies Inst. Med. Chem. Univ. Szeged. 2, 3-15.

Straub, F.B. (1943). Actin II. Studies Inst. Med. Chem. Univ. Szeged 3, 23-37.

Straub, F.B. & Feuer, G. (1950). Adenosine triphosphate, the functional group of actin. Biochim. Biophys. Acta 4, 455-470.

Strynadka, N.C.J. & James, M.N.G. (1989). Crystal structures of the helix-loop-helix calcium binding proteins. Annu. Rev. Biochem. 58, 951-998.

Szent-Gyorgyi, A. (1941-1942). The contraction of myosin threads. Studies Inst. Med. Chem. Univ. Szeged 1, 16-26.

Szent-Gyorgyi, A. (1945). Studies on muscle. Acta Physiol. Scand. 9, Supp. XXV, 1-115.

Szent-Gyorgyi, A. (1949). Free energy relations and contraction of actomyosin. Biol. Bull. 96, 140-161.

Szent-Gyorgyi, A.G. (1953). Meromyosins, the subunits of myosin. Arch. Biochem. Biophys. 42, 305-320.

Takahashi, K., Hiwada, K., & Koboku, T. (1986). Isolation and characterisation of a 34 kDa calmodulin and actin binding protein from chicken gizzard smooth muscle. Biochem. Biophys. Res. Commun. 141, 20-26.

Tokiwa, T. & Tonomura, Y. (1965). The pre-steady state of the myosin adenosine triphosphate system. J. Biochem. (Tokyo) 57, 616-623.

Trentham, D.R., Eccleston, J.E., & Bagshaw, C.R. (1976). Kinetic analysis of ATPase systems. Quart. Rev. Biophys. 9, 217-281.

Tsao, T-C. (1953). Fragmentation of the myosin molecule. Biochem. Biophys. Acta 11, 368-382.

von Muralt, A.L. & Edsall, J.T. (1930a). Studies on the physical chemistry of muscle globulin III. The anisotropy of myosin and the angle of the isocline. J. Biol. Chem. 89, 315-349.

von Muralt, A.L. & Edsall, J.T. (1930b). Studies on the physical chemistry of muscle globulin IV. The anisotropy of myosin and double refraction of flow. J. Biol. Chem. 89, 351-386.

Weber, A. (1959). On the role of calcium in the activity of adenosine 5 triphosphate hydrolysis by actomyosin. J. Biol. Chem. 234, 2764-2769.

Weber, H.H. (1925). Das kolloidale verhalten der muskeleiweisskorper II. Isoelektrischer punkt and loslichkeit des myosins. Biochim. Zeit. 158, 473-490.

Weber, H.H. (1934). Ergeb. Physiol. Exptl. Pharmakol. 36, 109-150.

Weber, H.H. & Portzehl, H. (1954). The transference of muscle energy in the contractile cycle. Progr. Biophys. Biophys. Chem. 4, 60-111.

Weeds, A.G. & Lowey, S. (1971). Structure of the myosin molecule II. The light chains of myosin. J. Mol. Biol. 61, 701-725.

Wilkinson, J.M., Perry, S.V., Cole, H.A., & Trayer, I.P. (1971). Characterisation of the components of inhibitory factor (troponin B) preparations of the myofibril. Biochem. J. 124, 55P.

Wilkinson, J.M., Perry, S.V., Cole, H.A., & Trayer, I.P. (1972). The regulatory proteins of the myofibril. Biochem. J. 127, 215-228.

Xie, X., Harrison, D.H., Schlichting, I., Sweet, R.M., Kalabokis, V.N., Szent-Gyorgyi, A.G., & Cohen, C. (1994). Structure of the regulatory domain of scallop myosin at 2.8 Å resolution. Nature (London) 368, 306-312.

Yount, R.G., Cremo, C.R., Crammer, J.C., & Kerwin, B.A. (1992). photochemical mapping of the active site of myosin. Phil. Trans. Roy. Soc. 336, 55-61.

Chapter 5

THREE HUNDRED YEARS OF BACTERIAL MOTILITY

Judith P. Armitage

INTRODUCTION

I realize that the title of my contribution is somewhat erroneous; bacteria have probably been swimming for 3 billion years, but we have only known about it for 300 years, and perhaps only realized in the past 10 years that it may be of fundamental importance to all aspects of our lives, from catching diseases to having enough food to eat!

Although this volume concentrates on great advances in biochemistry over the past 30 years I hope that readers will bear with the increased time scale of my overview by an order of magnitude! The purpose is to show the rate of progress, and also illustrate that at a phenomenological level bacterial behavior had been expertly described in the nineteenth century. A few very perceptive, enthusiastic men devoted a good deal of their lives to the phenomenon, and it then remained untouched for decades to follow. Perhaps

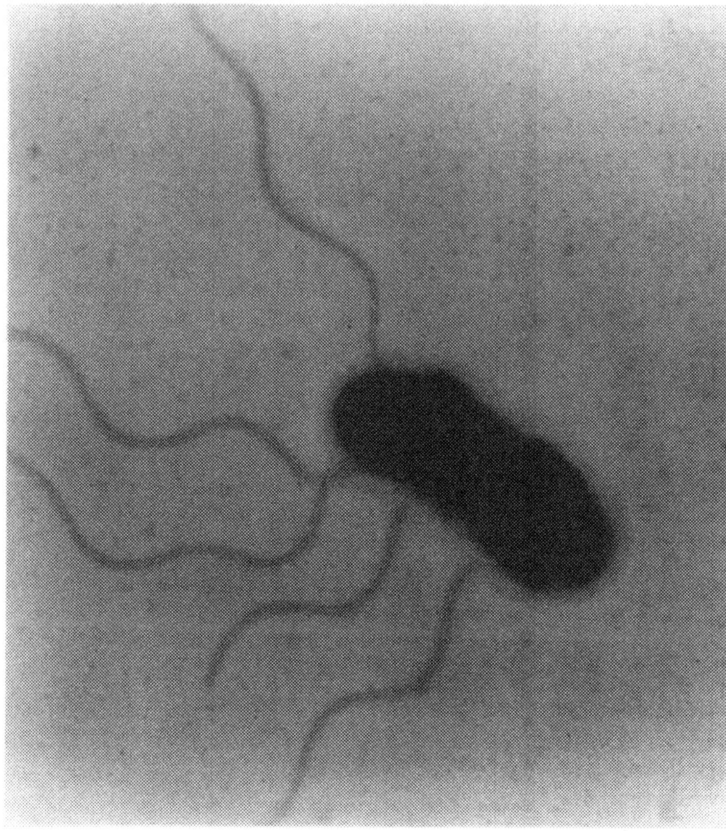

Figure 1. Negatively stained *E. coli* showing 5 flagella. (Adler, 1965 with permission).

five made any real contribution to the field before 1960, and this should not be forgotten in any article since their contributions were so very large. This illustrates the change in approach to research, from the enthusiastic amateur through the early professional scientist with a broad based view of biology to the modern teams dissecting problems at a molecular level, and perhaps sometimes losing sight of the foundation of their research—bacterial swimming—in the minutia of genes and proteins. What is apparent is that the majority of us work away producing little snippets of data to add to the pile, while every now and then someone comes along and looks at the problem from a different angle and, often by means of a simple experiment, makes that great leap forward that can only leave the majority of us feeling suitably humble.

THE BIRTH OF A SUBJECT

It is arguable that the birth of bacteriology as a subject was completely dependent on bacterial swimming. There may be upwards of a million different bacteria species, a large percentage of them motile. If they hadn't been motile, the discovery of bacteria, the oldest of the living domains, would have had to wait at least another century, rather as if Captain James Cook had missed Australia and it had taken another century for an explorer to reach the largest of the free continents. The deliberate swimming movements of tiny particles in pepper water seen by Antony van Leeuwenhoek[9] in 1676 convinced him he was looking at "very little animalcules." The detailed descriptions and measurements of larger protozoa makes it more or less certain that he was looking at motile bacteria, and in all his experiments he used active movement as the definition of life. His descriptions show that he was well aware of the differences between what we now know as Brownian motion and deliberate swimming. [All the descriptions of Leeuwenhoek's work comes from an excellent biography and translation of his letters written in 1932 by C. Dobell (Dobell, 1932).]

Leeuwenhoek was in many ways a very unusual man to be one of the founding fathers of a branch of science. He was a draper and solid citizen of Delft with no formal education and no attachment to any university. He carried out all his microscopical studies while working as a chamberlain and wine-gauger to the Sheriffs of Delft and continuing his commercial life. He never published a scientific paper or book, instead putting down all his observations in letters, written in Dutch, rather than the usual Latin or English, and sending them to the Royal Society in London, then a very new organization, where they were read to the Fellows. Although his eventual election to the Society in 1680 was, by his own admission, one of the most important honors of his life, his commercial life (and the intermittent war between Holland and England!) stopped him ever visiting and presenting his observations in person to the Society.

How did it come about that a draper rather than an academic made the first observations of bacteria? It is probably less surprising than it seems at first sight. Hand lenses had long been used to assess the density of the warp and weft in cloth and may have been one of his everyday tools during his apprenticeship in Amsterdam. In addition he made his one and only visit to London and Kent in 1668, the year after the second reprint of Robert Hooke's highly influential and popular book "Micrographia" in which Hooke describes the appearance of different textiles under a lens. Although published in English it is likely that a book so important for his commercial life could easily have been translated by a friend—indeed, Leeuwenhoek wrote many letters which were translated by friends into English before being sent. It seems too much of a coincidence that the year after his visit to the south of England

Leeuwenhoek made his first microscopic observations, many of them of the same type of object described by Hooke.

Initially Leeuwenhoek looked at inanimate objects but soon turned his interest to living things and in 1674 he made his first unmistakable identification of protozoa and in 1676 of bacteria. We probably have the enthusiastic first secretary to the Royal Society, Henry Oldenberg, to thank for the volume of material written by Leeuwenhoek because Oldenberg took it upon himself to bring the new Royal Society to the attention of anyone he thought might be interested all across Europe (to the extent that he, a German, was briefly imprisoned in the Tower of London as a possible spy because of his numerous correspondence with Europe). In 1668, *Philosophical Transactions* published an account of a new microscope made by Eustachio Divini through which could be seen "an animal lesser than any of those hitherto seen." This prompted one of Oldenberg's correspondents, Reiner de Graaf, a physician and friend of Leeuwenhoek, to write him that "a certain most ingenious person here, named Leeuwenhoek, has devised microscopes which far surpass those we have hitherto seen" and enclosing the first of Leeuwenhoek's letters describing his observations with his single lens microscope. The first letter is modest and suggests that he may have been looking down microscopes for about 5 years before that first letter. At almost the same time Constantin Huygens wrote to Robert Hooke, the President of the Royal Society, that Leeuwenhoek although "unlearned both in sciences and languages, but of his nature exceedingly curious and industrious—I trust you will not be unpleased with the confirmation of so diligent a researcher as this man is." The Royal Society obviously liked what they read and there followed continuous correspondence over 50 years, with Leeuwenhoek's last letters being sent from his deathbed at the age of 90.

His observations made him very famous in his lifetime with kings, queens, emperors, and tsars beating a path to his door to look down his single lens microscopes, to the extent that eventually he turned away anyone without an appointment. Oddly, however, the microscope was a curiosity of the age and by the late 1690s Hooke recorded that microscopes were generally out of use and repute for lack of "any inquisitive genius in the present Age." The field was left to Leeuwenhoek, and then abandoned for decades to follow.

What did Leeuwenhoek see with his single lens and how can we be so sure that what he saw included bacteria? In July 1676 he "incubated" an infusion of crushed pepper corns in water. When he examined this infusion he described some organisms smaller than any he had seen before and ... "their whole body appeared of one and the same thickness throughout, without my being able to make out a head or other bodily parts, and therewithall their body was very clear and it was thus very troublesome to succeed in seeing them alive in the water. They moved with bendings as an eel swims in the water; only with this difference, that whereas an eel always swims with its head in

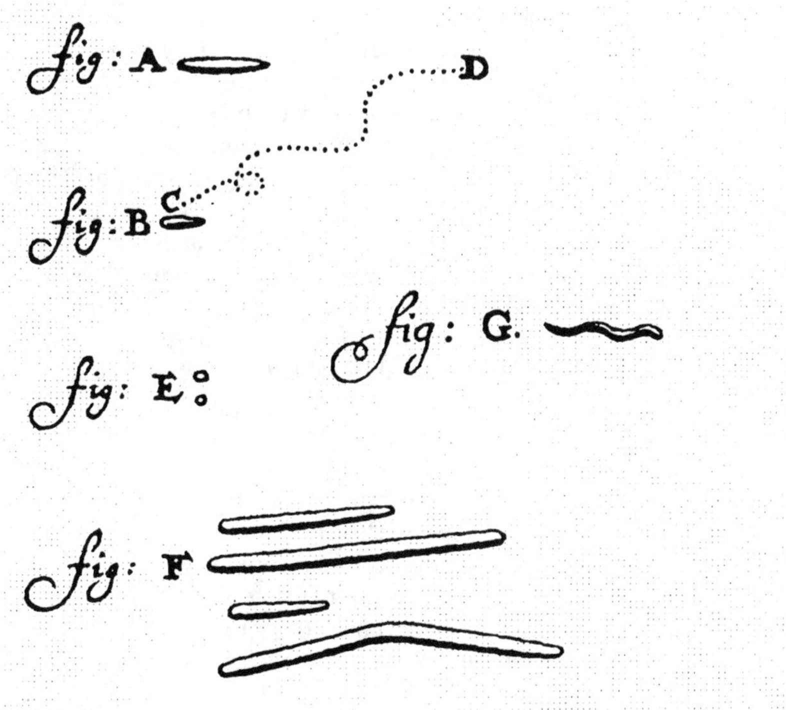

Figure 2. First description of bacterial swimming from Leeuwenhoek's letter to the Royal Society of 17th September, 1683.

front and never tail first, yet these animalcules swam as well backwards as forwards, though their motion was slow." This sounds very much like the swimming of spirilla which we now know swim using bipolar tufts of flagellar and swim equally well in either direction. He went on to say ... "that no more pleasant sight has yet come before my eye than these many thousands of living creatures seen all alive in a drop of water, moving among one another, each several creature having its own proper motion" (see Figure 2). When he added vinegar to the pepper water ... "very little animalcules which had been multifarious then lay without motion and after a time I concluded they were all dead." This tends to confirm that motility was his criterion for life. That they were bacteria can be concluded from their "exceedingly small" size relative to the little animalcules he describes with "tails" (which were clearly flagellate protozoa), their uniform thickness, and backward and forward swimming. When relating the "very small animalcule" to what was probably a ciliate he described it as a good 1000 times smaller, with the whole body no bigger than a "paw" of the larger animalcule.

There have been many doubters both during Leeuwenhoek's lifetime and since that he could really have seen something as small as a bacterium with a single lens microscope. This was not helped by his secrecy over his methods of observing them. It seems certain he used capillaries clipped to the lens which were rested against his forehead, but he had "a particular method for observing very small creatures " which he kept "for himself alone" along with his best microscopes, thus keeping himself ahead of the field (competition in science is therefore not new!). However in 1677 the minister to the English Congregation in Delft visited and wrote to the Royal Society validating Leeuwenhoek's observations when he saw swimming animalcules in water drawn into a capillary the thickness of a horsehair which stopped moving after the addition of a small amount of vinegar. Hooke himself confirmed Leeuwenhoek's observations in an unpublished letter to the Royal Society and at a meeting at which the Fellows examined Hooke's pepper water infusions. Much more recently Brian Ford has re-examined surviving microscopes and the probable specifications of many that were lost and shown that those still in existence can magnify up to ×266 (to 1.3 μm resolution) and his lost microscopes, particularly the so called "Utrecht" lens, may well have managed over ×400, making it more or less certain, combined with the detailed descriptions, that he was looking at swimming bacteria (Ford, 1991).

Over the years Leeuwenhoek made numerous observations of swimming bacteria isolated from all sorts of environments, from pepper water to tooth scrapings to fecal samples from all sorts of animals, including himself. After incubating ground pepper in rainwater in his bedchamber for 3 days he found "so great a number of such inconceivably small creatures that a man's mind may not contain them all and in my judgement the sort that were most plentiful were much more than one thousand times thinner than a hair of one's head and 3 to 4 times as long as thick." When looking at protozoa he could obviously see their flagella and cilia, describing them as their "paws" but with the "living animalcules which are yet more than 100x less" he could make out no "paws" "though from their structure and motion of their bodies I am persuaded that they too are furnished with paws withal: and if their paws are proportional to their body, like those of the bigger creatures on which I see paws, then, taking their measure at but 100× less it follows that a million of their paws together would but make up the thickness of a hair of my beard; while these paws, besides their organs of motion, must also be furnished with vessels whereby nourishment must pass through them." Later in a letter to Huygens, translated for Hooke, he states that "oft times my thoughts run to the extreme small vessels and sinews wherewith the very little animalcules are furnished withal."

Using his microscope he even discovered anaerobic bacteria, sealing a capillary of pepper water and incubating it for several days. His observations again confirm that he must have been watching bacteria as he was very puzzled by the fact that although their numbers increased dramatically he never saw any "young" but the cells were all always the same diameter and more or less the same size, suggesting binary fission. Leeuwenhoek's most famous observations came in 1683 when he drew a figure of about five species of bacteria isolated from scrapings from between his teeth. The descriptions of the behavior of the different species along with the drawings suggest he saw bacilli: "(they) shot through the water like pike—with very strong and swift motion"; cocci "moved ahead nimbly and hovered so together as you might imagine a big swarm of gnats or flies, flying in and out amongst each other," and spirochaetes: "an unbelievably great number of living animalcules a-swimming more nimbly than any I had seen up to this time. They bent their bodies in such curves in going forward—they were in such numbers that all the water seemed to be alive." Again the cells were of uniform thickness and they moved equally well with either end moving. It seems probable that he also was the first person to note gliding motility in bacteria as he describes another organism isolated from the mouth which again was of uniform diameter but longer than before and "these had so little motion that I had most times to confess they might not be living creatures at all yet when I kept my eyes on them without getting tired I could make out that they bend their bodies very slow in a very faint curve." He also realized that viscosity affected their ability to swim and compared the ability to move through the scraping of his teeth to when they were mixed with spittle and "the animalcules find themselves released and never stop moving."

His detailed observations and description of swimming appear to have attracted no contemporary disciples, perhaps because he had no academic appointment during his life and remained a gifted amateur. There were no students to carry the torch into the eighteenth century. We therefore have few records suggesting any real interest in bacteria until the beginning of the nineteenth century. Indeed, even Linnaeus doubted that the organisms seen by Leeuwenhoek and others presented anything other than a developmental stage of fungi. The gauntlet was not picked up until a group of German microbiologists began to use the compound microscopes that evolved from telescopes to examine bacterial behavior. This work was not now carried out in isolation, as it had been by Leeuwenhoek, but became part of the increasing momentum pushing microbiology forward as a major discipline.

THE NINETEENTH CENTURY

It is much easier to be sure of the bacterial species studied by the nineteenth century microbiologists, not only because their descriptions are much clearer that those of Leeuwenhoek and their systematic study better documented, but also because these studies were carried out in academic environments and generations of students carried the knowledge through to the present day. The first really clear account of bacterial swimming, and the first sighting of those "paws" which so fascinated Leeuwenhoek, came from the work of Christian Gottfried Ehrenberg from Berlin (Ehrenberg, 1883). Ehrenberg was a great traveller and wrote a large number of monographs which were eventually published in 1883 as a complete volume about 50 years after his original observations started. He was almost certainly looking at the photosynthetic bacterium *Chromatium* which swims using bundles of flagella at its poles. Interestingly, for the next century, studies on the motility of photosynthetic species were to run neck and neck with those on the ubiquitous *Escherichia coli* and it was only the advent of molecular genetics that caused the photosynthetic species to fall out of favor as the species of choice. Ehrenberg was the first person to record observing flagella *in vivo*, recording "wave-shaped" flagella which he thought were probably required for motility. Almost 50 years after Ehrenberg, in 1881, another German, Thomas W. Engelmann[10], isolated *Chromatium* from the Rhine, visible from his window, and carried out the first behavioral studies on bacteria.

Engelmann provides us with the earliest scientific account of bacterial behavior in a modern sense. By this time bacteria as a group had been recognized and modern classification begun, although there was disagreement as to whether they should be classified with plants or animals (a controversy which was, some might be surprised to note, continuing in my student days!). The study of bacterial behavior led Engelmann to write that "...they are only considered to be the most simple of all existing organisms because they are the smallest, which we happen to know. But small and simple is not the same thing...." In 1881 Engelmann looked at "putrefactive" bacteria and found that when the oxygen concentration started to fall they would move to places where there was more oxygen—the edge of the coverslip or around air bubbles given off from green algal cells under microscope illumination. As long as the oxygen remained high they sustained vigorous motility. He also noted that carbonic acid not only slowed them down but "great uneasiness is caused to them" which he suggested meant a capacity for sensing differences in oxygen and carbonic acid tension or "breathing sensitivity" similar to that seen in animals. He also saw that bacteria would accumulate around certain nutrients, suggesting a hunger response, a sensation he used to suggest that bacteria were undoubtedly animals and their behavior proved the "unity of organic nature."

Using the bacterium, "Bacterium photometricum", probably again a species of *Chromatium,* Engelmann carried out not only the first aerotaxis experiments but also the first phototactic ones. He placed an opaque disc with a circular hole between a light source (a burner) and a microscope and using a totally reflecting prism he could focus a light spot onto the otherwise dark, sealed slide. Bacteria accumulated in the light spot, after 10 min filling the entire illuminated space. He then carried out an experiment using a layer of bacteria and an intense source of light from a gas light and examined where in the spectrum the bacteria accumulated and compared this to the absorption spectrum of living cells. He found that the cells migrated to the red end of the spectrum, particularly in the far-red at 850 nm, with weaker accumulations between 810 and 570 nm and a third accumulation between 550 and 510 nm. However, the blue region and the visible red-to-orange region remained relatively empty. Watching the bacteria swimming he saw that they appeared "frightened" back at the light/dark boundaries of the spectrum, and also at certain wavelength boundaries, for example yellow/red and far-red/red, whereas moving in the opposite direction caused no response. This led to accumulation within certain spectral bands.

The response to far-red light particularly fascinated Engelmann since it obviously differed from that of other animals and from plants. He found that the pattern of accumulation matched perfectly the absorption pattern of the pigmented bacteria. Absorption could not be measured in the far-red region of the spectrum but the intensity of accumulation in this region led him to assume that the bacteria also absorbed far-red light and that this is probably used in bacterial photosynthesis. Hence a study on bacterial behavior led to the first real identification of the absorption spectrum of bacteriochlorophyll. Engelmann analyzed his data in detail trying to fit them into current understanding. It was still not realized that bacterial photosynthesis did not produce oxygen as a by-product, but Engelmann deduced from his experiments that the behavioral responses could not be caused by oxygen being produced (Engelmann, 1881, 1883). He thought instead that light was causing the production of one or more chemicals, the slow decay of which might be producing the power for the movement of the cells, perhaps producing something akin to starch in plants. In this way he could explain the phenomenon of photoinduction, whereby light caused nonmotile dark-incubated bacteria to swim, the response to specific wavelengths of light, and the duration of swimming after light was reduced or switched off.

We know today that he was watching the proton motive force (pmf, Δp) being developed by photosynthetic electron transfer, with swimming stopping as the pmf decays after light was switched off. It did not, however, explain why high light had what he described as a "soothing effect" on cells; that is they swam at a more uniform speed than cells in low light, a phenomenon we still cannot explain.

His description of *Chromatium* swimming is perfect. He described how they swam at a fairly uniform speed, between about 20 and 40 $\mu m\ s^{-1}$, rotating around their longitudinal axis at about 3–6 revolutions per second. The cells swam primarily in the same direction but occasionally briefly reversed. If incubated in the dark there were no motile cells in the morning, but after 5–10 min in light the majority were again motile, and after a period in light it took several hours for motility to be completely lost when put back into the dark—hence his suggestion of the formation of a food compound such as starch.

A major irritation to many working on eukaryotic behavior is the tendency for people (like me!) to refer to the boundary response shown by photosynthetic bacteria as "phototaxis". The responses seen by Engelmann were reversals caused by a drop in intensity or a change in wavelength—a positive photophobic response but not a change in orientation of the cell in response to the direction from which the incident light was coming. Engelmann recognized that "true" phototaxis, as defined by Strasburger, required a response to the direction of light. When he looked for a response to the direction of light he found neither a positive nor a negative response. He therefore suggested that light sensing in bacteria must depend on the even distribution of pigment over the bacterial surface, rather than specifically localized photoreceptors as seen in eukaryotic microbes. It is salutary to think that, with the exception of understanding the nature of the driving force for the flagella, we know little more about phototaxis than Engelmann knew in his days. His questions are still my questions!

At more or less the same time, Wilhelm Pfeffer[12] was working in Tubingen, looking at the behavior of a wide variety of bacterial species (Pfeffer, 1883; Weibull, 1960; Ford, 1991). He was particularly interested in chemotaxis and osmotaxis and developed a method for measuring tactic responses that is still the central tool for looking at taxis to this day. He drew out a fine glass capillary, filled it with a chemical solution and sealed it at one end. He then placed this into a culture of bacteria and watched to see whether the bacteria swam towards and into the capillary as the chemical diffused out, or swam away leaving a zone of clearing around the mouth of the capillary (Figure 3). Using this method he was able to measure the differences in sensitivity of bacteria to a range of chemicals and show that attractants varied from species to species. He found that most species were attracted to peptone and to K^+ while *Chromatium* was attracted to hydrogen sulfide, a repellent for most bacteria. Oxygen attracted most species but sulfur bacteria and spirilla were repelled.

Pfeffer was probably the first person to look at chemosensing in a quantitative rather than a simply qualitative way, using his capillary assay to measure threshold concentrations and the differences in sensitivity both of different species and the same species to different chemicals. He also showed that after a response the sensitivity was modifiable; in fact he identified adaptation, the central most important feature enabling any creature to sense a gradient. Pfeffer wrote that "the intensity of stimulus required to produce a perceptible reaction is proportional to the intensity of the existent stimulus. Until a threshold is reached the absolute difference between the contents of the liquid in the capillary and the liquid outside will increase as the amount of stimulant outside increases. It is concluded that a stimulant in homogeneous distribution does not direct the movement of the organism; however, it does influence the perception of the stimulus by the organism." In this way he linked bacterial behavior to the general phenomenon of perception and neuroscience since this is a restatement of "Weber's Law" which states that whether or not a signal will be detected depends on how much that signal differs from the background ($L \propto \log I/I^0$ where $L =$ sensation, $I =$ intensity of signal, and $I^0 =$ background).

Having established adaptation or "habituation," as he described it, Pfeffer calculated that if bacteria had a single system for "perception" then one substance added to a level where the response was "desensitized", i.e. adapted, it should cause the same desensitization to all other compounds. He showed that this was not the case and helped discover, with his colleague, Kniep, at least three different types of "perception"—to asparagine, protons, and hydroxyl ions—suggesting that there are specific sensitivies in bacteria similar to those of taste in humans. This recognition of different sensory pathways had to wait almost 100 years for detailed molecular confirmation.

Like Engelmann, Pfeffer presumed a uniformity in life and therefore tended to compare bacteria with eukaryotic microbes, for example by assuming that all species would respond to meat extracts. He confirmed Engelmann's data on aerotaxis and concluded that aerotaxis and chemotaxis may have different sensory systems (which we now think is correct) since high oxygen caused a repellent response whereas lower oxygen caused an attractant response in many species. Pfeffer also showed that the repellent response to acids was caused by protons by looking at the response under conditions that altered the degree of dissociation not the concentration—results confirmed again very recently.

Osmotaxis was also of great interest. This is particularly poignant since, as a study, it has remained untouched from those early experiments until recently reexamined by Julius Adler[11], the man who is almost single handedly responsible for the current studies on bacterial chemotaxis. Pfeffer exposed bacteria to mixed gradients of chemicals and osmolites and showed that positive chemoresponses reduced the repellent response of a highly osmotic compound, showing that the sensory pathways are different but the final response is the result of the integration of the two signals.

Therefore, at the end of the nineteenth century it had been established that bacteria were motile by means of flagella, lacked a head or tail, and therefore probably sensed their environment by receptors distributed all over their bodies. Photoresponses paralleled absorption by photosynthetic pigment and both motility and taxis in photosynthetic species were probably linked to some sort of metabolic process. It was known that bacteria responded to gradients of a wide range of chemicals, including oxygen, and the strength and type of response was species-dependent. As in other organisms, responses showed a threshold, saturation, and adaptation to stimuli. It was also recognized that bacteria responded simultaneously to a range of different stimuli and that they modified the overall response. There the subject lay, with a few minor observations, for over 50 years.

A

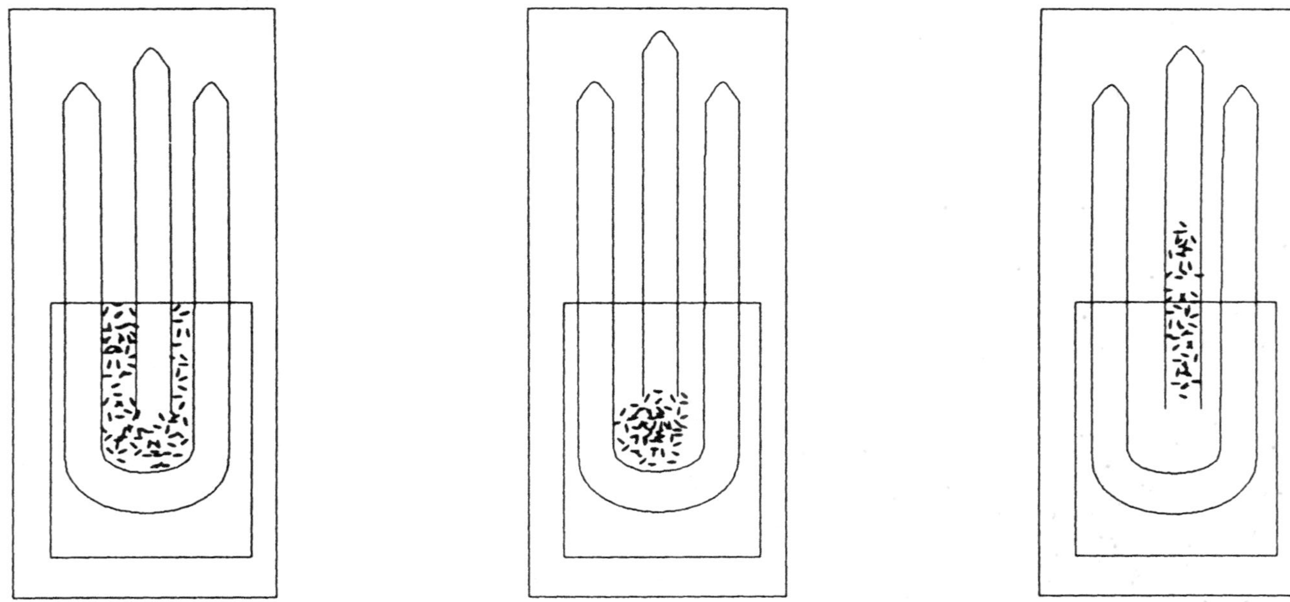

Figure 3. Capillary assay for bacterial chemotaxis devised by Pfeffer. (**A**) A blind ended capillary filled with an attractant and inserted into a suspension of bacteria. (Photograph of an original Adler capillary.) (**B**) After a set time the number swimming into the capillary is measured by washing the cells out into medium and spreading aliquots onto nutrient agar plates. (**C**) The graph shows the number of bacteria accumulating in the capillary with time, in response to 1 mM aspartate. (Adler, 1973 with permission).

B C

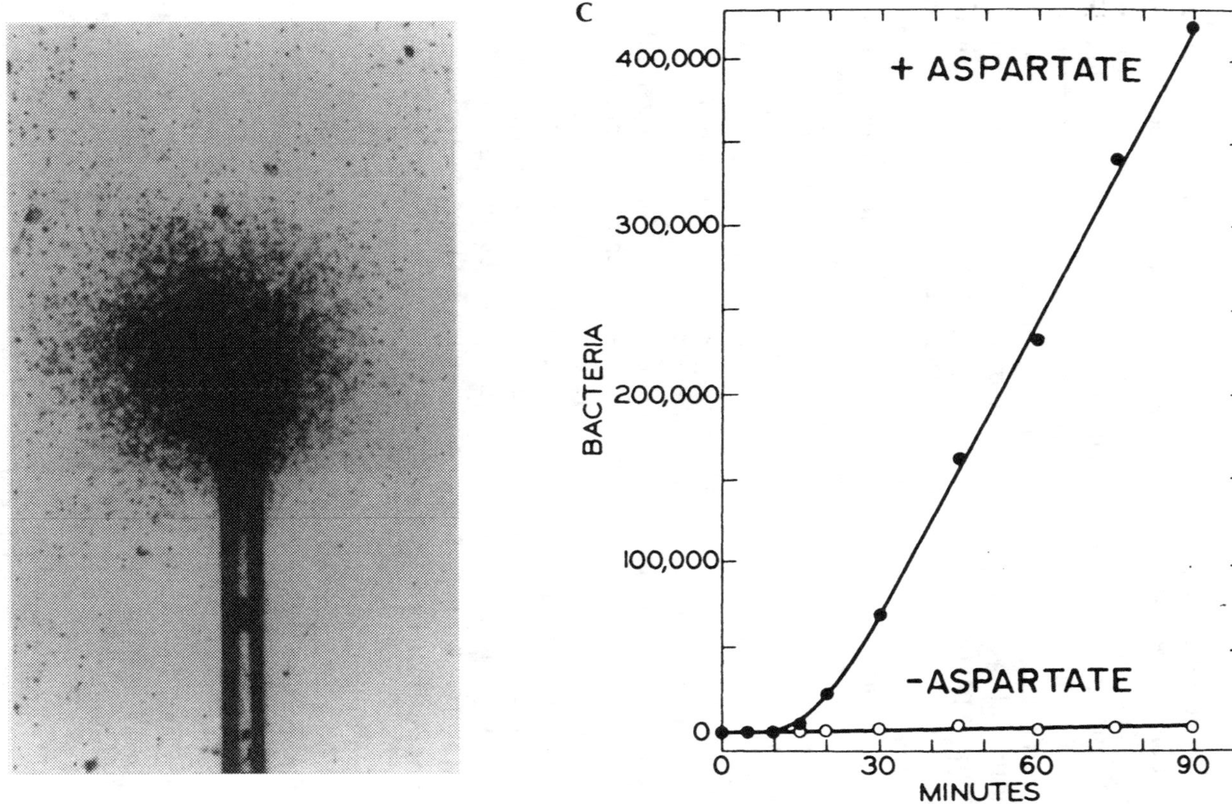

Figure 3. (continued)

THE 1950s

Behavior

In the early 1950s Roderick Clayton, a scientist who went on to make contributions to research in photosynthesis, looked at the behavior of *Rhodospirillum rubrum,* a purple non-sulfur photosynthetic bacterium (Clayton, 1953a,b,c, 1957, 1958). Manten had already reexamined phototaxis in this species and shown that the action spectrum of the behavioral response coincides with the photosynthetic spectrum (Manten, 1948). Clayton extended this and the early work of Engelmann, confirming the coincidence of the action spectrum of photosynthesis and quantifying the step-down behavioral response. He not only looked at the photoresponses but the interaction with chemoresponses and put the studies on a more quantitative footing. He quantified thresholds and the nature of the response, showing that the response depended not only on the strength of the signal but the background concentration (or light intensity). From this he formulated a hypothesis, again based on Weber's Law, but over a more limited range of 5–500%, and on recent advances in neurobiology. He suggested that changes in levels of ATP controlled behavior. By this time ATP was considered the central energy source of the cell, whereas Engelmann had thought it was starch.

Although the work of Adler with *E. coli* some 15 years later suggested that changes in energy levels are not involved in behavioral control of flagella, there is recent evidence that they may well be involved in some species. There is a salient lesson here: detailed and accurate descriptions of bacterial behavior were neglected for several decades because it was assumed that all bacterial species would behave in the same way. It now seems probable that the core of the sensing system has a very long ancestry (it is found in all three of the domains). Therefore, over perhaps 2 billion of years of evolution, bacterial species have adapted the pathway to optimize their survival in a particular environment, and there is more than one way of sensing and signaling to that central pathway. We must not be blinkered against the data from the past because more recent data appears on a different system.

The Flagellum

Motility and the number and organization of flagella on bacterial cells became primarily a tool of the systematic bacteriologist. The first edition of *General Microbiology* by Stanier, Doudoroff, and Adelberg in 1958 devotes most of the three pages on motility to describing how to stain the structures, while there is no mention at all of chemotaxis (Stanier et al., 1958). Whether or not a bacterium was motile and the number and arrangement of stained flagella were used to identify or classify a species, but nothing further.

There was still some controversy about whether or not flagellar filaments were really responsible for movement. As mentioned above, it had been possible since the nineteenth century to visualize flagellar bundles by using a high light intensity dark field. It was calculated that flagellar bundles rotated at about 40 rps when cells were swimming at a few micrometres per second, with the cell body counterrotating at about 12 rps. When the swimming speed increased, Pijper reported that although the cell body could still be seen to rotate, the flagellar bundle appeared to lose its helical shape (this we now know to be the result of the limitation of the microcinematography of the time). He therefore proposed that the helical bundles were simply extracellular slime trailing behind the cell, which took on the form of a helix because of the rotation of the cell body, and it was this rotation of the cell body which resulted in swimming (Pijper, 1948). In addition he could not always see flagella on all swimming cells by dark field, and he reported that rapid shaking could remove flagella without altering swimming behavior (Pijper, 1955).

There was a series of "sharp" letters to *Nature* between Pijper and Weibull (Pijper, 1951; Weibull, 1951) with Weibull being a major advocate of flagella as organelles of movement. Pijper was particularly worried about the cellular attachment of the filaments and how they came together as a bundle in peritricously flagellate species. He argued that if, as suggested, they arose in the cytoplasm this would require a hole (or several holes if there were large numbers of flagella) in the cell wall, and the flagella would then need to bend to come together behind the cell as the helical bundle. He considered this, with the frequent tumbling he had seen, to be far too awkward to work. Amazingly in the early papers using dark field microscopy he had observed the transitions in helical wave form, and the splitting of the bundle into individual filaments, but, because he continued to consider flagella to be products rather than causes of movement, it took a further two decades before subsequent workers would reinterpret his findings (Pijper, 1955). Pijper did some remarkable experiments and produced some outstanding pictures of flagella, but his interpretation remained limited by his resistance to the concept of flagella as organs of movement on the majority of species.

An excellent review has been written documenting the controversy between Pijper and his detractors, principly Conn, Elrod, and Morton (Strick, 1996). It admirably looks at the debate in the context not only of science, but of the personalities involved and the "battle" between the "old" visual morphological approach to science, as exemplified by Pijper, and the "modern" technology-based science, exemplified by the electron microscope used by van Iterson in Delft and Conn and Elrod in the U.S. It was also a battle between the "old-style" medical researcher and the "professional" scientists. Pijper, with some justification, felt he was the only one looking at living organisms; unfortunately his microscope, which depended on the bright light of the South African sun, was not readily reproducible elsewhere and was itself unreliable. The review is well worth reading as an insight into the sociology of a scientific debate, possibly only equalled recently in the debate over chemiosmosis. There is also a great deal more information about the development in understanding of flagellar structure and function in the 10 years between 1946 and 1956 than is possible to cover in this brief overview.

The bulk of the evidence presumed flagella were responsible for swimming and the majority of bacteriologists believed that flagella were the organs of motility. This was finally confirmed in experiments where chloramphenicol was added to cultures sheared of flagella, thus stopping subsequent protein synthesis. These cells did not regain motility, although motility was regained within 20 min if protein synthesis was allowed. These data suggested that in his experiments using sheared cells Pijper probably was looking at bacteria with short, regrown filaments.

The major regions of the flagellum had been identified but investigation remained hampered by the inability to observe individual flagella *in vivo* and by the assumption that they had to work in the same way as eukaryotic flagella, even though it had been recognized that the structures were very different. It was shown that the filaments remained attached to the bacterium after treatment with lysozyme to remove the outer membrane. In electron micrographs, basal granules, described as 100–500 Å membrane bound vesicles, could be seen at the base of the flagella, in the cytoplasm, and these were considered to be the equivalent of the blepharoplasts (or basal granules) of higher eukaryotes (Figure 4). It was known that a "hook" region passed from the basal body, through the membrane to the flagellum filament itself, and the stability of the hook was different from that of the filament, and this had been exploited to isolate the two. In suspension it was observed that the isolated filaments formed helices, which had a characteristic wavelength and amplitude when dried down into two dimensions.

Weibull did most of the excellent studies at this time and described the filament as a tube composed of fibers of a globular protein, flagellin. It was thought that these fibers probably coiled around each other, but the geometry was unclear. Flagella could be easily sheared from bacterial cells without disrupting the cell, and therefore a fair amount of chemistry and even X-ray diffraction studies had been carried out on flagellin (Ashbury and Weibull, 1949). The filaments disintegrated into their individual monomers when treated with acid or detergents. Koffler had used this property to analyse the amino acid composition of flagellin and shown that it lacked cysteine, and had very little histidine, tryptophan, proline, or tyrosine. Analysis of the amino acid composition led Weibull to put flagellin into the myosin group of proteins and develop a model by which flagella would function like contractile fibers (Wiebull, 1951).

A single bacterial species under any one set of growth conditions will usually produce only one type of flagellin protein. Following disintegration by incubating isolated flagella at acid pHs, Koffler and several other groups showed that raising the pH allowed the filaments to repolymerize at a rate dependent on temperature and protein concentration. However, despite this ability to disintegrate and repolymerize, and to form a stable helix when isolated, no research was carred out on the mechanisms involved in swimming. Instead the isolation and purification of flagellin was primarily used for the production of pure antiflagellin antibody and, because flagellin is unique for each species and even strain of bacteria, it provided (and still provides) an excellent tool for typing pathogenic strains.

The reasons for this lack of progress were clearly recognized by Weibull in the first edition of *The Bacteria* (Weibull, 1960), the limitations being the lack of suitable experimental techniques. The small diameter of the bacterial flagellum (15 nm) is below the resolution of the light microscope. The same limitations that had stopped Leeuwenhoek seeing the "paws" on his animalcules also limited twentieth century microbiologists. Although it was possible to see the bundles of flagella, as had been seen by Ehrenberg, it was difficult to interpret the behavior of a bundle.

Weibull realized that one of the major problems involved in working out how bacteria swim was not only the small diameter of the flagellum but also that hydrodynamic forces acting on microscopic organisms are very different from those acting on macroscopic organisms and therefore models designed for large objects may not apply. He recognized that, using the relationship between size and viscosity calculated by Reynolds, for a macroscopic object moving through liquid the major force is inertia, but as an object gets smaller the inertial forces get less important and viscosity becomes the dominant force.

By this time there had been numerous observations of bundles of flagella, and several workers had noted that they appeared to rotate "like a rotating corkscrew", but it was still assumed that the apparent helical rotation of the filaments was an artifact of observation, as illustrated by this statement by Weibull: "Flagellar bundles had the appearance of rotating helices i.e., the appearance characteristic of filaments down which series of helical waves travel continuously ... on the *reasonable assumption that the base of the flagellar bundle is unable to rotate relative to the cell body,* a wave motion would tend to cause the cell body to counterrotate." The slow movement of the cell body as the result of the viscous drag would then make movement through the medium possible by applying a torque to the flagellar helix. It was assumed that the helical wave was propagated by sequential contraction of subfibrils within the flagellar filament.

The energy source for flagellar activity was controversial; some workers reported contraction and increased motility in the presence of ATP, whereas others found no ATPase activity in isolated filaments (as would be found for eukaryotic flagella) and several species continued to swim in the presence of cyanide, azide, or fluoride. Fleming reported that *Proteus vulgaris* swam faster

A

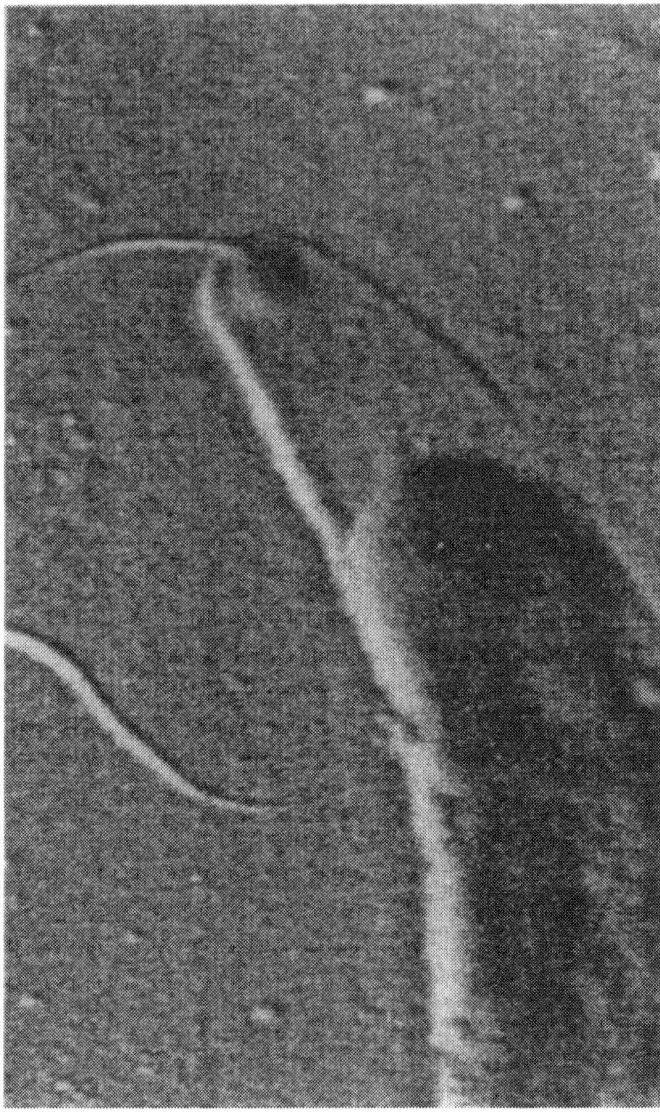

Figure 4. Evolution of the image of the flagellar motor. (**A**) Electron micrograph of a flagellum of a *Spirillum* species showing a basal granule. (Weibull, 1960 with permission). (**B**) Electron micrograph of the *E. coli* basal body showing 4 rings thought to correspond to the outer membrane (L and P rings) and the inner membrane (S and M rings). (De Pamphilis and Adler, 1971c with permission). (**C**) Electron micrograph of a bacterial basal body showing the cytoplasmic component of the motor (the C-ring). (Khan et al., 1992 with permission).

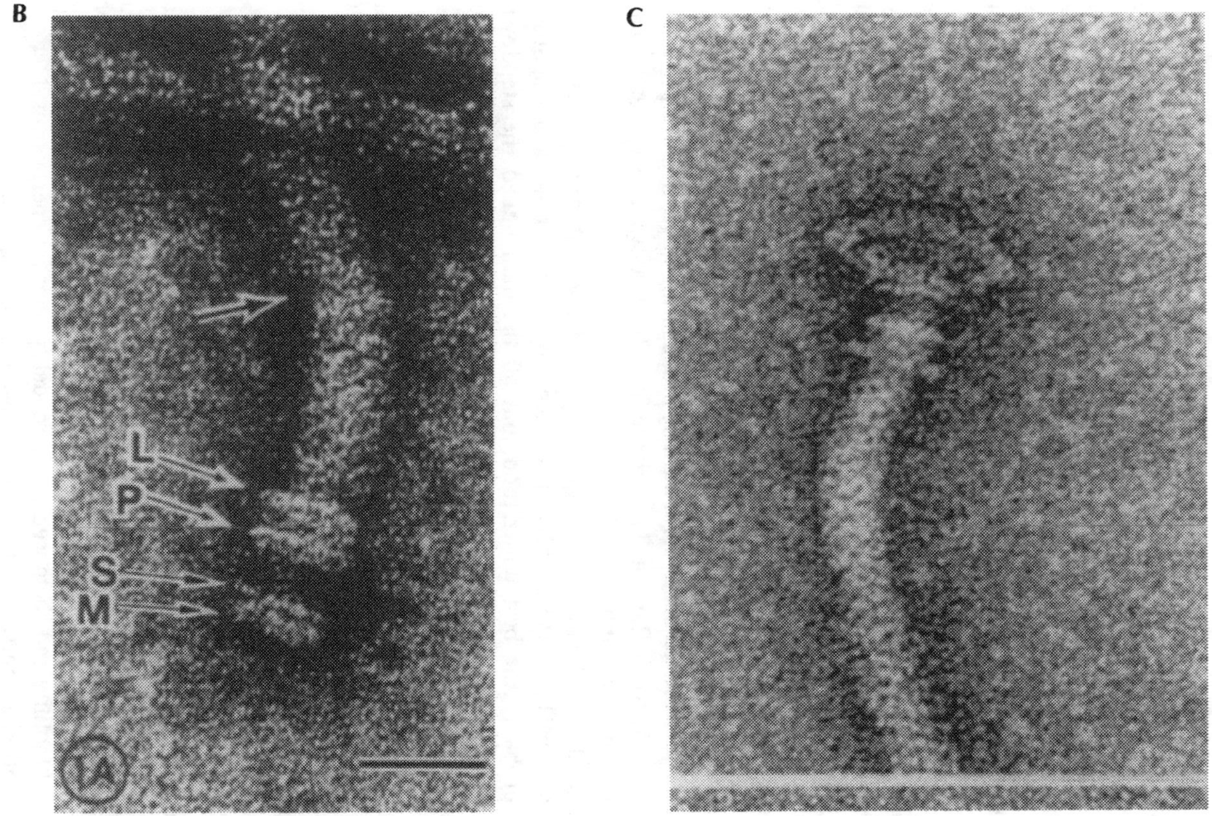

Figure 4. (continued)

when incubated at high temperature, but eventually became "exhausted", and he suggested this was similar to the behavior of a muscle fiber, and the flagellum might be considered an extracorporeal muscle fiber (Fleming, 1950). The small diameter of the filament led many to consider it impossible to include a complex enzyme system in the structure, although Weibull considered enzymes might be attached to the surface of the flagellum and be washed off during preparation.

A novel suggestion for driving bacterial flagella came from Peter Mitchell (Mitchell, 1956). He proposed that the heat generated by metabolism could pass through the hollow flagellar filaments creating a temperature gradient which could cause a streaming of water along the flagella and the subsequent movement of the bacterium in the opposite direction. He calculated that this thermokinetic process could easily drive a cell at 10 $\mu m\ s^{-1}$. His alternative suggestion was an electrokinetic mechanism whereby the cell produced a charge gradient by secreting positive charge at one end of the cell and absorbing it at the other end. The electric forces between the cell and the medium would cause them to move against each other (a suggestion transiently revived recently to try to explain free swimming in a nonflagellate cyanobacterium).

Therefore by the end of the 1960s the accepted position was that bacterial flagella were hollow structures made of protein, attached in the cytoplasm of the cell, which moved bacteria forward by propagating waves in a way probably very like eukaryotic flagella. The driving force for that wave propagation was unknown, but probably ATP.

The understanding of tactic responses was little advanced since Pfeffer and Engelmann. The study was still a phenomenological one. Phototactic responses were probably the best understood, but all the data linked the stepdown response to a change in metabolism, i.e.: "Taxis in purple bacteria is mediated through the development of an excitatory state which is transmitted to the locomotory areas, causing a coordinated motor response. The development of excitation is triggered by a decrease in the concentration of a compound whose rate of synthesis parallels the rates of anabolic activities which support growth." (Weibull, 1960).

ADLER AND THE GREAT LEAP FORWARD

The real breakthrough came with the experiments of Julius Adler, who is now considered one of the founding fathers of modern studies of bacterial behavior, an area which has grown from a means of species identification to a major area of bacterial research. There are now many laboratories worldwide using bacterial motility and chemosensing as model systems for energy transduction, transcriptional control, and sensory transduction, many of them connected directly by a scientific lineage to Adler since, unlike Leeuwenhoek, he did have students who were so excited by the possibilities of the area they went on and developed their own research laboratories, and these are now producing the next generation of research. The majority of research groups can claim a direct lineage from four laboratories, those of Daniel Koshland Jr. in Berkeley, Howard Berg in Boulder/CalTech/Harvard, Melvin Simon in CalTech, and Julius Adler in Wisconsin.

Our understanding of both the mechanisms of bacterial motility and environmental sensing advanced more or less simultaneously, and both areas owe a great debt to Julius Adler. Dr. Adler started looking at the behavior of bacteria as a more tractable system than his real love, butterflies. He used the same basic methodology as Pfeffer 80 years before, a blind-ended capillary filled with a compound and placed in a suspension of bacteria, but he used *E. coli.* He took the technique one stage further and filled the capillaries with buffer and a metabolizable attractant and watched bands of bacteria travelling down the capillary as they generated a gradient by metabolizing the chemoeffector. This led directly to the second major tool for studying chemotaxis; swarm plates (Armstrong et al., 1967). They consist of soft agar plates containing one or more metabolic chemoeffector. Bacteria are inoculated in the center of the plate and as they metabolize the nutrient they create a gradient going out from the colony. Motile chemotactic bacteria swim outwards following the gradient, forming a ring of moving bacteria. If the swarm plate contained a complex medium three rings appeared. If only two carbon sources were present, e.g. galactose and serine, there were two bands and a single carbon source produced one band (Figure 5). Using this method, Adler was able to screen for and isolate a range of nonchemotactic mutants (Armstrong et al., 1967) which he characterized and mapped onto the early *E. coli* genetic map (Armstrong and Adler, 1969a,b). The great strength of Adler's work is that he took relatively simple tools and used them to carry out exceptionally illuminating experiments; soft agar plates had been used for years as a diagnostic tool to identify whether an isolate was motile and therefore help classification, but their potential use in pure research had not been realized.

By the end of the 1960s well-characterized metabolic mutants were becoming available. Adler showed that mutants unable to transport or metabolize certain chemoattractants still showed wild-type chemotaxis towards those compounds; chemotaxis and transport were therefore separate biochemical systems (Adler, 1969; Ordal and Adler, 1974). Conversely, chemotaxis mutants were isolated that still showed normal transport of the chemoeffectors. This strongly suggested that, contrary to the idea that taxis depends on metabolic change as had been advocated by Clayton (but see later), there were specific chemosensory receptors independent of metabolism. This was supported by a series of experiments using swarm plates in which different attractants were added in a background of one or more chemoeffectors. These competition experiments identified a number of chemoreceptors specific for groups of

A

B

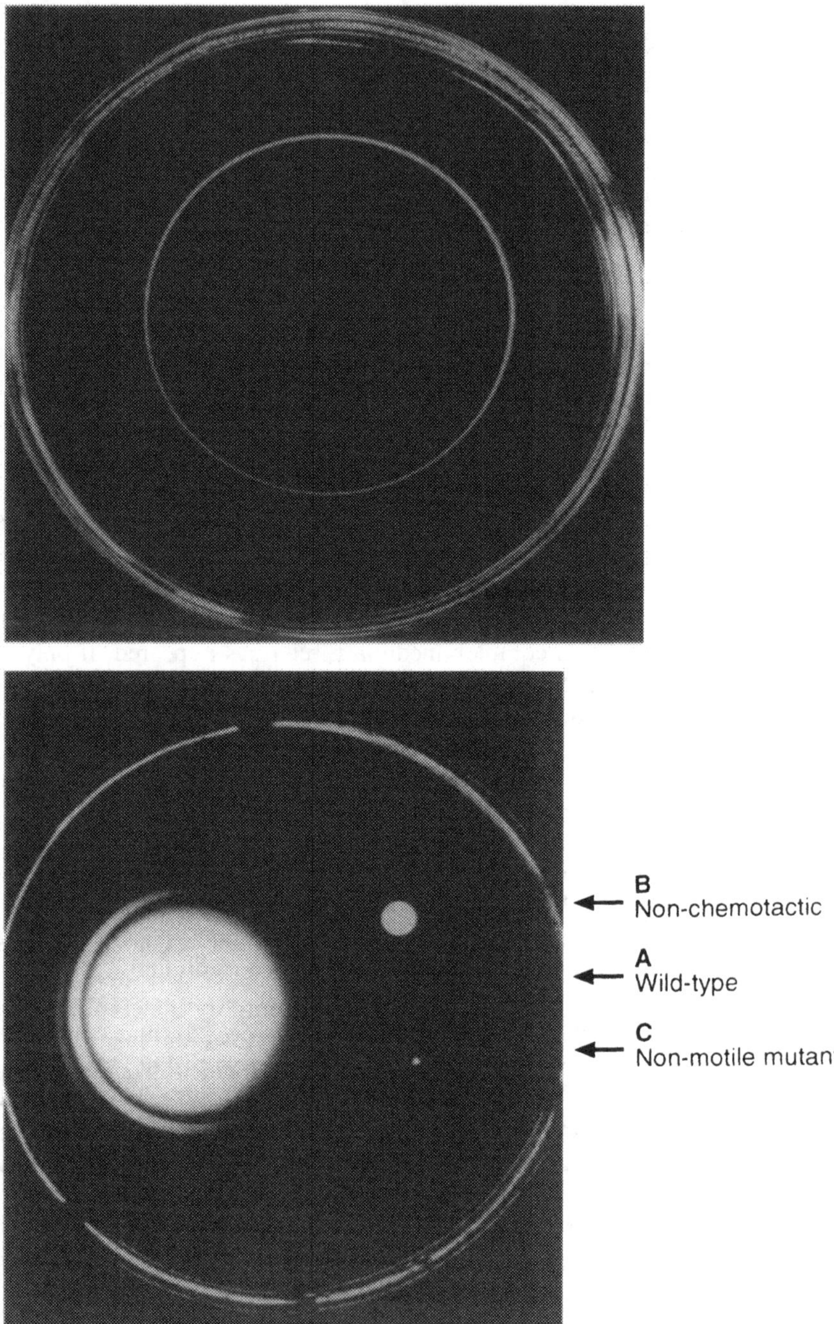

C

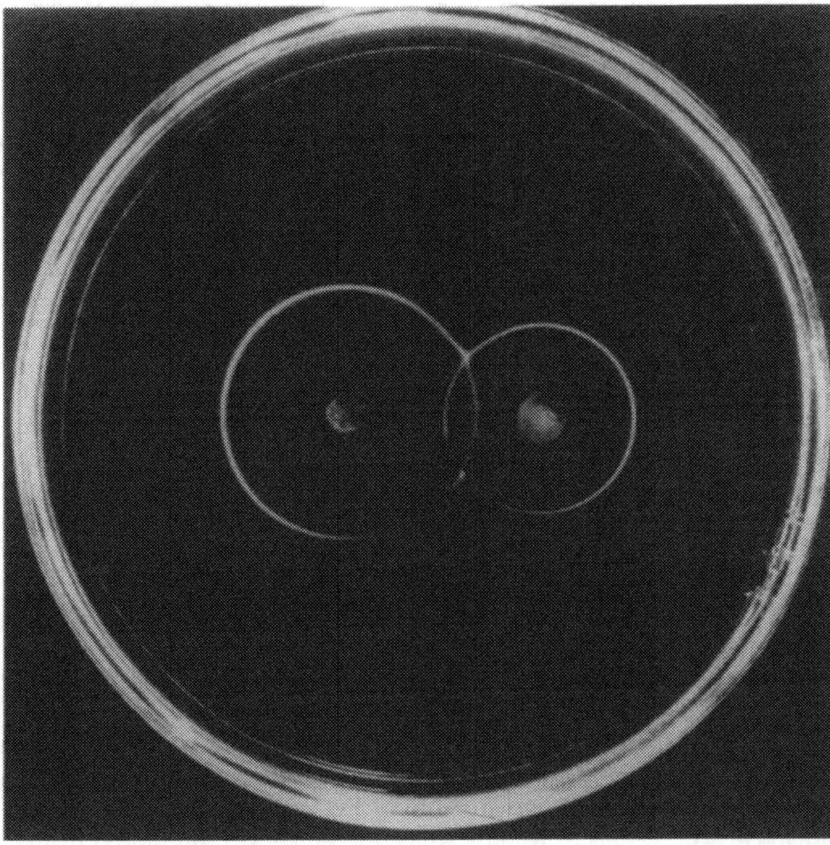

Figure 5. Swarm plates showing the behavior of *E. coli* in soft agar containing different chemoattractants (from Adler, 1976 with permission). (**A**) The use of swarm plates to isolate different motility and chemotactic mutants. On a plate containing galactose cells consume galactose where they are deposited, proliferate, and follow the gradient they create, forming an expanding ring at the edge of the attractant-rich zone. (**B**) Behavior of *E. coli* on nutrient soft agar swarm plates. A is the spreading of wild-type *E. coli*, moving out as a gradient of nutrients is created by metabolism. B is a nonchemotactic mutant which although motile, cannot respond to a gradient, and C is a nonmotile mutant which does not spread from the site of inoculation. (**C**) The response of *E. coli* inoculated at two different sites to a single metabolic chemoattractant. Two mutant strains are plated on agar containing ribose and galactose. One strain cannot metabolize ribose and the other cannot metabolize galactose. Each consumes the attractant it can metabolize, forming a gradient and an expanding ring. The two rings expand independently.

attractants and repellents on *E. coli* (Mesibov and Adler, 1972). If mutants in individual chemoreceptors were used, the number of rings on swarm plates was reduced. It was apparent that the first ring contained cells following the gradient created by metabolizing the dominant chemoeffector; the second band was of cells following a different nutrient sensed by a different chemoreceptor, and the third yet a different metabolic chemoeffector. In this way it was shown that one receptor, named Tsr, was responsible for sensing serine, Tar for aspartate and maltose, and Trg for galactose and ribose. Swarm plates and flow chambers using tethered cells (see below) have remained the two most important basic techniques in studying bacterial behavior.

Interestingly, mutants in the periplasmic binding proteins necessary for the transport of some chemoeffective sugars lost taxis, but mutants in the transport proteins themselves were normal for taxis (Hazelbauer and Adler, 1971; Adler et al., 1973). This strongly suggested that the periplasmic domain of the chemoreceptors could interact both with some individual amino acids and with sugar-bound periplasmic binding proteins to produce the same response.

The simple swarm plate method was used to identify mutants in specific chemosensory pathways and also to show that some mutants lost all chemotactic responses, while remaining motile. One of the strengths of the method was that it separated not only chemotaxis mutants, but also nonmotile from nonchemotactic mutants since the nonmotile grew as a tight colony while the nonchemotactic grew as a diffused colony. In this way Adler identified a possible sensory pathway very rapidly with a series of receptors feeding information to a common sensory pathway. There was one particular group of attractants that did require transport to produce a response. These were the sugars using the phosphotransferase transport system. However, although these compounds required transport to produce a response, the general chemotactic mutants had lost their responses to these compounds, showing that whatever signal they caused fed into the common pathway (Adler and Epstein, 1974). The mechanism by which the signal was transduced from the membrane receptor to the flagellum remained, however, a mystery.

The Rotary Motor

At about the same time several groups were looking at the flagellum itself. Julius Adler made membrane preparations from *E. coli* and *Bacillus subtilis* and produced some excellent electron micrographs of negatively stained flagellar basal bodies (DePamphilis and Adler, 1971a,b,c) (Figure 6). Interestingly the same structures had been identified in *Rhodospirillum rubrum* a few years earlier, but the observations were taken no further (Cohen-Bazire and London, 1967). The basal body appeared as a series of rings, two apparently coinciding with the outer membrane and two with the inner membrane, very reminiscent of an electric motor (Larsen et al., 1974b)!

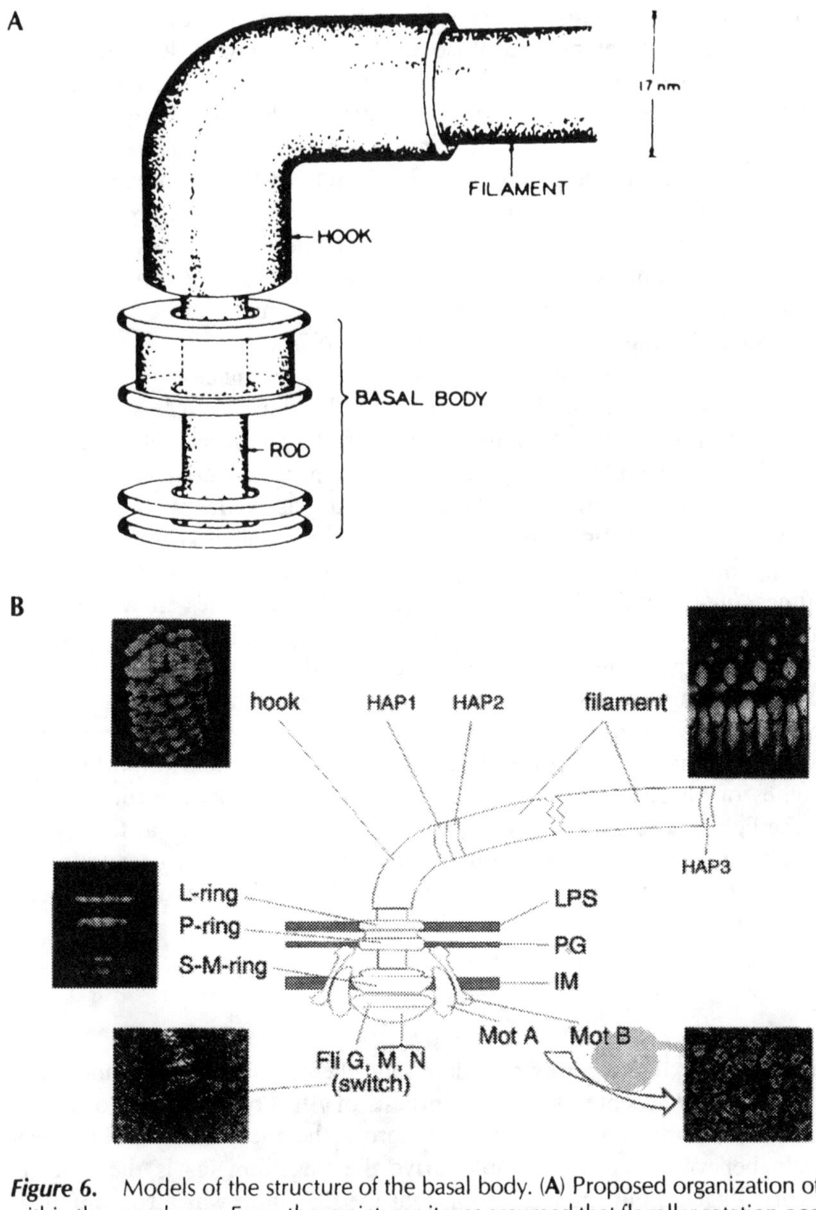

Figure 6. Models of the structure of the basal body. (**A**) Proposed organization of rings within the membrane. From these pictures it was assumed that flagellar rotation occurred as a result of the two rings in the cytoplasmic membrane rotating against each other. (DePamphilis and Adler, 1971 with permission). (**B**) Current model of a bacterial motor with electron micrographs of the individual structures. It is now realized that the S and M rings are in fact one protein and are passive structures; rotation occurs between the Mot proteins and the switch proteins in the cytoplasm. (Armitage, 1992 with permission).

Antibodies raised to purified flagella had been used for several years as a diagnostic tool for different clinical isolates. Now, however, Berg and Simon independently used these antibodies for a completely different purpose and serendipidously discovered bacterial motor rotation. Howard Berg at Boulder coated small polystyrene beads with antibody raised against *E. coli* flagella and these were then bound to the bacterial flagella. If flagella moved because of a propagated wave the antibody bound beads would move up and down, but basically stay in the same place. If, however, there was rotation of a rigid corkscrew (as it appeared in earlier studies but had been dismissed as impossible!) the beads would move in and out of focus in a set pattern. The beads rotated (Berg and Anderson, 1973; Berg, 1974).

E. coli has about six flagella and they therefore tangled and clumped if polyclonal antibody was added. However, if the flagella were sheared and allowed to grow to short stumps they did not interfere with each other. Simon's group at CalTech took advantage of this and tethered sheared *E. coli*, which had been allowed to regrow short filaments, to glass slides with antiflagella antibody. If the flagellum move as a whiplash the cell bodies might show random jiggling motion but no more, however the cell bodies rotated smoothly, sometimes in a counterclockwise direction and sometimes in a clockwise direction (Silverman and Simon, 1974) (Figure 7). This again suggested a rotating motor and showed once and for all that the driving force came from the base of the flagellum, not from enzymatic activity within the filament itself.

This all pointed to a flagellum which was a rigid helix being rotated at its base—but by what? This also provided the basis of the next important tool for studies of bacterial motility. Since individual filaments are too small to see in the light microscope it was impossible to study behavior at the level of single cells, but if the cell body could be made to rotate (and because of the drag at a much slower rate) the behavior of the flagellum could be inferred from the rotational behavior of the cell body.

Energetics

Peter Mitchell[1] had recently postulated the chemiosmotic theory and it had become generally accepted that the synthesis of ATP required the formation of an electrochemical proton gradient across the membrane. It had been generally believed that ATP would drive the flagellum, as is the case for eukaryotic flagella, but the realization that the flagellum was a rotating helix led to a major rethink. Three groups independently showed that it was indeed the electrochemical proton gradient (Δp) that drives flagellar rotation, not ATP (recently this has been expanded to show that the sodium motive force can drive rotation in alkalophiles). Adler and Berg independently showed that while uncouplers of oxidative phosphorylation, such as dinitrophenol and CCCP (carbonylcyanide-*m*-chlorophenylhydrazone), caused a loss of motility, inhibitors of ATP synthesis such as arsenate did not (Larsen et al., 1974a; Manson et al., 1977). Skulachev and co-workers also showed that motility was driven by the Δp, but extended it to show that, like the ATP synthase, the motor could be driven by either the $\Delta\Psi$ (the electrical component of the Δp) or the ΔpH (the chemical component of the Δp). Valinomycin plus potassium had no effect on bacterial swimming, although they collapsed the $\Delta\Psi$, while weak acids which collapsed the ΔpH also had no effect when added alone. However if valinomycin plus K^+ and a weak acid were added together to collapse both components of the proton gradient, motility was lost (Glagolev and Skulachev, 1978).

The first electric motor in biology had been identified! Bacteria had invented the wheel! There was a proton driven rotary motor at the base of the flagellum and everything suggested that the rings seen in electron micrographs might serve the purpose of a true motor, one anchored and one rotating as the proton flowed through, rotating a passive helix.

Observations *In Vivo*

Taking the earlier method of high intensity light microscopy one stage further, Macnab at Yale used 500-W xenon arc illumination (rather than focused sunlight as had been used a century earlier) to follow the behavior of individual flagella on swimming *S. typhimurium* (Macnab, 1976). He observed that, when swimming smoothly, all the flagella rotated together as a bundle with a certain fixed helix, but when the cells changed direction, or tumbled, the flagella flew apart and some showed a shorter wavelength with a different handedness. Experiments showed that a cell tethered by a single flagellum rotated part of the time in a counterclockwise (CCW) direction, but every now and then it switched and rotated in a clockwise (CW) direction. Together this suggested that CCW rotation resulted in a certain helical waveform which allowed the flagellar to come together in a functional bundle for swimming, but if the motor rotated CW the handedness of the helix and the wavelength changed, forcing the bundle apart and the cell to tumble. This random switching in flagellar rotation explained the three-dimensional swimming pattern identified by Berg using a sophisticated tracking system which could follow the behavior of individual cells (Berg and Brown, 1972). The technique was also used to analyse the sensing mechanisms involved in chemotactic accumulation (see below). There were now three major problems: (1) how can a structure made of a single protein produce different helical forms? (2) how can the motor rotate in two directions when the proton motive force is always directed inwards? and (3) how is the direction of rotation controlled by environmental signals?

A

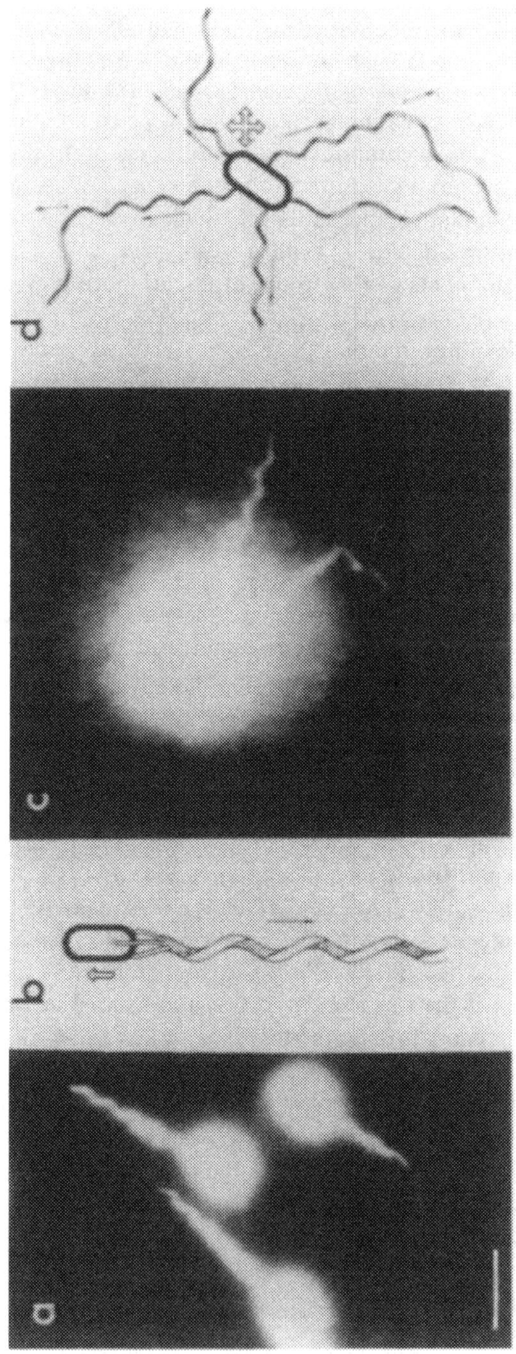

B

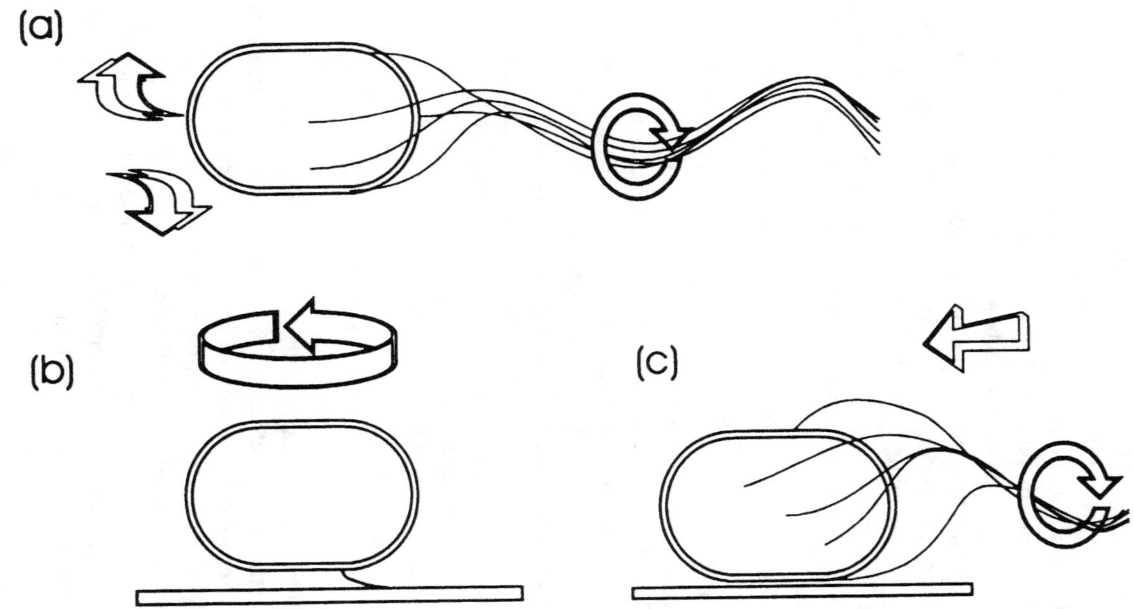

Figure 7. Methods for examining the behavior of the flagellar motor *in vivo*. (**A**) High intensity dark field microscopy of free swimming enteric bacterium. Light scattering allows individual flagella to be visualized. (a) Shows a smooth swimming cell pushed by a flagellar bundle, diagrammatically represented in (b). (c) Shows a tumble where a majority of flagella switch the direction of rotation causing the bundle to fly apart, represented diagrammatically in (d). (Macnab, 1987 with permission). (**B**) Other microscopic methods for examining the activity of the flagellar motor. (a) Although it is not possible to see the bacterial flagella using normal light microscopy, the counter rotation of the cell body, seen as body wobble, can be used to measure the activity of the motor. (b) Bacteria can be tethered to microscope slides by antiflagellar antibody and the rotation of the cell body in response to motor rotation measured. (c) The cell body itself can be tethered to a glass slide and the behavior of the flagella measured by using strobed laser dark field microscopy to analyze the rotation of the filaments.

The computer-controlled tracking system designed by Howard Berg, a physicist turned biologist, is the only one of its kind ever constructed as far as I am aware. It was (and still is!) able to track the swimming pattern of a single bacterial cell in three dimensions. Using this system, Berg followed the pattern of an unstimulated cell and compared this to that of a stimulated cell (Berg and Brown, 1972). This could then be correlated with the rotational behavior of cells tethered to glass slides by an anti-flagellar antibody seen by Simon and others. Under unstimulated conditions *E. coli* swam in an almost perfect three-dimensional random pattern, but a gradient of an attractant caused the tumbling frequency to be reduced when the cell swam towards the attractant, but remain unchanged when swimming away. In contrast, an increase in a repellent chemical, sensed through the same receptor, caused the cell to increase tumbling when moving towards the chemical (Berg and Brown, 1972; Tsang at al., 1973). Thus the random swimming pattern is biased, moving the bacterium towards a favorable environment, either towards an attractant or away from a repellent.

The tethering experiments also showed that after the initial response, whether an increase in tumbling in response to a repellent or a decrease in response to an attractant, the cells went back to their prestimulus behavior within a few minutes, the response period depending on the strength of stimulus (Berg and Tedesco, 1975). As had been suggested in the nineteenth century, bacteria did not sense the direction of a stimulus, they have neither head nor tail, but detected changes with time (Parkinson and Revello, 1978). That bacterial chemotaxis depends on temporal sensing of the environment rather than spatial sensing became one of the central tenets of bacterial chemosensing. Howard Berg did a series of modelings calculating the result of different forms of behavior, and showing that accumulation could occur with a biased random walk as long as there was adaptation (Berg, 1983).

Identification of the Receptors

Julius Adler had shown in his early studies of *E. coli* behavior that methionine auxotrophs starved of methionine were motile but did not show chemotaxis. They could respond to a stimulus but did not return to normal behavior after a period of time, instead remaining "excited", i.e. there was no adaptation and therefore no gradient sensing (Goy et al., 1978). Armstrong showed that ethionine resistant mutants of *E. coli* were also nonchemotactic and had very low *s*-adenosyl methionine intracellular pools. There therefore seemed to be a role for *s*-adenosyl methionine in bacterial chemotaxis (Armstrong, 1972; Silverman and Simon, 1977b).

Several proteins ranging in apparent molecular mass from about 55 to 60 kD had been identified by SDS-polyacrylamide gel electrophoresis in the cytoplasmic membrane of *E. coli.* These proteins were posttranslationally methylated in wild-type cells but not methylated in nonchemotactic mutants (Kort et al., 1972; Aswad and Koshland, 1975). The breakthrough in understanding the link between the methylated membrane proteins and sensory signaling was made more or less simultaneously in several laboratories (Toews et al., 1979; Rollins and Dahlquist, 1980; Hazelbauer and Engstrom, 1981; Sherris and Parkinson, 1981; Boyd et al., 1983). It was shown that what appeared to be a number of different proteins on the gel were in fact single receptor proteins whose apparent molecular weight shifted substantially when methylated. The addition of an attractant caused a change in the extent of methylation of one set of membrane proteins but would not change the methylation pattern of the other proteins. The apparent increase in methylation, and molecular weight, remained at the new level until the attractant was removed and the molecular weight and pattern of labeled bands again changed, with the methylation level being reduced and the apparent molecular weight of the remaining labeled proteins falling. These membrane proteins were renamed the "methyl-accepting chemotaxis proteins" (MCPs). It became apparent that these were the receptors for attractants and repellents. Methylation levels increased after addition of an attractant (and this caused the apparent increase in MW) while the addition of a repellent caused a decrease in methylation.

Parkinson in Utah was responsible for taking the initial mutant work of Adler further. He produced most of the chemotaxis mutants used around the world, and he continued analyzing the genetic organization of the system (Parkinson, 1980). Two nonchemotactic mutants were identified (then called CheX and CheB mutants) which like methionine auxotrophs could respond to a stimulus but did not adapt, and therefore were nonchemotactic (Yonekawa et al., 1983). These mutants also showed abnormal methylation of the MCPs. CheB mutants had MCPs which were always fully methylated and did not change the extent or apparent molecular weight on stimulation, whereas no methylation could be measured in CheX mutants. (CheX was later renamed CheR). This, with the data on methionine auxotrophs, suggested a role for methionine in adaptation, and adaptation involved the MCPs (Kehry et al., 1983). Isolated CheB was shown to be a methyl esterase, whereas CheX(R) was shown to be a methyl transferase. The two proteins competed to add or remove methyl groups from three or four glutamate residues on the cytoplasmic side of the membrane spanning MCPs.

It was realized the methylation of the membrane proteins whose change had been measured on SDS-PAGE gels was the process of adaptation to the chemotactic stimulus. Methylation of the membrane protein increased after addition of an attractant and decreased when it was removed. The apparent change in molecular weight of the MCPs on SDS gels was the result of the different number of methyl groups added to the receptor proteins, up to four per receptor with each methyl group resulting in an apparent increase in molecular weight of the protein (Chelsky and Dahlquist, 1980; Kehry and Dahlquist, 1982; Kehry et al., 1984). Dahlquist studied the kinetics of the adaptation process using labeled methionine added to cells in a flow assay, and followed the release of radiolabeled methanol after addition and removal of an attractant, thus providing the first quantitative studies on adaptation.

Sequencing and site directed mutagenesis of the MCPs revealed four glutamate residues on the cytoplasmic side (identified by tryptic digestion and hydropathy profiles) of the MCP (Kleene et al., 1977). These glutamate residues are carboxymethylated by the methyl transferase, CheR, in response to an attractant binding, and demethylated by the methyl esterase, CheB, when the attractant binding is reduced or there is an increase in repellent concentration (Parkinson and Hazelbauer, 1983). Interestingly it might be expected that bacteria that swim in straight lines would spread furthest in soft agar, but Berg showed that in fact smooth swimming cells "hit" agar molecules and cannot escape, remaining trapped at the sight of inoculation. Smooth swimming and tumbly mutants therefore have the same phenotype on soft "swarm" plates (Wolfe and Berg, 1989).

Molecular genetics became the great tool for analyzing the chemosensory system, as with all other systems at this time. However, analysis of the mutant phenotypes depended primarily on two of the simple techniques developed 20 years earlier: swarm plates and tethered cells. Swarm plates determined whether the cells were chemotaxis or motility mutants and observation of tethered cells down a microscope gave a detailed picture of the behavior of individual cells. This last method was taken a stage further when Berg and Block designed a simple flow cell for the microscope in which the environment of tethered cells could be changed rapidly and the behavior of the cells followed (a process initially done manually but now computerized) (Block and Berg, 1984). These mutant studies identified six genes whose interruption led to the loss of chemotaxis, while analysis of the behavior of tethered cells of these mutants identified different behavior patterns (Silverman and Simon, 1977a) (Figure 8). Mutations in four of the genes led to smooth swimming, no tumbling, and therefore a loss of chemotaxis.

The Signal

By the early 1980s it was known that increased binding to the cytoplasmic side of the membrane-spanning MCPs caused a signal to be sent via the four cytoplasmic gene products to the flagellar motor, and the receptors were "reset" by controlling the level of methylation. But how was the signal sent, what was the signal, and how did the signal control the flagellar motor and adaptation processes? The answers to these questions were provided by a series of different experimental approaches. Koshland's group had been very active looking at the structural aspects of signaling through the MCPs. It was shown by chemical cross-linking and tryptophan fluorescence that MCPs operated as dimers (Mowbray et al., 1985; Milligan and Koshland, 1988). There was also some suggestion that, like many eukaryotic sensors, the individual monomers came together in a signaling dimer state. Koshland showed that this was not the case. MCPs were always dimers, but their conformation changed on attractant binding. Cystine cross-linking of the membrane-spanning domains showed that the interaction of the four membrane-spanning domains of the MCP dimers was important for transducing a signal from the periplasmic to the cytoplasmic side of the membrane and, if the ability to move relative to each other was inhibited by cross-linking, signaling was also lost.

Howard Berg in another of many elegant experiments made long non-septate cells from a polyhook mutant of *E. coli.* He then attached dead *E. coli* cells to the polyhooks by an antihook antibody and could thus see the motors of the very long *E. coli* filament rotating individually. He used this system to confirm what had been initially noticed by Macnab using dark field microscopy (Macnab and Han, 1983), namely that the individual motor operate asynchronously; that is switching their direction of rotation is relatively independent (Ishihara et al., 1983). However, Berg took the observation further and iontophoresed an attractant at one end of the long cell and followed the switching of the motors down the filament (Segall et al., 1985). If any part of the sensory signal was electrical rather than chemical the signal should travel rapidly down the length of the long bacterial cell, and probably lose little of its strength. If it was a diffusing molecule it would travel a limited distance and the strength weaken as it diffused, the distance traveled being at least partly dependent on the size of the signal. He found that only the motors close to the site of stimulation responded, i.e.the signal decayed. He suggested the signaling molecule was small, in the region of 10 to 15 kD. One of the gene products known to be involved in signaling, CheY, happened to be 12 kD in size. This experiment also confirmed that the individual flagellar motors operated independently, with each having its own spontaneous reversal frequency.

Michael Eisenbach, in Israel, developed a very tricky technique he had started when working with Adler (Eisenbach and Adler, 1981). He made cell envelopes which still retained a rotating motor, allowing the roles of the different potential Che proteins and substrates in controlling motor activity to be studied directly. The outer membrane is needed for flagellar rotation, probably as an anchor for the stator of the motor, making studies *in vitro* difficult. This is such a painstaking technique with such a high level of failure that few others have successfully managed it. He was able to show that clockwise rotation of the motor depended on CheY, supporting the view that CheY is indeed the signal from the receptor to the motor. However, in his early experiments the motors of the spheroplasts appeared to be locked, he could not get them to change their direction of rotation and in the presence

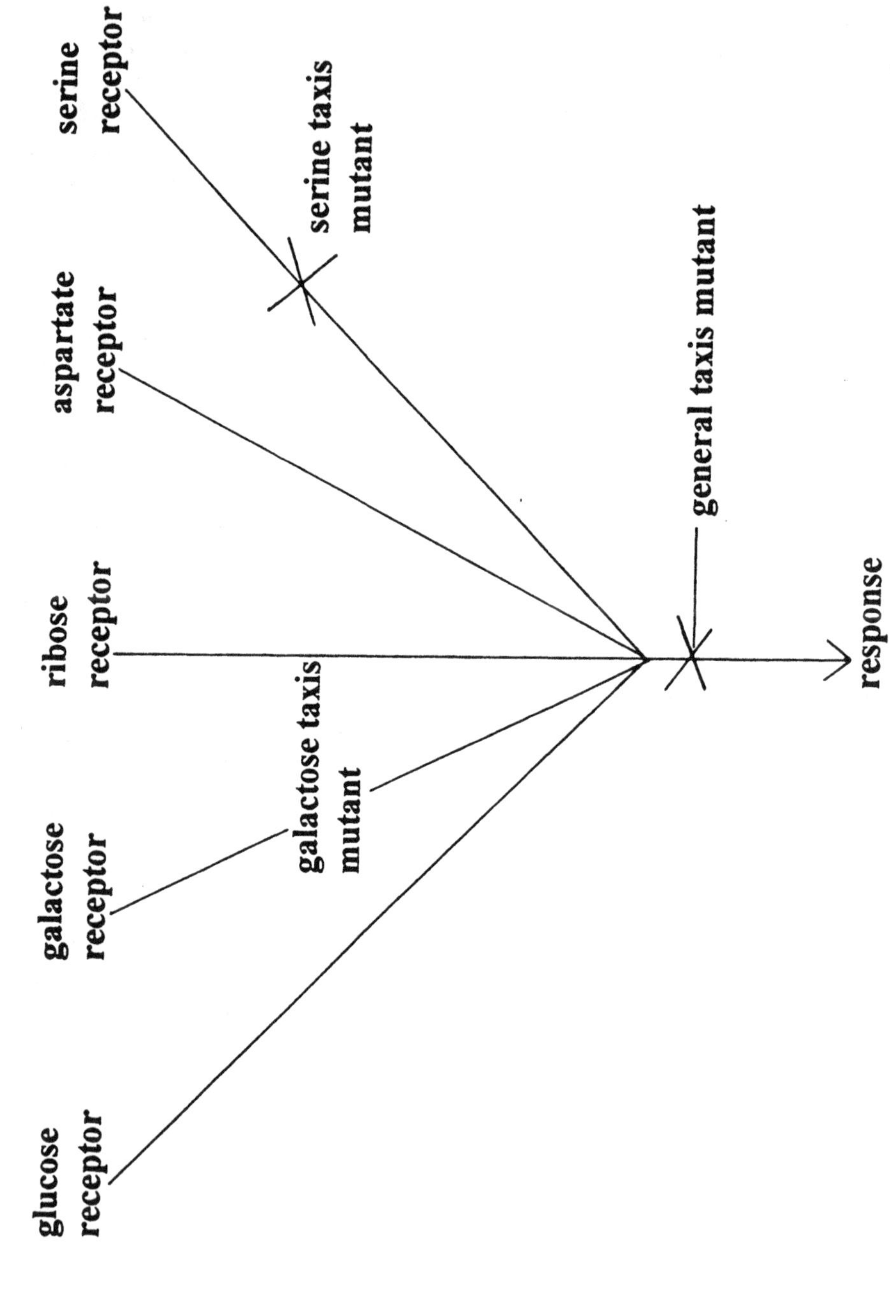
A
glucose receptor
galactose receptor
ribose receptor
aspartate receptor
serine receptor
galactose taxis mutant
serine taxis mutant
general taxis mutant
response

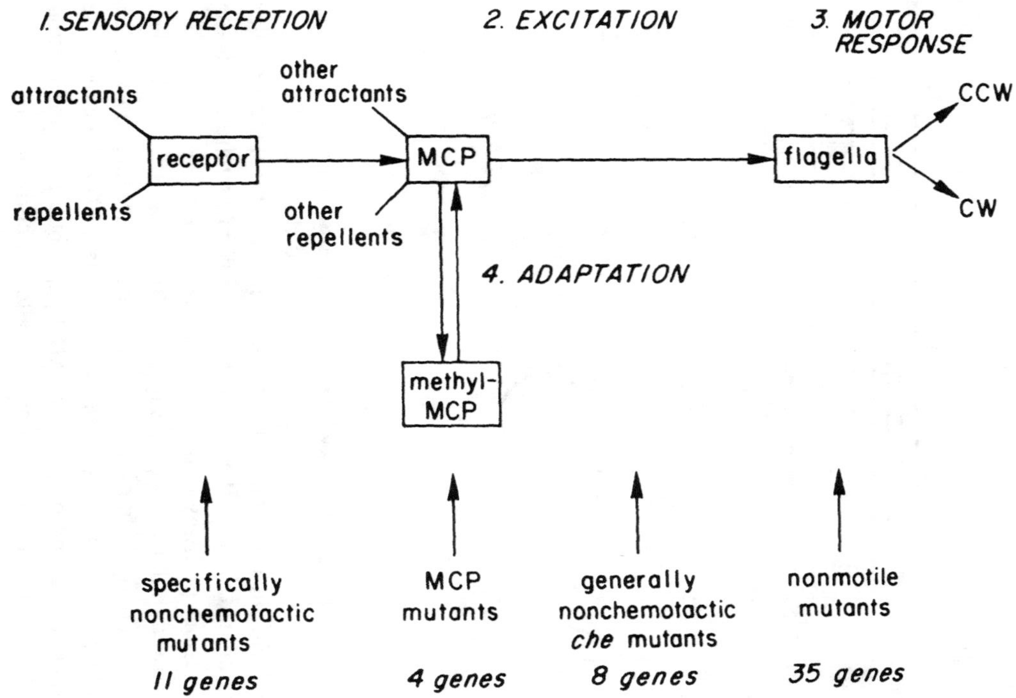

Figure 8. Evolution of the understanding of the chemosensory pathway of *E. coli*. (**A**) The view in 1969 when it was understood that there were dedicated receptors signaling to the motor, but no molecular details. (**B**) The view in 1984 after methyl accepting chemotaxis protein had been identified and the number of genes involved in each stage of the sensory pathway had been identified. The molecular details were however still not understood. (**C**) The current view of chemosensing. The individual components have been identified, sequenced, and in many cases crystallized, protein–protein interaction studies have been carried out and the path of phosphotransfer is understood.

C

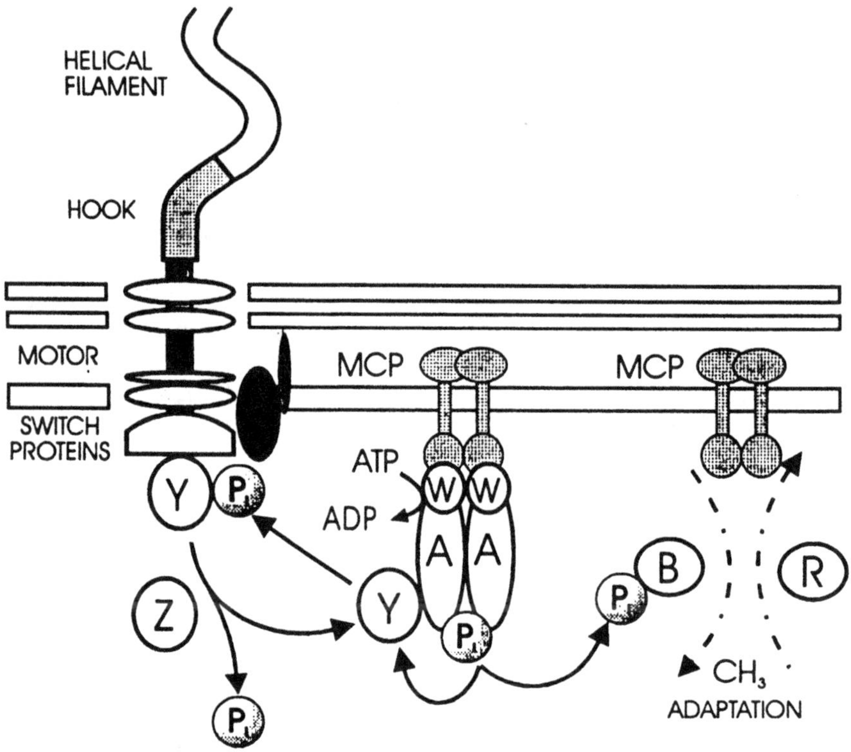

Figure 8. (continued)

of CheY they rotated exclusively clockwise (Ravid and Eisenbach, 1984; Ravid et al., 1986). Further technical refinement have allowed the roles of different Che proteins and chemically modified Che proteins in controlling flagellar rotation to be examined. This is the closest to an *in vitro* preparation of the complete chemosensory pathway that exists (see Figure 8).

Yasuo Imae showed that, as had been suggested 100 years earlier by Pfeffer, bacteria really do respond to changes in temperature, swimming towards the temperature at which they were grown (Maeda and Imae, 1979). The response was thought initially to depend only on one MCP, Tsr, but it was then found that this caused the negative response, while Trg allowed both the positive and negative response, and both are needed for a complete temperature response (Nara et al., 1991). It was, and is, assumed that temperature alters the cytoplasmic conformation of the MCPs and this is responsible for signaling in the same way as a chemical signal.

The molecular biology continued apace, and homologues of the MCPs were identified in a wide range of species by Western and Southern blot analysis (Nowlin et al., 1985). The genes were all sequenced and subjected to numerous site-directed mutagenesis. The nomenclature for the chemotaxis and flagellar genes was unified between *E. coli* and *S. typhimurium;* therefore any papers from before 1988 need to be read in conjunction with Iino et al. (1988). The kinetics for the methylation/demethylation reactions were calculated and the chemistry worked out, but the mechanism of sending the signal and its attenuation remained elusive. In the mid-1980s people in eukaryotic sensory transduction had secondary messengers, but little receptor chemistry, while the prokaryotic researchers had a lot of molecular detail about the receptor and gene organization, but understood little about the process of signal transduction.

Histidine Protein Kinase

This changed when Parkinson in Utah showed that CheA (previously CheD) was essential for signal transduction and it was a kinase, but not a "usual" transducing kinase in the eukaryotic sense, but a histidine kinase (HPK). By the mid-1980s Mel Simon's group had managed to overexpress and purify the sensory transduction proteins in concentrations high enough to carry out some classical biochemistry (Hess et al., 1987, 1988). They were able to show that CheA was both an autophosphorylase, phosphorylating at a histidine residue in the presence of ATP, and a kinase, phosphorylating CheY. Attempts at phosphorylation *in vivo* had not revealed any labeling, although evidence had suggested that phosphorylation was important in chemosensing as arsenate-inhibited taxis but not motility. The kinetics identified *in vitro* showed that the concentration *in vivo* combined with the turnover rate of CheA-P was such that detecting CheA-P *in vivo* would be very difficult. However, the recognition of phosphorylation *in vitro* allowed a massive advance in the field, particularly when it was realized that CheA, a histidine protein kinase, was one of a major family of bacterial sensory proteins.

All HPKs identified previously in prokaryotes had been involved in controlling gene transcription in response to environmental change. For example, if the concentration of available cytoplasmic glutamate falls, this is sensed by a cytoplasmic protein which alters the activity of a specific HPK which in turn phosphorylates a DNA binding protein which activates transcription of the genes coding for glutamate synthetase. Other HPK-dependent systems control responses to osmotic stress, phosphate limitation, and oxygen level.

The ability to isolate the chemotaxis proteins and study phosphorylation *in vitro* moved both the field of chemotaxis and of HPK-dependent environmental sensing forward at a remarkable pace (Kuo and Koshland, 1987;

Stock and Stock, 1987; Wolfe et al., 1987; Borkovich et al., 1989; Stock et al., 1989). It was soon shown that CheZ, one of the other essential proteins, was involved in increasing the rate of dephosphorylation of CheY (Borkovich and Simon, 1990). It turned out that the form of CheY which interacted with the flagellar motor was probably CheY-P. This was supported both biochemically and physiologically using Eisenbach's flagellate vesicles and by genetic suppressor studies. Thus CheY-P causes tumbling and CheY, which probably does not bind to the flagella motor, does not. The change in the switching frequency of the flagellum seen in chemotaxis is therefore the result of altering the concentration of CheY-P in the cell.

The rate of phosphotransfer for CheA-P to CheY was found to depend crucially on the presence of MCPs, added as membrane vesicles in the phosphorylation experiments *in vitro*, and the fourth of the essential proteins,CheW (Parkinson et al., 1983; Lui and Parkinson, 1989). Second-site suppressor studies and site-directed mutagenesis showed that an MCP dimer binds two molecules of CheW on its cytoplasmic side (Lui and Parkinson, 1989). Little was, or is, known about CheW, but the CheW proteins appear to transduce the conformational change from the MCP dimer to a dimer of CheA attached to these two CheW molecules (Liu and Parkinson, 1991). CheA also operates as a dimer, each histidine phosphorylating in response to the addition of a repellent and transferring the phosphate to a conserved aspartate on the CheY molecule. CheA transfers its phosphate to the CheY bound to the other partner of the dimer pair. CheY-P is then released and transmits the signal to the motor, causing a tumble. The whole process takes about 100 ms. The kinetics and numbers of proteins *in vivo* suggests that CheY-P formation also serves to amplify the signal coming from the MCP.

Adaptation

So how do bacteria adapt to sensory signals? Adler had shown that methanol was produced from MCPs *in vivo* (Toews and Adler, 1979) and this was taken further by Dahlquist using a combination of classical biochemistry and molecular genetics. Using flow analysis he showed that the activity of the methyl esterase involved in removing methyl groups from the MCPs was also altered on chemotactic stimulation, and controlled by CheA. This biochemical approach was used to look at the effect of adding opposing stimuli simultaneously, and the effect of adding stimuli using different MCPs. The results were important, showing that a signal through one MCP could alter, to some extent, the methylation state of another MCP, suggesting that there is communication between MCPs probably by the overall control of the esterase activity. These experiments also showed that a combination of positive and negative stimuli produced no overall net change in methanol release and no behavioral response. There is therefore an "integration" signal within the cell controlling the methylesterase activity.

The pattern of methylated proteins had shown that the majority of methyl groups placed on MCPs after a stimulus were on the MCP specific for that sensory signal, but the flow cell experiments showed that the control of the activity of the methyl esterase was global and not specific. Thus some other component must control the specificity while a global regulator controls the activity of the methyl esterase (Parkinson, 1988).The histidine protein kinase, CheA, therefore controls both signaling and adaptation. However, although both CheY and the esterase, CheB, are activated by the CheA, the turnover rate and amplification of the signal to CheY is much greater than that of CheB. The half-life of both CheY-P and CheB-P *in vivo* has been shown to be less than 10 s, which probably contributes to the short latency of the signal (Block et al., 1982).

The CheY signal is relatively fast, about 100 ms, as shown by iontophoresis and the response kinetic experiments of Berg, while adaptation takes longer—between tens of seconds and several minutes dependent on the strength of the signal and therefore the extent of CheB phosphorylation. It appears that adaptation depends on the resetting of the MCP into a nonsignaling state. If there is an increase in attractant, the level of CheA-P is reduced and the activity of the esterase relative to the stable transferase falls. The conformational changes in the MCPs that cause the sensory signal probably expose glutamate residues on the cytoplasmic signaling domains of the MCP which become methylated and reset the signaling state. Binding a repellent causes an increase in CheY-P, and thus tumbling. It also increases the activity of the esterase relative to the transferase, thus removing the methyl residues and releasing them as methanol. This again resets the signaling state and the bacterium returns to prestimulus behavior. However, the control of adaptation is still not completely understood, nor is the sensitivity of the system; only a few receptors need to change their level of occupancy to cause a response, implying a high level of signal amplification.

CheY has now been crystallized, unfortunately not in the phosphorylated state, but it shows an active acidic pocket containing the conserved aspartate residues essential for signaling and the possible face of the protein required to bind the flagellar switch (Stock et al., 1989). This should lead to a better understanding of the molecular interactions involved in signaling.

NONENTERIC SYSTEMS

All the above advances were made using *E. coli* or its sister species *S. typhimurium*, primarily as a result of the molecular techniques worked out for these species. These are two closely related species, normally living in a fairly specialized niche, but even they have sensory transducers which differ. *E. coli* has the MCP Tap which senses dipeptides while *S. typhimurium* has Tcp which senses citrate (a metabolite for *S. typhimurium* but not *E. coli*) and phenol. The different effectors for Tap and Tcp showed that there was some variation even between closely related species and it should have come as no surprise that other more distantly related species showed variation of the central signaling scheme. There are after all probably well over a million different species specialized for almost every niche on Earth with about 3.8 billion years of evolution behind them. Is it realistic to assume that other species work in exactly the same way? During the time of the great advances being made with *E. coli* a few groups were beavering away with other species. It appears that the "Che"-based sensory system probably evolved very early in biological history since some form is found in all bacterial groups and in the archaea. Recently homologues of histidine protein kinases have been found even in eukaryotes (Alex and Simon, 1994). As would be expected for a sensory system with such a long history, there have been changes and adaptations reflecting the complex environments and metabolisms of other species. Interestingly, although the sensory system has been carried through to higher organisms, no homologue of the motor has been found, although it has been suggested that part of the proton utilizing membrane-spanning enzyme, ATP synthase, rotates.

George Ordal has worked for many years on the Gram-positive species, *Bacillus subtilis*. Early on in this period of rapid advance he showed that *B. subtilis* has MCPs, and that it has the methyl transferase and esterase enzymes (Ullah and Ordal, 1981; Burgess-Cassler et al., 1982). However, *B. subtilis* appeared to respond to a much wider range of attractants and repellents than *E. coli*, but competition experiments suggested a limited number of common binding proteins that interacted with the MCPs (Nettleton and Ordal, 1989). Homologues were found to the enteric Che proteins, but CheY-P was found to cause smooth swimming rather than tumbles and, although the MCPs were methylated, the level of methylation did not change on stimulation. Ordal showed that the rate of turnover of methyl groups altered on stimulation, and rather than being released as methanol, methyl groups were transferred to a cytoplasmic protein and could be used for relabeling. Ordal has been the only researcher looking at *B. subtilis* for several decades and his constant output of good data has served to emphasize that, although great strides have been made in the understanding of enteric chemotaxis, we must not assume that all species operate identically. Billions of years of evolution have certainly modified the systems for different species. If there are a million species there may well be tens of variations of a central theme.

Similarly David Zusman looking at the gliding social bacterium *Myxococcus xanthus* found homologues of the *che* genes (Blackhart and Zusman, 1986). This organism glides at a maximum speed of 5 $\mu m\ s^{-1}$ which leads to an argument about whether a bacterium could sense a gradient if it were going this slowly. However, deletion of the genes leads to changes in the reversal frequency of the gliding cells and a loss of normal pattern formation. Homologues of all the *che* genes have been identified, but intriguingly the *mcp* gene appears to code for a cytoplasmic protein, suggesting that the bacterium is moving along a "metabolic" gradient using a cytoplasmic signal (McBride et al., 1989).

This relates to the finding of scientists investigating members of the α-group of bacteria which comprise a wide range of free living, environmentally important species including the Gram negative photosynthetic eubacteria, rhizobial species (*Rhizobium*, *Agrobacterium*, *Azospirillum*) and the differentiating species such as *Caulobacter crescentus*. MCPs were identified in *C. crescentus* by Lucy Shapiro, but little work has been done on their chemotactic role. Instead she used them as a marker in differentiation. *C. crescentus* has a fascinating life cycle with a sedentary stalked phase which carries out cell division budding off a motile swarmer cell from the nonstalked pole of the cell (Shapiro, 1976; Shapiro et al., 1989). The swarmer cell does not carry out cell division, but swims until it finds a suitable environment when it loses its flagellum and develops a stalk, initiating DNA replication and cell division to produce more daughter cells. Shapiro found that only swarmer cells contained MCPs and that although there is no DNA synthesis in swarmers, mRNA coding for the proteins required for motility and taxis are segregated into the future swarmer cell, not the stalked parent cell (Gomes and Shapiro, 1984; Alley et al., 1991).

In the other species of α-bacteria more attention has been paid to chemosensing, but until recently with little success (Gotz et al., 1982). MCPs have not been found in these species although they are chemotactic to a wide range of compounds, primarily sugars and organic acids. Detailed work by the group led by Judith Armitage showed that *R. sphaeroides* requires transport and limited metabolism to show a chemotactic response, although recent work has identified at least two cytoplasmic MCP (Sockett et al., 1987; Poole and Armitage, 1988, 1989).

This takes the work of Clayton in the 1950s full circle. The very elegant experiments of Adler with *E. coli* had shown that transport and metabolism were not required for chemotaxis, and this led to the "energy"-based theory of chemotaxis developed by Clayton being abandoned. Clayton however worked with the photosynthetic relative of *R. sphaeroides*, an α-group bacterium, not an enteric species which belongs to the γ-subgroup. It seems probable that the system so completely worked out for the enteric species may not be the complete story for all bacterial species, and the variations probably include a metabolically based sensing system as first suggested by Clayton. A metabolically based system would guide bacteria under growth-limiting environments to areas where metabolic activity could increase, perhaps with dedicated MCPs only coming in after growth limitation has been overcome.

Oxygen and Light

The early workers on bacterial behavior had almost all looked at taxis towards oxygen and light; aero- and phototaxis. This area has been amazingly neglected over the past 30 years while chemotaxis has leapt from discovery to discovery. Perhaps the principle reason for this is that very early experiments showed that respiratory or photosynthetic electron transport were required to sense oxygen or light, and these are also essential to create the pmf (proton gradient) needed to drive flagellar rotation under those conditions. Any change in electron transport changes the pmf and thus directly effects flagellar behavior, making it difficult to separate a change in pmf from a signal. In addition the whole process has remained remarkably intractable to molecular genetic techniques. All mutants isolated with altered phototactic and aerotactic responses so far have proved to be altered in the electron transport chain and none have been isolated in the system which one assumes must signal changes in the electron transport chain to the flagellum. This suggests that the signal is either the pmf directly altering the behavior of the flagellum, a so called "protometer", perhaps an active part of the motor sensing changes in the pmf, or the signal is an essential part of cellular metabolism, mutations in which are lethal. All that is known is that the signal depends not on a pigment but on active electron transfer. Whether the primary signal is a redox change in a component or a change in pmf is currently unknown (Harayama, 1977; Armitage and Evans, 1981; Armitage et al., 1985).

Much more is known about the photoresponses of the archaebacterium *Halobacterium halobium* (renamed recently *H. salinarium*). *H. salinarium* is particularly famous for its "purple" membrane identified by Dieter Oesterhelt and Walter Stokenius (Oesterhelt and Stockenius, 1971). Bacteriorhodopsin (BR), to give the purple membrane its correct name, has created a whole research field of its own as the first identified light-driven ion pump in biology. The proton pumping activity of bacteriorhodopsin, a homologue of eukaryotic rhodopsin, uses red light under anaerobic conditions to generate the pmf required for ATP synthesis. In the early 1980s John Spudich looked at the photoresponses in this organism. *H. salinarium* lives in a very extreme environment, in salt lakes with an optimum requirement of 3.5 M salt. These lakes tend to be very turbid, often semicrystalline. Therefore to maximize the light intensity and thus allow BR to operate *H. salinarium* needs to maintain itself as close to the surface as possible, but salt lakes are in a very hot, high light intensity environment where UV irradiation is also high and the bacterium risks damaging effects of UV.

Spudich found that *H. salinarium* was repelled by blue light, but attracted to orange light of wavelengths close to the absorbance maximum of bacteriorhodopsin. Oddly, incubating the cells in a background of orange light made them more sensitive to blue light. He identified a new retinal-containing pigment protein (the sodium pump halorhodopsin had also been identified) required for phototaxis and called it "sensory rhodopsin" (SRI) (Spudich and Bogomolni, 1984). This protein turned out to have fascinating photobehavior. The chromophore can undergo either a one or a two photon photocycle. The orange-absorbing form of SRI goes through a fast photocycle when illuminated with light at 587 nm changing into the blue absorbing form, SR_{373}. If there is no light at 373 nm, SR_{373} will slowly decay back to SR_{587}, but if it absorbs a photon of blue light at 373 nm it will rapidly transform back to SR_{587}. Thus in damaging blue light there is a high concentration of SR_{587} and this induces a repellent signal, whereas a high concentration of SR_{373} means high levels of orange light (and low levels of blue light) and causes a positive signal. The single photopigment can thus maintain *H. salinarium* in the optimum environment for bacteriorhodopsin function without risking major cellular damage.

Unexpectedly it was found that when this sensory rhodopsin was deleted *H. salinarium* was still repelled by blue light. A second retinal-containing sensory photopigment was identified, SRII (Wolfe et al., 1986; Takahashi et al., 1990). This is a constitutive protein present at low levels and unlike SRI only undergoes a simple single photocycle, responding to blue light at 487 nm. This pigment therefore keeps *H. salinarium* out of damaging blue light under all growth conditions including when it is growing aerobically or fermentatively, and therefore lacking bacteriorhodopsin. If oxygen or the fermentative substrate arginine falls, BR and SRI are induced directing cells to the right environment for their new growth state.

How do SRI and SRII signal to the flagellum of *H. salinarium*? Methylation has been shown to be involved and recent experiments have indicated that MCP-like proteins interact closely with the SR proteins, probably transforming the change in SR caused by light absorbance into a signal through the MCP-like transducer protein. This in turn interacts with proteins related to the Che protein of enterics since a CheA homologue has been identified (Spudich et al., 1989). This processing sequence may help to explain how a signal operating in the fast ms time scale, as the photocycle does, can produce a behavioral response on a timescale of seconds.

Flagellar Structure

The flagellum is a semirigid rotated helix made of polymerized monomers of a single protein. How can a single protein form a helix, and how is this helix deformed to produce the shapes seen during tumbling? Iino's group in Japan had identified a range of mutants which swam abnormally and appeared to have flagella with different waveforms compared to wild-type cells (Fujita et al., 1981). Hotani showed, by using high intensity dark field microscopy, that different physical environments including high and low pH and increased viscosity in flow chambers could change the waveform of the flagellum. Using these methods Hotani identified 12 different flagellar filament states, 2 straight forms isolated from nonmotile but flagellate mutants, and 10 forms with differing pitches and handedness (Kamiya et al.,1982).

At the same time several groups, including that of Calladine in Cambridge, had been studying negative stained electron micrographs of isolated flagella (Calladine, 1978, 1982). A very regular array of subunits could be seen along a filament, and there were signs that there was a central hole through this filament. Modeling this structure suggested that the filament was made of a cylinder of 11 strands of flagellin subunits tilted at a slight angle to the axis of the filament. The curved hook region, which joined the flagellar filament to the cell, showed a very similar pattern. Several models were suggested for the formation of the different waveforms, the most likely being that the flagellin monomers could produce one of two different lengths of flagellin strands or protofilaments. If one or more strands were shorter that the others then the whole structure would deform into a helix. The different helical forms could all be explained by changing numbers of long and short strands in the filament.

The above was eventually confirmed by David DeRosier using very elegant computer imaging of multiple images of the mutants with straight flagella, which turned out to have all the flagellin subunits interacting in the same way. However the straight flagella fell into two groups, one with all the subunits interacting in one conformation and the other group with a different pattern of interaction (Trachtenberg and DeRosier, 1987). He was able to show that the flagellin monomers interact with their neighbor above and below in one of two ways, but the interaction, dictated from the motor via the hook, once formed at the base propagated up the full length of the filament. The short straight filament had all the strands interacting in the shorter confirmation, whereas the long straight filament mutants had all the subunits interacting in the alternative way. The interactions were determined by the amino- and carboxy-terminal regions of the flagellin, the regions conserved between flagellins of diverse species of bacteria and tucked into the center of the filament, while the highly variable region, which provides cells with their antigenicity, was exposed on the outside and not involved in forming the structure. The electron microscopy also revealed that there was indeed a hole down the center of the filament, although its diameter was controversial, and, as we shall see later, this provides another of the unexpected findings associated with the flagellum.

Genetics

There is only one structure in a bacterial cell which is more complex than its flagellum, and that is the ribosome. However, a ribosome is completely cytoplasmic whereas the flagellum is composed of cytoplasmic, membrane spanning, and extensive extracellular domains. It is still unclear what is the exact number of genes involved in making a flagellum, but the rough number is about 40 or about 5% of a bacterial genome! Even when not being expressed (and making a flagellum is a serious commitment in energy for a cell), just maintaining that amount of DNA takes measurable energy and there is good evidence that motility and sensing are advantageous in its search for food. In very well mixed chemostats, nonmotile mutants rapidly outgrew motile wild-type species, but in poorly mixed environment more closely related to natural environments the motile strains dominated (Pilgram and Williams, 1976). The high percentage of motile bacteria in nutrient-poor environments, from salt lakes to hyperthermal vents, supports the importance of motility in competition for limited resources.

The flagellum has also provided an amazing model system for several different aspects of biology. The protein structure is so complex that it must be synthesized in a set order for correct assembly; it has therefore provided a model for transcriptional control. Although complex, flagellation is not essential for survival in the laboratory which makes mutant isolation and analysis easier than with an essential protein complex such as a ribosome. Each cell tends to have a limited number of flagella often located at the same position, e.g. the pole, proving a model for targeting studies. Since they have domains which are intra- and extracellular, they have been used to investigate protein transport into, across, and outside the cell wall/membrane.

The most comprehensive early study of assembly was carried out by the group of Tetsuo Iino in Japan (Suzuki et al., 1978; Stader et al., 1986). He took a range of nonflagellate mutants, with mutations identified in different alleles, and then carried out a basal body purification procedure on large volumes of cells and negatively stained the extracts (Iino, 1985). Given the limited number of motors per cell, the results from this painstaking study are remarkable. He was able to identify the first visible part of the motor in the membrane, the ring structure in the cytoplamic membrane, and the genes necessary for its formation. He identified the next structure, the outer membrane-associated rings plus the transmembrane rod and so on until he had a sequential model of gene expression and the physical motor and flagellar structure which requires those genes.

Robert Macnab did a very comprehensive analysis of the genes and their organization in *S. typhimurium,* linking this with exceptionally good preparations of basal bodies (Aizawa et al., 1985; Yamaguchi et al., 1986) and producing a volume of work far too extensive to reproduce here. Investigation of the tight control of the expression of the 14 or so operons localized in 4 regions of the chromosome in *E. coli* and *S. typhimurium* is still ongoing and has led to the identification of both a novel sigma factor (Arnosti and Chamberlin, 1989) and an anti-sigma factor. If the motor proteins do not assemble correctly they feed back onto the later operons, and often their own operons, inhibiting expression. This is an essential, tightly controlled process since the final operon, coding for FliC, the flagellin, cannot be expressed before the motor onto which it must be located is correctly assembled because the 20,000 or so copies that make up a full length flagellum would be toxic if they accumulated intracellularly (Macnab, 1990).

The mechanism of assembly is fascinating. Early researchers showed that flagellin monomers would assemble onto stubs of sheared flagella *in vitro*, but under normal conditions assembly is limited with added flagellin monomers failing to add to fully grown flagella *in vivo*. In 1956 Iino showed that the flagellin of *S. typhimurium* had phase variation as a result of gene inversion, producing a second flagellin (possibly to evade the immune system).The flagellins were not only antigenically different but also produced a flagellar filament with a different wavelength. He exploited this to look at flagella growth. By changing environments he caused a phase shift and showed that the flagellum grows not from the proximal but from the distal end of the filament (Iino, 1969). It was generally assumed that the length of the filament was finite, and there was a signal controlling length. However, expression studies showed that the length is not finite, but the filament continues to grow with length being limited only by breakage and limits of flagellin transport. The filament can therefore grow to lengths of 10 μm on a cell of only 2 μm, and a single peritrichous cell may have flagella of varying lengths.

Macnab identified a group of cytoplasmic proteins required for flagellar formation. These appeared to be located on the cytoplasmic surface of the switch proteins. Sequence analysis showed that although not involved in driving rotation, one of these proteins was homologous to the β-subunit of the ATPase (Dreyfus et al., 1993). When mutated, the flagellin monomers accumulated in the cytoplasm. When Macnab sequenced the basal body and filament genes he demonstrated that, although the proteins for the components of the motor had classical signal sequences which directed them through the membrane, the hook, the hook associated proteins (HAPs) and flagellin proteins had no recognizable signal sequence. Deletion of a region at the N-terminal end however stopped export and assembly (Kuwajima et al., 1989). This led to the theory that the flagellin monomers are transported to the basal body, perhaps by a chaperone-type protein and then exported, using the energy of ATP hydrolysis, through a hole in the center of the motor and then up the center of the filament to be polymerized at the tip with the aid of one of the hook-associated proteins (HAP2). The length of filament may therefore simply be limited by steric hindrance of the unfolded monomers blocking the central hole as the flagellum gets longer. This type of transport system has now been identified in other systems and may be a pathway for the release of some toxins.

Motors

The model of the motor as four rings described by Adler remained for well over a decade since all basal body preparations had the same, or closely related, images in the electron microscope. However, mutations were found that produced a paralyzed but flagellate phenotype. These mutants had normal looking basal bodies but could not drive flagellar rotation, although the pmf was normal. Howard Berg showed that one gene, *motA,* coded for a membrane-spanning protein which is almost certainly part of the proton pore essential for rotation. Overexpression of the protein in *E. coli* led to poor growth and the isolated membranes were "leaky" to protons, whereas if the *motA* gene from a paralyzed mutant was overexpressed cells grew normally and had proton-tight membranes (Blair and Berg, 1990, 1991). A second gene, *motB*, in which mutations led to a paralyzed phenotype was identified and found to be located next to *motA*. The *motB* gene, however, had only one membrane-spanning domain and did not behave like a proton pore. It did, however, have a sequence that could interact with the peptidoglycan in the cell wall, and it therefore became the candidate protein for the "anchor" or stator.

All motors require a stator as well as a rotor to develop torque, and it seemed possible that the Mot complex provided the stator half of the flagellar motor. Howard Berg then carried out yet another elegantly simple experiment. *E. coli* mutants missing either *motA* or *motB* were tethered via antiflagella antibody. The mutants contained the missing gene on a plasmid behind a *lac* promoter and expression was induced by the addition of a β-galactoside and the rotational behavior of the tethered cell followed. He found a stepwise increase in the rotation rate of the tethered cells, with each step increasing the speed of rotation by a fixed rate (Block and Berg, 1984). In the initial experiments there seemed to be about eight steps in rotation rate until the rate became maximal. The same results were found whether it was *motA* or *motB* which was being expressed (Blair and Berg, 1988). This suggested two important properties of the motor: (1) the MotA/B complex forms the force generating unit of the flagellar motor, and (2) the units can operate independently, each unit giving a set amount of force.

Using freeze fracture studies of bacterial membranes, Shahid Khan showed that there were rosettes of protein studs in the wild-type cells. These patterns of proteins had already been seen in electron micrographs of *Aquaspirillum serpens*, but it had been assumed that they were related to localization of the polar tuft (Coulton and Murray, 1978). The dimensions of the rosette corresponded to the calculated diameter of a basal body plus the Mot complexes (Khan et al., 1988). Paralyzed mutants did not have these rosettes. The number of particles in these rosettes rarely was exactly 8 but tended to vary between 10 and 13. Whether this difference in number is an artifact of freeze fracture or in the statistical problems involved in analyzing the rotation rates of the tethered cells (which have a larger load on their cell bodies than free swimming cells) has not been resolved.

None of the paralyzed mutants, however, were affected in the "switch" responding to the chemotaxis signal. The site of interaction of the Mot complex with the motor and the CheY-P signal transducer with the motor remained elusive. It was assumed that the M/S ring, so obvious in electron micrographs, was the site of action of CheY-P and involved in rotation (see models later). However, no paralyzed mutants mapped to genes thought to code for the M/S ring. Electron microscopy by DeRosier complemented by the molecular genetics of by Macnab and Yamanuchi have recently shown that the "two" rings seen in electron micrographs are really one protein, the product of a single gene (Ueno et al., 1992). This fits with structural studies, suggesting that the other half of the motor and the CheY-P interactive site are probably on the cytoplasmic side of the motor, not in the membrane. Several genes were identified which when knocked out resulted in a nonflagellate phenotype but could not be "fitted" into the classical flagellar structure.

Some very elegant second-site suppressor studies by Parkinson started to identify the role of the protein products and the protein–protein interactions within the motor. To overcome the problem of inhibition of flagellar synthesis when these genes were mutated, a number of laboratories used site-directed mutagenesis and in-frame mutation of these genes, which allowed a complete flagellum to be synthesized but with mutated motor protein. These techniques combined with antibody studies and electron microscopy confirmed that the M/S ring is not an active part of the motor. The second half of the motor, taking protons from the MotA protein, is the FliG complex located under the M/S ring in the cytoplasm.

The proteins FliM and FliN form part of the switch complex, or C-ring, with FliG. Mutations in these two genes led to a range of altered swarming patterns and mutants that could not change the direction of flagellar rotation. It was also found that mutations in CheY that led to a smooth swimming mutant could be compensated by mutations in FliM or FliN, now known as the switch complex (Parkinson et al., 1983). The data now all add up to a passive structure,the M/S ring, in the cytoplasmic membrane driven by the movement of proton between multiple Mot complexes surrounding the membrane structure, the stator half of the motor, and the ring of FliG proteins attached to the cytoplasmic face of the motor, the rotor. Switching between counterclockwise and clockwise rotation depends on the CheY-P protein interacting with the FilM and FliN switch complex on the cytoplasmic face, but how the direction of rotation is switched is unknown.

Motor Rotation

Since the realization that the bacterial flagellum did indeed rotate, the possible mechanism involved in transforming a gradient of protons—electrochemical energy into mechanical rotation—has held a persistent fascination for biophysicists, chemists and theoreticians alike. New models appear at regular intervals trying to accommodate new data. One major problem with getting to grips with motor rotation is our mental concept of motors. In a macroscopic world the major force resisting movement is inertia, with viscosity playing a minor role. However in the microscopic world there is little inertia, but viscous forces are very high. This means that displacement is not the primary driving force, and when a bacterium stops swimming it stops in a distance equivalent to the diameter of a proton—there is no gliding. It was recognized as early as the 1960s that macroscopic models would not be accurate representations of the bacterial motor.

Howard Berg has been largely responsible for the biophysical data and Macnab for the structural data currently used to build models of flagellar rotation but there have been a large number of models over the past 30 years. Early structural studies showed the central membrane-spanning multiring complex and from this it was assumed that the M-ring rotated against the S-ring as protons moved freely between the rings. One of the earliest models had proton interactive sites along the faces of the M-ring and S-ring at angles to each other (Lauger, 1977). Protons flowed down the proton gradient interacting with the proton binding sites on each ring and causing one to move against the other. Assuming one ring is fixed in the membrane, as a stator, and the other is attached to the flagellum and free to rotate, the filament will rotate as protons move down the gradient. This is still an acceptable model and has currently been adapted to cope with the identification of the Mot and switch complexes (Kleutsch and Lauger, 1990).

Skulachev showed that, as with ATP synthase, the motor could be driven by either the electrical potential difference or the pH difference caused by the electrochemical proton gradient across the membrane (Glagolev and Skulachev,1978; Khan et al., 1985).Temperature had no major effect on rotation rates of tethered cells, implying that the driving force does not involve covalent forces (Khan and Berg, 1983). A great deal of the data on rotation has come from comparing tethered cells, where the load on the cell body is high, with free-swimming cells, where the loads are low. Berg has looked at both systems and shown that the efficiency of energy conversion is different under the two conditions. By looking at such parameters as rotation rate under artificially energized conditions, and the "stall" torque in controlled viscous environments, he has calculated the number of protons per rotation to be about 1000, with the motor rotating at about 100 Hz. The efficiency of tethered cells is close to 100%, the rate of rotation increasing linearly as the pmf increased therefore suggesting that all protons do work as they moved down the gradient. Free swimming cells, however, reached a maximum swimming speed as the pmf increased, suggesting that the efficiency was much lower at high values of pmf (Lowe et al., 1987; Meister and Berg, 1987; Meister et al., 1987). This also led to the idea that the motor is tightly rather than loosely coupled (Meister et al., 1989).

The identification of the Mot complexes as the stator component of the motor meant that models had to be modified. Berg developed a model in which the Mot complexes, anchored by an elastic linkage (probably MotB) to the cell wall, are located in a ring around the rotor, acting as half-proton channels (Berg and Khan, 1983). The Mot complexes interact with proton binding sites on the rotor and are only free to move (in either direction) a limited distance by thermal movement. If a proton is bound between the MotA stator and FliG rotor site this constrains the direction of movement and the Mot complex "pulls" the FliG and rotor complex and the proton is now free to move into the cytoplasm. The direction of proton gradient is the only constraint in this model on the direction of rotation and as long as it stays in one direction the motor will rotate one way. The model should be freely reversible and Berg showed that artificially energized cells with reversed direction of pmf rotated their flagellar backwards. Obviously the motor does change its direction of rotation under normal conditions, and the direction of the proton gradient does not change. It is assumed that the binding of CheY-P to the switch complex must change the conformation of the FliG–MotA interaction sites and allow reversal.

One problem with all models of the motor has always been the lack of visible "steps" in the motor rotation; if there are limited Mot proteins then some steps should be visible during rotation—however, this is probably a limit in technology. It is now thought that the Mot complexes may interact with as many as 200 FliG proteins and this would make steps in rotation below the current limits of detection.

Several other models have been produced. Peter Mitchell produced a model (after his earlier thermal motor) based on the "proton well" idea of the ATPase in which the pmf is converted to a pH gradient, and this used to drive rotation in a well-and-turnstile model (Mitchell, 1984). Wagenknecht proposed a model where the motor does not physically rotate, but there is twisting within the motor which translates into rotation at the flagellar level (Meister et al., 1989).

The large number and variety of models reflects both the interest in this problem, and the distance we are from an answer to it. While molecular biology and behavioral studies have made the chemosensory system of *E. coli* the best understood in biology, and the molecular studies have led to a great understanding of the control of the synthesis of probably the most complex protein structure in bacteria, the mechanisms involved in driving rotation are still not understood and will probably provide food for heated discussions at conferences for a few years to come.

SUMMARY

Where have we come since the early descriptions of bacterial swimming and behavior? Some extremely detailed and perceptive descriptions at the turn of the last century have hardly been improved upon. However the advent of molecular genetics and improved microscopy has led to a detailed molecular description of the sensory-transducing processes leading to a behavioral response. We know that the flagellum is a complex structure, the genes being transcribed in a strict hierarchy and their products assembled in order. Understanding this has helped identify a new bacterial export system possibly important in virulence. We know now that the flagellar motor rotates—the only known rotary structure in biology—driven by protons (sodium ions are used to drive rotation in bacteria living in high salt or alkaline environments). Where do we go from here? There are still a series of answers needed ranging from "How does the molecular motor work?" to "What is the role of motility in the natural environment?"

Perhaps the most obvious change over the years has been that early data came from a very few individuals quietly working, usually alone and more or less in isolation, whereas today the research is being carried out by large research groups producing a constant flow of data. However, even with the number of large groups, major advances still tend to come from a few perceptive individuals (with the help sometimes of a major advance in technology).

ACKNOWLEDGMENTS

I would like to thank Rudiger Schmitt at the University of Regensberg for help with locating and translating many of the early German articles, and the librarians at the University of Regensberg for their patience. I would like to thank Julius Adler for the translation of the early papers by Engelmann and for numerous figures from his early papers and photographs of Pfeffer, Engelmann, and himself. I would like to emphasize that this a purely personal view of the subject and I am unable to do justice to all the more recent players in this fast moving and exciting field.

REFERENCES

Adler, J. (1965). Chemotaxis in *E. coli.* Cold Spring Harbor Symp. 30, 289-293.

Adler, J. (1966). Chemotaxis in bacteria. Science 153, 708-716.

Adler, J. (1969). Chemoreceptors in bacteria. Science 166, 1588-1597.

Adler, J. (1973). A method for measuring chemotaxis and the use of the method to determine optimum conditions for chemotaxis by *E. coli.* J. Gen. Microbiol. 74, 77-91.

Adler, J. (1976). The sensing of chemicals by bacteria. Sci. American 234, 40-47.

Adler, J. & Epstein, W. (1974). Phosphotransferase-system enzymes as chemoreceptors for certain sugars in *Escherichia coli* chemotaxis. Proc. Nat. Acad. Sci. USA 71, 2895-2899.

Adler, J., Hazelbauer, G.L., & Dahl, M.M. (1973). Chemotaxis toward sugars in *Escherichia coli*. J. Bacteriol. 155, 824-847.

Aizawa, S., Dean, G.E., Jones, C.J., Macnab, R.M., & Yamaguchi, S. (1985). Purification and characterization of the flagellar hook-basal body complex of *Salmonella typhimurium*. J. Bacteriol. 161, 836-849.

Alex, L.A. & Simon, M.I. (1994). Protein histidine kinases and signal transduction in prokaryotes and eukaryotes. Trends Genet. 10, 133-138.

Alley, M.R.K., Gomes, S.L., Alexander, W., & Shapiro, L. (1991). Genetic analysis of a temporally transcribed chemotaxis gene cluster in *Caulobacter crescentus*. Genetics 129, 333-342.

Armitage, J.P. (1992) Bacterial motility and chemotaxis. Science Progress 76, 451-477.

Armitage, J.P. & Evans, M.C.W. (1981). The reaction centre in the phototactic and chemotactic responses of *Rhodopseudomonas sphaeroides*. FEMS Microbiol. Lett. 11, 89-92.

Armitage, J.P., Ingham, C., & Evans, M.C.W. (1985). Role of the proton motive force in phototactic and aerotactic responses of *Rhodopseudomonas sphaeroides*. J. Bacteriol. 163, 967-972.

Armstrong, J.B. (1972). An S-adenosyl methionine requirement for chemotaxis in *Escherichia coli*. Canad. J. Microbiol. 18, 1695-1701.

Armstrong, J.B. & Adler, J. (1969a). Location of genes for motility and chemotaxis on the *Escherichia coli* genetic map. J. Bacteriol. 97, 156-161.

Armstrong, J.B. & Adler, J. (1969b). Complementation of nonchemotactic mutants of *Escherichia coli*. Genetics 61, 61-66.

Armstrong, J.B., Adler, J.. & Dahl, M.M. (1967). Non-chemotactic mutants of *Escherichia coli*. J. Bacteriol. 93, 390-398.

Arnosti, D.N. & Chamberlin, M.J. (1989). Secondary sigma factor controls transcription of flagellar and chemotaxis genes in *Escherichia coli*. Proc. Nat. Acad. Sci. USA 86, 830-834.

Ashbury, W.T. & Weibull, C. (1949). X-ray diffraction study of the structure of bacterial flagella. Nature (London) 163, 280-282.

Aswad, D.W. & Koshland, D.E.J. (1975). Evidence for an S-adenosyl methionine requirement in the chemotactic behavior of *Salmonella typhimurium*. J. Mol. Biol. 97, 207-223.

Berg, H.C. (1974). Dynamic properties of bacterial flagellar motors. Nature (London) 249, 77-79.

Berg, H.C. (1983). Random Walks in Biology. Princeton, NJ.

Berg, H.C. & Anderson, R.A. (1973). Bacteria swim by rotating their flagellar filaments. Nature (London) 245, 380-382.

Berg, H.C. & Block, S.M. (1984). A miniature flow cell designed for rapid exchange of media under high-powered objectives. J. Gen. Microbiol. 130, 2915-2920.

Berg, H.C. & Brown, D.A. (1972). Chemotaxis in *Escherichia coli* analysed by three-dimensional tracking. Nature (London) 239, 500-504.

Berg, H.C. & Khan, S. (1983). A model for the flagellar rotary motor. In: Mobility and Recognition in Cell Biology (Anonymous), pp. 485-497. Walter de Gruyter, Berlin.

Berg, H.C. & Tedesco, P.M. (1975). Transient responses to chemotactic stimuli in *Escherichia coli*. Proc. Nat. Acad. Sci. USA 72, 3235-3239.

Blackhart, B.D. & Zusman, D.R. (1986). Analysis of the Products of the *Myxococcus xanthus frz* Genes. J. Bacteriol. 166, 673-678.

Blair, D.F. & Berg, H.C. (1988). Restoration of torque in defective flagellar motors. Science 242, 1678-1681.

Blair, D.F. & Berg, H.C. (1990). The MotA protein of *E. coli* is a proton-conducting component of the flagellar motor. Cell 60, 439-449.

Blair, D.F. & Berg, H.C. (1991). Mutations in the MotA protein of *Escherichia coli* reveal domains critical for proton conduction. J. Mol. Biol. 221, 1433-1442.

Block, S.M. & Berg, H.C. (1984). Successive incorporation of force generating units in the bacterial rotary motor. Nature (London) 309, 470-472.

Block, S.M., Segall, J.E., & Berg, H.C. (1982). Impulse responses in bacterial chemotaxis. Cell 31, 215-226.

Borkovich, K.A., Kaplan, N., Hess, J.F., & Simon, M.I. (1989). Transmembrane signal transduction in bacterial chemotaxis involves ligand-dependent activation of phosphate group transfer. Proc. Nat. Acad. Sci. USA 86, 1208-1212.

Borkovich, K.A. & Simon, M.I. (1990). The dynamics of protein phosphorylation in bacterial chemotaxis. Cell 63, 1339-1348.

Boyd, A., Kendall, K., & Simon, M. (1983). Structure of the serine chemoreceptor of *Escherichia coli*. Nature (London) 301, 623-626.

Burgess-Cassler, A., Ullah, A.H., & Ordal, G.W. (1982). Purification and characterization of *Bacillus subtilis* methyl-accepting chemotaxis protein methyltransferase II. J. Biol. Chem. 257, 8412-8417.

Calladine, C.R. (1978). Change of waveform in bacterial flagella: the role of mechanics at the molecular level. J. Mol. Biol. 118, 457-479.

Calladine, C.R. (1982) Construction of bacterial flagellar filaments, and aspects of their conversion to different helical forms. In: Prokaryotic and Eukaryotic Flagella (Amos, W.B. & Duckett, J.G., Eds.), pp. 33-51. Cambridge University Press, Cambridge.

Chelsky, D. & Dahlquist, F.W. (1980). Structural studies of methyl-accepting chemotaxis proteins of *Escherichia coli*: evidence for multiple methylation sites. Proc. Nat. Acad. Sci. USA 77, 2434-2438.

Clayton, R.K. (1953a). Studies in the phototaxis of *Rhodospirillum rubrum*. II. The relation between phototaxis and photosynthesis. Archiv fur Mikrobiologie 19, 125-140.

Clayton, R.K. (1953b). Studies in the phototaxis of *Rhodospirillum rubrum*. I. Action spectrum, growth in green light and Weber-law adherence. Archiv fur Mikrobiologie 19, 107-124.

Clayton, R.K. (1953c). Studies in the phototaxis of *Rhodospirillum rubrum*. III. Quantitative relationship between stimulus and response. Archiv fur Mikrobiologie 19, 141-165.

Clayton, R.K. (1957). Patterns of accumulation resulting from taxes and changes in motility of microorganisms. Archiv fur Mikrobiologie 27, 311-319.

Clayton, R.K. (1958). On the interplay of environmental factors affecting taxis and mobility in *Rhodospirillum rubrum*. Archiv fur Mikrobiologie 29, 189-212.

Cohen-Bazire, G. & London, J. (1967). Basal organelles of bacterial flagella. J. Bacteriol. 94, 458-465.

Coulton, J.W. & Murray, R.G.E. (1978). Cell envelope associations of *Aquaspirillum serpens* flagella. J. Bacteriol. 136, 1037-1049.

DePamphilis, M.L. & Adler, J. (1971a). Attachment of flagellar basal bodies to the cell envelope: specific attachment to the outer, lipopolysaccharide membrane and the cytoplasmic membrane. J. Bacteriol. 105, 396-407.

DePamphilis, M.L. & Adler, J. (1971b). Purification of intact flagella from *Escherichia coli* and *Bacillus subtilis*. J. Bacteriol. 105, 376-383.

DePamphilis, M.L. & Adler, J. (1971c). Fine structure and isolation of the hook-basal body complex from flagella of *Escherichia coli* and *Bacillus subtilis*. J. Bacteriol. 105, 384-395.

Dobell, C. (1932). Antony van Leeuwenhoek and His "Little Animals". John Bale, Sons and Danielsson, London.

Dreyfus, G., Williams, A.W., Kawagishi, I., & Macnab, R.M. (1993). Genetic and biochemical analysis of *Salmonella typhimurium* FliI, a flagellar protein related to the catalytic subunit of the F_0F_1 ATPase and to virulence proteins of mammalian and plant pathogens. J. Bacteriol. 175, 3131-3138.

Ehrenberg, G.S. (1883). Die infusionstrierchen als vollkommene organismen. In: (Engelmann, T.W., Ed.), p. 15. Leipzig.

Eisenbach, M. & Adler, J. (1981). Bacterial cell envelopes with functional flagella. J. Biol. Chem. 256, 8807-8814.

Engelmann, T.W. (1881) *Bacterium photometricum*. Arch. Ges. Physiol. Bonn. 30, 95-124.

Engelmann, T.W. (1883) *Bacterium photometricum*: Ein beitrag zur vergleichenden physiologie des licht und fabensinnes. Pfug. Arch. Gesamte Physiol. Menschen Tiere 42, 183-186.

Fleming, A. (1950). Further observations on the motility of *Proteus vulgaris* grown on penicillin agar. J. Gen. Microbiol. 4, 457-463.

Ford, B.J. (1991). The Leeuwenhoek Legacy. Biopress and Farrand Press, Bristol and London.

Fujita, H., Yamaguchi, S., Taira, T., & Iino, T. (1981). A simple method for the isolation of flagellar shape mutants in *Salmonella*. J. Gen. Microbiol. 125, 213-216.

Glagolev, A.N. & Skulachev, V.P. (1978). The proton pump is a molecular engine of motile bacteria. Nature (London) 272, 280-282.

Gomes, S.L. & Shapiro, L. (1984). Differential expression and positioning of chemotaxis methylation proteins in *Caulobacter*. J. Mol. Biol. 178, 551-568.

Gotz, R., Limmer, N., Ober, K., & Schmitt, R. (1982). Motility and chemotaxis in two strains of *Rhizobium* with complex flagella. J. Gen. Microbiol. 128, 789-798.

Goy, M.F., Springer, M.S., & Adler, J. (1978). Failure of sensory adaptation in bacterial mutants that are defective in a protein methylation reaction. Cell 15, 1231-1240.

Harayama, S. (1977). Phototaxis and membrane potential in the photosynthetic bacterium *Rhodospirillum rubrum*. J. Bacteriol. 131, 34-41.

Hazelbauer, G.L. & Adler, J. (1971). Role of the galactose binding protein in chemotaxis of *Escherichia coli* towards galactose. Nature (London) 230, 101-104.

Hazelbauer, G.L. & Engstrom, P. (1981). Multiple forms of methyl-accepting chemotaxis proteins distinguished by a factor in addition to multiple methylation. J. Bacteriol. 145, 35-42.

Hess, J.F., Oosawa, K., Matsumura, P., & Simon, M.I. (1987). Protein phosphorylation is involved in bacterial chemotaxis. Proc. Nat. Acad. Sci. USA 84, 7609-7613.

Hess, J.F., Bourret, R.B., & Simon, M.I. (1988). Histidine phosphorylation and phosphoryl group transfer in bacterial chemotaxis. Nature (London) 336, 139-143.

Iino, T. (1969). Polarity of flagellar growth in *Salmonella*. J. Gen. Microbiol. 56, 227-239.

Iino, T. (1985). Genetic control of flagellar morphogenesis in *Salmonella*. In: Sensing and Response in Microorganisms (Eisenbach, M. & Balaban, M., Eds.), pp. 83-92. Elsevier, Amsterdam.

Iino, T., Komeda, Y., Kutsukake, K., et al. (1988). New unified nomenclature for the flagellar genes of *Escherichia coli* and *Salmonella typhimurium*. Microbiol. Rev. 52, 533-535.

Ishihara, A., Segall, J.E., Block, S.M.. & Berg, H.C. (1983). Coordination of flagella on filamentous cells of *Escherichia coli*. J. Bacteriol. 155, 228-237.

Kamiya, R., Hotani, H., & Asakura, S. (1982). Polymorphic transition in bacterial flagella. In: Prokaryotic and Eukaryotic Flagella (Amos, W.B. & Duckett, J.G., Eds.), pp. 53-76. Cambridge University Press, Cambridge, U.K.

Kehry, M.R., Bond, M.W., Hunkapiller, M.W., & Dahlquist, F.W. (1983). Enzymatic deamidation of methyl-accepting chemotaxis proteins in *Escherichia coli* catalyzed by the *cheB* gene product. Proc. Nat. Acad. Sci. USA 80, 3599-3603.

Kehry, M.R. & Dahlquist, F.W. (1982). Adaptation in bacterial chemotaxis: *cheB*-dependent modification permits additional methylations of sensory transducing proteins. Cell 29, 761-772.

Kehry, M.R., Doak, T.G., & Dahlquist, F.W. (1984). Stimulus-induced changes in methylesterase activity during chemotaxis in *Escherichia coli*. J. Biol. Chem. 259, 11828-11835.

Khan, I.H., Reese, T.S., & Khan, S. (1992). The cytoplasmic component of the bacterial flagellar motor. Proc. Nat. Acad. Sci. USA 89, 5956-5960.

Khan, S. & Berg, H.C. (1983). Isotope and thermal effects in chemiosmotic coupling to the flagellar motor of *Streptococcus*. Cell 32, 913-919.

Khan, S., Dapice, M., & Reese, T.S. (1988). Effects of mot gene expression on the structure of the flagellar motor. J. Mol. Biol. 202, 575-584.

Khan, S., Meister, M., & Berg, H.C. (1985). Constraints on flagellar rotation. J. Mol. Biol. 184, 645-656.

Kleene, S.J., Toews, M.L., & Adler, J. (1977). Isolation of a glutamic acid methyl ester from an *Escherichia coli* membrane protein involved in chemotaxis. J. Biol. Chem. 252, 3214-3218.

Kleutsch, B. & Lauger, P. (1990). Coupling of proton flow and rotation in the bacterial flagellar motor: stochastic simulation of a microscopic model. Eur. Biophys. J. 18, 175-191.

Kort, E.N., Goy, M.F., Larsen, S.H., & Adler, J. (1972). Methylation of a membrane protein involved in bacterial chemotaxis. Proc. Nat. Acad. Sci. USA 72, 3939-3943.

Kuo, S.C. & Koshland, D.E.J. (1987). Roles of *cheY* and *cheZ* gene products in controlling flagellar rotation in bacterial chemotaxis in *Escherichia coli*. J. Bacteriol. 169, 1307-1314.

Kuwajima, G., Kawagishi, I., Homma, M., Asaka, J.-I., Kondo, E., & Macnab, R.M. (1989). Export of an N-terminal fragment of *Escherichia coli* flagellin by a flagellin-specific pathway. Proc. Nat. Acad. Sci. USA 86, 4953-4957.

Larsen, S.H., Adler, J., Gargus, J.J., & Hogg, R.W. (1974a). Chemomechanical coupling without ATP: the source of energy for motility and chemotaxis in bacteria. Proc. Nat. Acad. Sci. USA 71, 1239-1243.

Larsen, S.H., Reader, R.W., Kort, E.N., Tso, W., & Adler, J. (1974b). Change in direction of flagellar rotation is the basis of the chemotactic response in *Escherichia coli*. Nature (London) 249, 74-77.

Lauger, P. (1977). Ion transport and the rotation of bacterial flagella. Nature (London) 268, 360-361.

Liu, J. & Parkinson, J.S. (1991). Genetic evidence for interaction between the CheW and Tsr proteins during chemoreceptor signalling by *Escherichia coli*. J. Bacteriol. 173, 4941-4951.

Lui, J. & Parkinson, J.S. (1989). Role of CheW protein in coupling membrane receptors to the intracellular signaling system of bacterial chemotaxis. Proc. Nat. Acad. Sci. USA 86, 8703-8707.

Lowe, G., Meister, M., & Berg, H.C. (1987). Rapid rotation of flagellar bundles in swimming bacteria. Nature (London) 325, 637-640.

Macnab, R.M. (1976). Examination of bacterial flagellation by dark-field microscopy. J. Clin. Microbiol. 4, 258-265.

Macnab, R.M. (1977). Bacterial flagella rotating in bundles: a study in helical geometry. Proc. Nat. Acad. Sci. USA 74, 221-225.

Macnab, R.M. (1987). Biology of the chemotactic reponse. In: *E. coli* and *Salmonella typhimurium* (Neidhardt, F.C., Ed.), pp. 732-759. Amer. Soc. Microbiol., Washington, DC.

Macnab, R.M. (1990). Genetics, structure, and assembly of the bacterial flagellum. In: Biology of the Chemotactic Response (Armitage, J.P. & Lackie, J.M., Eds.), pp. 77-107. CUP, Cambridge, UK.

Macnab, R.M. & Han, D.P. (1983). Asynchronous switching of flagellar motors on a single cell. Cell 32, 109-117.

Maeda, K. & Imae, Y. (1979). Thermosensory transduction in *Escherichia coli*: inhibition of the thermosensory response by L-serine. Proc. Nat. Acad. Sci. USA 76, 91-95.

Manson, M.D., Tedesco, P., Berg, H.C., Harold, F.M., & van der Drift, C. (1977). A proton motive force drives bacterial flagella. Proc. Nat. Acad. Sci. USA 74, 3060-3064.

Manten, A. (1948) Phototaxis in the purple bacterium *Rhodospirillum rubrum* and the relationship between phototaxis and photosynthesis. Ant. van Leeuw. 14, 65-86.

McBride, M.J., Weinberg, R.A., & Zusman, D.R. (1989). "Frizzy" aggregation genes of the gliding bacterium *Myxococcus xanthus* show sequence similarities to the chemotaxis genes of enteric bacteria. Proc. Nat. Acad. Sci. USA 86, 424-428.

Meister, M. & Berg, H.C. (1987). The stall torque of the bacterial flagellar motor. Biophysical J. 52, 413-419.

Meister, M., Caplan, S.R., & Berg, H.C. (1989). Dynamics of a tightly coupled mechanism for flagellar rotation. Biophysical J. 55, 905-914.

Meister, M., Lowe, G., & Berg, H.C. (1987). The proton flux through the bacterial flagellar motor. Cell 49, 643-650.

Mesibov, R. & Adler, J. (1972). Chemotaxis towards amino acids in *Escherichia coli.* J. Bacteriol. 112, 315-326.

Milligan, D.L. & Koshland, D.E.J. (1988). Site-directed cross-linking: Establishing the dimeric structure of the aspartate receptor of bacterial chemotaxis. J. Biol. Chem. 263, 6268-6275.

Mitchell, P. (1956). Hypothetical thermokinetic and electrokinetic mechanisms of locomotion in microorganisms. Proc. Roys. Phys. Soc. (Edinburgh) 25, 32-34.

Mitchell, P. (1984). Bacterial flagellar motors and osmoelectric molecular motion by an axially transmembrane well and turnstile mechanism. FEBS Lett. 176, 287-294.

Mowbray, S.L., Foster, D.L., & Koshland, D.E.J. (1985). Proteolytic fragments identified with domains of the aspartate chemoreceptor. J. Biol. Chem. 260, 11711-11718.

Nara, T., Lee, L., & Imae, Y. (1991). Thermosensing ability of Trg and Tap chemoreceptors in *Escherichia coli.* J. Bacteriol. 173, 1120-1124.

Nettleton, D.O. & Ordal, G.W. (1989). Functional homology of chemotactic methylesterases from *Bacillus subtilis* and *Escherichia coli.* J. Bacteriol. 171, 120-123.

Nowlin, D.W., Nettleton, D.O., Ordal, G.W., & Hazelbauer, G.L. (1985). Chemotactic transducer proteins of *Escherichia coli* exhibit homology with methyl-accepting proteins from distantly related bacteria. J. Bacteriol. 163, 262-266.

Oesterhelt, D. & Stoeckenius, W. (1971). Rhodopsin-like protein from the purple membrane of *Halobacterium halobium.* Nature (London) 233, 149-152.

Ordal, G.W. & Adler, J. (1974). Isolation and complementation of mutants in galactose taxis and transport. J. Bacteriol. 117, 509-516.

Parkinson, J.S. (1980). Novel mutations affecting a signalling component for chemotaxis of *Escherichia coli.* J. Bacteriol. 142, 953-961.

Parkinson, J.S., Parker, S.R., Talbert, P.B., & Houts, S.E. (1983). Interaction between chemotaxis genes and flagellar genes in *Escherichia coli.* J. Bacteriol. 155, 265-274.

Parkinson, J.S. (1988). Protein phosphorylation in bacterial chemotaxis. Cell 53, 1-2.

Parkinson, J.S. and Hazelbauer, G.L. (1983). Bacterial chemotaxis: molecular genetics of sensory transduction and chemotactic gene expression. In: Gene Function in Prokaryotes (Beckwith, J., Davies, J.E., & Gallant, J.A., Eds.), pp. 293-318. Cold Spring Harbor, NY.

Parkinson, J.S. & Revello, P.T. (1978). Sensory adaptation mutants of *E. coli.* Cell 15, 1221-1230.

Pfeffer, W. (1883). Locomotorische richtungsbewegungen durch chemische rieze. Berichte der Deutschen Botanischen Gesellschaft 1, 524-533.

Pijper, A. (1948). Bacterial flagella and motility. Nature (London) 161, 200-201.

Pijper, A. (1951). Bacterial flagella. Nature (London) 168, 749-750.

Pijper, A. (1955). Shape of bacterial flagella. Nature (London) 175, 214-215.

Pilgram, W.K. & Williams, F.D. (1976). Survival value of chemotaxis in mixed cultures. Canad. J. Microbiol. 22, 1771-1773.

Poole, P.S. & Armitage, J.P. (1988). Motility response of *Rhodobacter sphaeroides* to chemotactic stimulation. J. Bacteriol. 170, 5673-5679.

Poole, P.S. & Armitage, J.P. (1989). Role of metabolism in the chemotactic response of *Rhodobacter sphaeroides* to ammonia. J. Bacteriol. 171, 2900-2902.

Ravid, S. & Eisenbach, M. (1984). Direction of flagellar rotation in bacterial cell envelopes. J. Bacteriol. 158, 222-230.

Ravid, S., Matsumura, P., & Eisenbach, M. (1986). Restoration of flagellar clockwise rotation in bacterial envelopes by insertion of the chemotaxis protein CheY. Proc. Nat. Acad. Sci. USA 83, 7157-7161.

Rollins, C.M. & Dahlquist, F.W. (1980). Methylation of chemotaxis-specific proteins in *Escherichia coli* cells permeable to S-adenosyl methionine. Biochemistry 19, 4627-4632.

Segall, J.E., Ishihara, A., & Berg, H.C. (1985). Chemotactic signalling in filamentous cells of *Escherichia coli*. J. Bacteriol. 161, 51-59.

Shapiro, E., Shapiro, B., & Shapiro, L. (1989). The molecular genetics of differentiation. Genetics 123, 427-429.

Shapiro, L. (1976). Differentiation in the *Caulobacter* cell cycle. Annu. Rev. Microbiol. 30, 377-407.

Sherris, D. & Parkinson, J.S. (1981). Posttranslational processing of methyl-accepting chemotaxis proteins in *Escherichia coli*. Proc. Nat. Acad. Sci. USA 78, 6051-6055.

Silverman, M. & Simon, M. (1974). Flagellar rotation and the mechanism of bacterial motility. Nature (London) 249, 73-74.

Silverman, M. & Simon, M. (1977a). Identification of polypeptides necessary for chemotaxis in *Escherichia coli*. J. Bacteriol. 130, 1317-1325.

Silverman, M. & Simon, M. (1977b). Chemotaxis in *Escherichia coli*: methylation of the *che* gene products. Proc. Nat. Acad. Sci. USA 74, 3317-3321.

Sockett, R.E., Armitage, J.P., & Evans, M.C.W. (1987). Methylation-independent and methylation-dependent chemotaxis in *Rhodobacter sphaeroides* and *Rhodospirillum rubrum*. J. Bacteriol. 169, 5808-5814.

Spudich, E.N., Takahashi, T., & Spudich, J.L. (1989). Sensory rhodopsin I and II modulate a methylation/demethylation system in *Halobacterium halobium*. Proc. Nat. Acad. Sci. USA 86, 7746-7750.

Spudich, J.L. & Bogomolni, R.A. (1984). Mechanism of colour discrimination by a bacterial sensory rhodopsin. Nature (London) 312, 509-513.

Stader, J., Matsumura, P., Vacante, D., Dean, G.E., & Macnab, R.M. (1986). Nucleotide sequence of the *Escherichia coli motB* gene and site limited incorporation of its product into the cytoplasmic membrane. J. Bacteriol. 166, 244-252.

Stanier, R.Y., Doudoroff, M., & Adelberg, E.A. (1958). General Microbiology. Macmillan, London.

Stock, A.M., Mottonen, J.M., Stock, J.B., & Schutt, C.E. (1989). Three-dimensional structure of CheY, the response regulator of bacterial chemotaxis. Nature (London) 337, 745-749.

Stock, A.M. & Stock, J.B. (1987). Purification and characterization of the CheZ protein of bacterial chemotaxis. J. Bacteriol. 169, 3301-3311.

Stock, J.B., Ninfa, A.J., & Stock, A.M. (1989). Protein phosphorylation and regulation of adaptive responses in bacteria. Microbiol. Rev. 53, 450-490.

Strick, J. (1996). Isis 87, 274-305.

Suzuki, T., Iino, T., Horiguchi, T., & Yamaguchi, S. (1978). Incomplete flagellar structures in non-flagellate mutants of *Salmonella typhimurium*. J. Bacteriol. 133, 904-915.

Takahashi, T., Yan, B., Mazur, P., Derguini, F., Nakanishi, K., & Spudich, J.L. (1990). Color regulation in the archaebacterial phototaxis receptor phoborhodopsin (sensory rhodopsin II). Biochemistry 29, 8467-8474.

Toews, M.L., Goy, M.F., Springer, M.S., & Adler, J. (1979). Attractants and repellents control demethylation of methylated chemotaxis proteins in *Escherichia coli*. Proc. Nat. Acad. Sci. USA 76, 5544-5548.

Toews, M.L. & Adler, J. (1979). Methanol formation *in vivo* from methylated chemotaxis proteins in *Escherichia coli*. J. Biol. Chem. 254, 1761-1764.

Trachtenberg, S. & DeRosier, D.J. (1987). Three-dimensional structure of the frozen-hydrated flagellar filament : The left-handed filament of *Salmonella typhimurium*. J. Mol. Biol. 195, 581-601.

Tsang, N., Macnab, R.M., & Koshland, D.E.J. (1973). Common mechanism for repellents and attractants in bacterial chemotaxis. Science 181, 60-63.

Ueno, T., Oosawa, K., & Aizawa, S.-I. (1992). M ring, S ring and proximal rod of the flagellar basal body of *Salmonella typhimurium* are composed of subunits of a single protein, FliF. J. Mol. Biol. 227, 672-677.

Ullah, A.H.J. & Ordal, G.W. (1981). *In vivo* and *in vitro* chemotactic methylation in *Bacillus subtilis*. J. Bacteriol. 145, 958-965.

Wagenknecht, T. (1986). A plausible mechanism for flagellar rotation. FEBS Lett. 196, 193-197.

Weibull, C. (1951). Movement of bacterial flagella. Nature (London) 167, 511-512.

Weibull, C. (1960). Movement. In: The Bacteria: A Treatise on Structure and Function (Gunsalus, I.C. & Stanier, R.Y., Eds.), pp. 153-205. Academic Press, New York and London.

Wolfe, A.J., Conley, M.P., Kramer, T.J., & Berg, H.C. (1987). Reconstitution of signalling in bacterial chemotaxis. J. Bacteriol. 169, 1878-1885.

Wolfe, A.J. & Berg, H.C. (1989). Migration of Bacteria in Semisolid Agar. Proc. Nat. Acad. Sci. USA 86, 6973-6977.

Wolfe, E., Bogomolni, R., Scherrer, P., Hess, B., & Stoeckenius, W. (1986). Color discrimination in halobacteria: spectroscopic characterisation of a second sensory receptor covering the blue-green region of the spectrum. Proc. Nat. Acad. Sci. USA 83, 7272-7276.

Yamaguchi, S., Fujita, H., Ishihara, A., Aizawa, S.-I., & Macnab, R.M. (1986). Subdivision of flagellar genes of *Salmonella typhimurium* into regions responsible for assembly, rotation, and switching. J. Bacteriol. 166, 187-193.

Yonekawa, H., Hayashi, H., & Parkinson, J.S. (1983). Requirement of the *cheB* function for sensory adaptation in *Escherichia coli*. J. Bacteriol. 156, 1228-1235.

DANIEL ARNON[1]

ROBIN HILL[2]

EDWIN KREBS[3]

H. H. WEBER[4]

HUGH HUXLEY[5]

JEAN HANSON[6]

ANDREW HUXLEY[7]

KENNETH BAILEY[8]

LEEUWENHOEK[9]

THEODOR ENGELMANN[10]

JULIUS ADLER[11]

WILHELM PFEFFER[12]

LOWELL HOKIN[13]

MABEL HOKIN[14]

EARL SUTHERLAND[15]

PHILIP COHEN[16]

MARJORY STEPHENSON[17]

ANDRÉ LWOFF[18]

FRANCOIS JACOB[19]

JACQUES MONOD[20]

PAUL ERLICH[21]

KARL LANDSTEINER[22]

MACFARLANE BURNET[23]

ELVIN KABAT[24]

GERALD EDELMAN[25]

RODNEY PORTER[26]

CESAR MILSTEIN[27]

SUSUMU TONEGAWA[28]

Chapter 6

TALKING TO CELLS–CELL MEMBRANE RECEPTORS AND THEIR MODES OF ACTION

Robin F. Irvine

INTRODUCTION

This chapter is essentially about the field of research that we now know as "signal transduction" or "cellular signaling". Currently this field comprises a significant proportion of the world's total research in the life sciences. This is not surprising if one thinks about it. The cells of our tissues are under the constant control of hormones, neurotransmitters, and growth factors, which are telling the cells to do this, do that, stop doing this, do that instead, etc. The great majority of these outside influences—"agonists" is a useful all-embracing term—are water-soluble. They have to be because they move and work in an aqueous environment. So, when they come up against the hydrophobic cell membrane (plasma membrane) of any cell, they must either be taken up into the cell by an active process (e.g. endocytosis, active transport) which is necessarily slow, or they must bind to a specific recognition site (a receptor) in the plasma membrane, which then registers their presence by sending a chemical signal into the cell. Signal transduction, therefore, is all about the nature of these chemical messages, how they are generated after the receptor has bound its ligand (the agonist), and how the cell uses them to alter its function.

Because these receptor-generated signals "plug-into" and modulate the homeostatic control mechanisms of a cell's functions, it is inevitable that in understanding receptor-mediated signal transduction we will understand the fundamentals of cellular function. Even the exceptions to the "water-soluble agonist" paradigm (e.g. steroid hormones, thyroxine, retinoic acid) follow the same principles, but, because of their hydrophobic nature (which in turn necessitates a slow action as they have to move around the body while bound to plasma proteins), these molecules can cross the plasma membrane unaided and be recognized by receptors inside the cell. Because they act on intracellular receptors, steroids are therefore outside the remit of the chapter, so for further information the interested reader should consult an excellent summary by Evans (1988). This is to be particularly recommended because the concept that a specific protein recognizes a hormone and transduces its signal owes much to the early studies of steroids.

These are extraordinary times for scientists working on signal transduction, times with which I can find parallels only with physics in the 1920s. Between 1960 and (perhaps?) 2020, we will have identified most of the key players in the processes that govern eukaryotic cell function, and at least the principles of what they do. The (loose) temporal scope of this chapter, 1960–1990, obviously covers the first half of this "golden age", and although it goes without saying that all I describe rests absolutely on the work of earlier pioneers, it is nevertheless true that virtually all of the concepts, paradigms, and molecular species that are now used as a basis for our present research spring from these 30 years.

Finally, I must raise the inevitable waiver clause: I cannot be totally comprehensive. I will be picking specific examples to illustrate general principles, and personal opinion and selection from my own perspective will inevitably influence what I put in and leave out. This is not a scholarly review, but is intended as a readable story that gives a flavor of some of the breakthroughs; to give a full story would take more than this entire book. Besides, as the historian A.L. Rowse (1955) has pointed out, any fool can generalize, but only using specific examples can give us true insight. And following that philosophy to a practical endpoint, a review like this is a very poor substitute for the real thing. Only reading the original papers can truly give a feel for how things were perceived and done "in those days", and I hope that at least some of the most important papers referred to here will be sought out in the dusty shelves of libraries. The reader will find his or her search will be made considerably easier by a forthcoming publication (Burgoyne and Petersen, 1997), which reprints some of them in original form.

MEMBRANE RECEPTORS

Introduction

The idea that receptors—specific recognition sites for agonists such as acetylcholine, adrenaline, and histamine—may exist goes way back into the early history of pharmacology. In the 1940s, pharmacologists distinguished two functions of adrenaline (α and β types) and of acetylcholine (muscarinic and nicotinic). These discoveries can be taken as the beginnings of the idea that a particular tissue must have a specific way of recognizing an extracellular agonist. But it is only in the last 30 or so years that we have clearly formulated the idea of a protein specifically designed to recognize an agonist in a "lock-and-key" manner, involving allosteric control of the receptor's structure as an integral part of the transduction process. Allostery and its history are outside the scope of this review, but it is important to remember the absolutely central role of allostery when considering signal transduction. Simply visualize how small an acetylcholine molecule is compared to a nicotinic acetylcholine receptor, and then think of the profound and rapid change in the receptor's structure that results from the interaction of the two molecules (so beautifully demonstrated recently by Unwin (1995)). The analogy of a bee stinging an elephant is a compelling one in terms both of relative size and of effect. The mechanism of protein structure and function that underlies this phenomenon, which permeates every single process that I describe below, is truly one of the fundamentals of life.

Receptor Structure

The idea that specific recognition sites for agonists exist on the surface of cells began to take on a new veracity with the appearance over the 1950s and 1960s of radiolabeled ligands of high specific activity. These ligands could be incubated with cells, or membrane preparations from cells, and then separated by filtration or centrifugation so that the extent, saturability, and specificity of binding could be determined. As many effects of hormones or neurotransmitters on tissue metabolism had already had their pharmacology mapped out fairly extensively (i.e. which antagonists could oppose an effect of a hormone and with what potency, and which could mimic the effect), the pharmacology could now be correlated with displacement of a radioactive ligand. The better the correlation, the stronger the indication that the binding was to a receptor rather than to something else. That something else could be, for example, (a) an enzyme that was there to metabolize the agonist, or (b) a protein that just happened to bind the ligand under the conditions and sensitivity of the assay.

Cuatracasas (1974) gives a very clear and detailed summary of what was done in the 1960s using this approach, and of the controversies that surrounded the data. Indeed, controversies were in plentiful supply. For example, a radioactive ligand might bind to several different proteins (as mentioned above), thus not all of the binding observed was to what the investigators were seeking ("the" receptor). Even when the overall specificity of a binding site looked encouraging, there often was still a contradiction between the apparent affinities observed in the binding of ligands to membranes or cells, as compared with the much lower concentrations of the same compounds needed to exert a metabolic effect. This became known as "receptor reserve" (i.e. you need to bind agonist to only a small proportion of receptors to saturate the physiological response of the cell). We can now account for receptor reserve in molecular terms (it is because signal transduction pathways amplify as they progress), but the results worried greatly those who were hoping that their radioactive ligands were indeed binding to true receptors.

Another source of concern was whether these binding sites for hydrophilic ligands were truly located within the plasma membrane. Subcellular fractionation and early immunocytochemical studies (e.g. using antibodies prepared to partially purified receptor protein; Ehrenfreis, 1962) in general supported a cell surface localization, but uncertainties inevitably accompany interpretation of such complex fractionations. The doubts were resolved only when the various receptors were fully purified and antibodies to them generated, and then, with the advent of molecular biology (where would signal transduction be without it?), the receptors cloned and characterized molecularly. Only with these advances, which all happened between 1960 and 1990, did receptors become a reality. I cannot possibly describe the whole history of these "quantum leaps", but I will take three examples (Figure 1): the nicotinic acetylcholine receptor (a ligand-gated ion channel); the β-adrenergic receptor (a G-protein-linked receptor), and; the insulin receptor (a tyrosine–kinase-linked receptor). These were in each case the first (or nearly so) of their kind, but I choose them principally because they are so well known, and even many nonscientists have heard of acetylcholine, adrenaline, and insulin.

Nicotinic Acetylcholine Receptor: A Ligand-Gated Ion Channel

Sometimes one is inclined to think that the only reason why a particular type of creature exists is that God put it there so humankind could study whatever process that creature takes to an extreme. Certainly such thoughts are provoked by the various "electric" eels and fishes (*Electrophorus, Torpedo, Narcine*, and *Narke*) which derive their electrogenic systems in part by having *huge* amounts of the nicotinic acetylcholine receptor. This made (comparatively) easy the purification, using ligand or inhibitor binding as an assay, of large quantities of nicotinic acetylcholine receptors from various

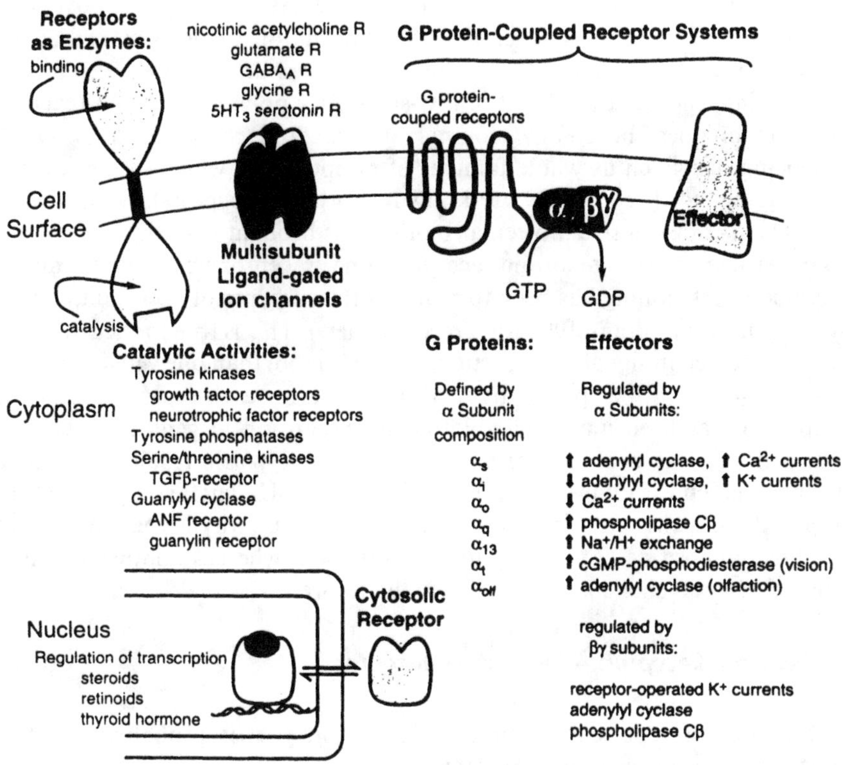

Figure 1. Structural motifs of physiological receptors and their relationships to signaling pathways. (Reproduced with permission from Ross, 1996).

electrogenic organisms in the 1960s and 1970s by several groups, principally those of Changeux, Raftery, Cohen, Polanic, Lindström, and others (see Popot and Changeux, 1984, for a review). Indeed, it is important to remember that in those days, before protein microsequencing and molecular biology, this was one of the few eukaryotic receptors that could be purified in sufficient quantity to be studied: we learned a great deal by characterization of its shape, size, subunit structure, and mechanism (i.e. how it functions as a ligand-gated Na^+ channel) before any cloning took place.

From these heroic studies emerged the picture of a large oligomeric protein with two α-subunits, and one each of β, γ, and γ, which formed a pore that spanned the membrane to act as a sodium channel (see Popot and Changeux, 1984). When the cloning and sequencing of these subunits was achieved (initially mostly in Raftery's laboratory, and then to a major degree in Numas';

for a review, see McCarthy et al., 1986), it put molecular detail into what was already a reasonably well-characterized structure. As I mentioned above, this sequence of events is not a generality; the majority of receptors cloned at the time of writing this review (1996) were more or less unknown entities before their cloning. Indeed, the advent of low-stringency polymerase chain reaction (PCR) techniques, homology cloning, and random complementary DNA sequencing has given us whole families of receptors for which we do not yet know the ligands (hence, they are known as orphan receptors); what a *volte face* 20 years of extensive molecular biology has brought about!

To return to the nicotinic acetylcholine receptor, the other major consequence of cloning has been to multiply the subtypes of each subunit (see Sivilotti and Colquhoun, 1995 for a recent review). This is true for nearly every protein involved in signal transduction—there is more than one version of each protein, usually with a specific tissue distribution for each type. This level of complexity has been made possible by the expansion in gene number that coincided with the evolution of vertebrates, probably as a result of an extra level of gene repression by methylation (Bird, 1994). Understanding *why* it has happened—i.e. answering a specific question such as, why is this member of the phosphoinositide C β-family found in tissue *a* whereas another is found in tissue *b*?—is a challenge now and for the future.

β-Adrenergic Receptor: A Serpentine Receptor

This receptor represents a family of receptors that is probably the most numerous: the seven-transmembrane domain, or serpentine, receptors. The amino terminus is exterior to the cell and the protein crosses the membrane seven times so that the carboxyl end is intracellular. All these receptors (as far as we know) interact with trimeric G-proteins (see below), and recent cloning counts estimate their number to be several hundred.

The β-adrenergic receptor had been extensively studied by ligand binding and by its physiological effects (including its ability to couple via G-proteins to adenylyl cyclase—see below) for many years, with, as one might expect, considerable controversy and argument (for a comprehensive review see Stiles et al., 1984). The tissue that had enough β-adrenergic receptors and was obtainable in sufficient quantities to enable purification (i.e. the nearest equivalent of the electroplax tissues and the nicotinic acetylcholinergic receptor) was the frog erythrocyte, and purification from this source (and later from guinea pig lungs) led to the first cloning (Dixon et al., 1986).

There is a suggested difference between a biochemist and a molecular biologist: when a protein is cloned, a biochemist hopes it has not been discovered before, whereas a molecular biologist hopes that it has. Another reason I have focused on the β-adrenergic receptor is that it illustrates particularly well this general principle of molecular cloning; sometimes you learn more from cloning the second or third examples of a receptor type than you do from the first. For the β-adrenergic receptor, a remarkable leap in knowledge stemmed from its cloning (Dixon et al., 1986) because it was closely related to the photo-sensitive protein rhodopsin. This observation was not in itself entirely unexpected (as they both couple to G-proteins), but was nevertheless of great importance because so much was already known about rhodopsin.

Rhodopsin is an unusual serpentine receptor in that it is not in a plasma membrane, but is instead located in an intracellular membrane (the membrane of photoreceptors); its "agonist" is a photon of light. The availability of rhodopsin in large quantities in bovine retinae (from the slaughterhouse, that ubiquitous and vital source of animal material) led Jeremy Nathans and Daniel Hogness to purify and clone it in 1983, and from its hydrophobicity profile (see Steitz et al., 1982) to deduce that it had seven transmembrane domains. This in turn was consistent with information on the closely related bacteriorhodopsin, a protein about which much was already known from the elegant work of Walther Stoeckenius and co-workers. For them, the evolutionary gift, again equivalent to the *Torpedo* electroplax for the acetylcholine receptor, was the purple membrane from *Rabobacterium halobinium*. This extraordinary beast lives under highly saline conditions and generates its energy by a light-driven proton extruder. Stoeckenius and Kunau (1968) isolated the "purple" membrane responsible for this process and showed that it contained a rhodopsin-like pigment. Indeed, the purple membrane contained this protein, an H^+-channel, in an essentially pure state: the membrane is almost pure bacteriorhodopsin in highly ordered arrays (Blaurock and Stoeckenius, 1971). It was therefore the ideal material for Henderson and Unwin (1975) to use for their classic study in which, by quantitative electron microscopy of the membrane from various angles, they showed bacterial rhodopsin to have seven transmembrane domains and a central pore.

Thus it was that when Dixon et al. (1986) determined the primary sequence of a β-adrenergic receptor, in one fell swoop the link through rhodopsin and bacteriorhodopsin gave us a large amount of information about its structure, topography, and probable functional domains. Moreover, as additional members of the serpentine receptor family were cloned, the conserved, part-conserved, and variable sequence domains gave us more and more insight into the functions of the different parts. Extensive site-directed mutagenesis has since led to an enormously detailed picture of how this protein family works.

Thus, the progression in less than 20 years from a binding site of debatable physiological significance to a protein of whose molecular behavior we have some intimate knowledge results from a typical mixture of luck, evolution having given us suitable tissues for study, and inspired science.

Insulin Receptor: Tyrosine Kinases

I have chosen insulin because it is the best-known hormone; indeed, it was in fact something of an embarrassment to those working in the insulin transduction field that more than 50 years after the spectacular discovery of insulin, we still had no idea how it worked! For the insulin receptor, too, we find a leap connected it with a research area that had been going on in parallel for years. In fact the receptor was not cloned until 1987, and only in the 1990s are we beginning to unravel its signal transduction pathway. However, following early purification attempts, and the discovery that it had tyrosine kinase activity, it became a part of the tyrosine kinase story, one of the most important signal transduction discoveries of the 1970s and 1980s.

In the wake of extensive studies that investigated binding of radioactive insulin to membranes and tissues (which included the elegant use of insulin attached to beads too large to be endocytosed, Cuatrecasas, 1969; for a review see Cuatrecasas, 1974), the first purification of insulin receptors to homogeneity was by Pedro Cuatrecasas in 1972 using affinity chromatography on concanavalinA–agarose and insulin–agarose columns; affinity chromatography is a technique now widely employed in biology, and it was introduced for this particular purpose. The final purification, from placenta, was 250,000-fold—there is no *Torpedo* electroplax or purple membrane equivalent for the insulin receptor!

During the following 10 years, investigations by a number of groups (particularly those of Mike Czech and Ron Kahn) led to a detailed picture of a heterotetrameric protein with two α- (155 kDa) and two β- (95 kDa) subunits, but its signal transduction pathway was still unknown. The key observation came from Kahn's, Oren's, and others' laboratories (see Kasugua et al., 1983, for references): the purified receptor was a protein kinase, specifically a tyrosine kinase. This provided a link with another set of observations running in parallel in a completely unrelated area. (How many times do unexpected links of this sort have to occur before those who direct science realize that science cannot always be directed?).

In 1970, Martin published a classic paper in which he described a temperature-sensitive mutant of the Rous sarcoma virus that, under restrictive conditions, could grow but could not transform cells. This was really the starting point for the recognition of cancer-related genes (oncogenes) because the gene responsible, *src*, cloned much later, was the first oncogene. Elegant work in many laboratories (principally those of Vogt, Hanafusa, and Erickson) identified genetically and biochemically the existence of this protein which transformed cells. In 1980, Hunter and Sefton showed that the *src* gene product was a protein with enzymic activity that enabled it to phosphorylate itself and other proteins, but the phosphate moiety was transferred from ATP to a tyrosine, rather than to a serine or threonine residue as previously found.

During the following decade a large number of growth factor receptors, e.g. those for epidermal growth factor (EGF) (Ushio and Cohen, 1980) and platelet-derived growth factor (PDGF) (Ek et al., 1982), were shown to contain tyrosine kinase activity. Since protein kinases, which play a central role in signal transduction, are to be covered in Chapter 7, I will take that story no further, but return to the insulin receptor which was then (in 1983–1984) plugged into the emerging picture. In fact, the insulin receptor has proved to be something of an exception among those receptors that transduce their signal by tyrosine phosphorylation. The latter are now classified broadly into (a) those that have a tyrosine kinase domain as an intrinsic part of the receptor (for a recent review see Van der Geer et al. 1994), or (b) those that couple to nonreceptor tyrosine kinases (for recent reviews see Irvine et al., 1996). Most of these are monomeric proteins that dimerize after ligand binding, but the insulin receptor is a heterotetramer (see above). Also, whereas those receptors with intrinsic kinase activity recruit their "targets" mostly by direct interaction of their phosphorylated tyrosines with SH2 domains, the insulin receptor appears to phosphorylate a "docking protein" IRS-1 (insulin receptor-substrate 1; White et al., 1985), which subsequently recruits the target proteins to it for activation.

Mention of SH2 (*src* homology 2) domains should not be passed by without some explanation. These were first identified by Tony Pawson's group (Sadowski et al., 1986) as a domain widespread among proteins that are activated by/associated with tyrosine phosphorylation. The full story of their specificity and function is still to be elucidated, but I have mentioned them here because they are perhaps the first clear example of a domain that (a) is used in many proteins, (b) has been shuffled around with its precise characteristics changed (for specificity) by evolution, and (c) is involved in protein-protein interaction and recognition as part of many intracellular signaling pathways. The identification of such domains is now proliferating at a great rate (the first clue is often sequence similarity), and the ways in which proteins talk directly to each other to transduce signals is currently, and in the future surely to be, one of the most exciting areas of signal transduction.

The insulin receptor was finally cloned in 1985 by Ullrich et al. and Ebina et al.; although we have still much to learn about how it works, we have again seen an extraordinary advance in our knowledge. From diverse sources in the last 20 years, our picture of the receptor has moved from a controversial set of binding data using radiolabeled insulin to a molecular entity of which we know at least the basic working principles.

COUPLING OF RECEPTORS TO INTRACELLULAR SIGNALS

Introduction and Principles

In the above sections I have outlined with three selected examples some of the profound advances during the years 1960 to 1990 that have led to our present understanding of the nature of membrane receptors. All these receptors follow the basic principle of having a structure that specifically recognizes a ligand (agonist), and this recognition results in an allosteric change in receptor conformation. Now we must consider the consequences of that change—intracellular signaling.

Most receptors are transmembrane proteins but there are some proteins now recognized as having a receptor function that do not span the membrane—for example those that are joined to the outer surface of the plasma membrane by glycosylphosphatidylinositol links. However, these polymerize when they bind their ligands, and as a result interact with other proteins that do not bind extracellular ligands, but are transmembrane proteins. In these instances—important examples being the receptors for recognition of B-cells by T-cells—the "receptor" is therefore actually a complex of many different proteins, yet it functions as a transmembrane entity. The change in receptor structure following any agonist binding therefore causes a change in the conformation of the intracellular domains, and it is this change that generates the intracellular signal. We now know there is a wide range of "cross-talk" in agonist–intracellular signal interaction. By this I do not refer to downstream cross-talk between intracellular signaling pathways—there is, of course, plenty of that. There are three, more immediate, mechanisms:

1. One agonist can activate a wide range of different receptors and signaling pathways in different tissues. For example, acetylcholine activates nicotinic receptors as described above, or the family of muscarinic receptors of which there are five or more. The muscarinic receptors, however, are entirely different from the nicotinic since they are serpentine (seven transmembrane domain) receptors which in turn can couple to different signaling systems depending on which member of the family is involved: (e.g. M_1 and M_3 receptors couple to the phosphoinositidase C pathway; M_2 and M_4 do not).
2. One signaling system (e.g. phosphoinositidase C) can be activated even in one cell type by several different receptors. This convergence into one signal can occur in one of two ways. (a) A group of very similar receptors that recognize very diverse agonists couple to the phosphoinositidase C (PIC) pathway in the same way. For example, in rat parotid glands, acetylcholine, adrenaline, and the peptide substance P all couple to the PIC pathway by their serpentine receptors (M_1, α_1 and substance P_1,

respectively). (b) Very different types of receptor coupling may occur in different ways to produce a similar endpoint. Bombesin and PDGF both activate phosphoinositidase C in mouse Swiss 3T3 cells, but the former works through a serpentine receptor via PICβ, and the latter through tyrosine kinase activation and acting on PICγ.

3. One agonist can interact with more than one signaling system in the same cell either (a) by one receptor coupling to multiple signaling systems or (b) by one agonist binding to more than one receptor (e.g. adrenaline applied to rat parotid glands activates adenylyl cyclase by β-adrenergic receptors and phosphoinositidase C (PIC) via α_1-receptors).

Thus, even leaving aside cross-talk between signaling systems, there is clearly the most enormous flexibility in which agonist will activate what and in which cell. This stems from a combination of (a) the similarities in the basic pathways used (e.g. the tyrosine kinase receptor family, the serpentine receptors, G-proteins, etc.), (b) the differences that nevertheless exist within these pathways (e.g. different serpentine receptors couple to different G-proteins), and (c) the ability of cells to feed different signals into common endpoints (e.g. PDGF and bombesin activate PIC by different pathways). An example of convergence even further "downstream" is that glucagon and vasopressin both activate glycogen breakdown via serpentine receptors in hepatocytes, but the former works via the cyclic AMP pathway, and the latter via Ca^{2+}, in turn mediated by activation of PIC.

All of this seems obvious now, but it was far from obvious in 1960. It is most informative to read reviews written about that time (indeed, the review by Cuatrecasas (1974) is a good example), when cyclic AMP was essentially the only understood and reliable second messenger. Protein kinase cascades, Ca^{2+}, inositides, diacylglycerol, and cyclic GMP were still unknown or poorly understood, and it is extraordinary what convolutions were employed to explain everything in terms of cyclic AMP because that was all there was. The very concept, so obvious to us now, that there might be other signaling systems, entirely distinct from cyclic AMP (albeit with some downstream cross-talk), that could be coupled to entirely distinct receptors (including different receptors for a common agonist) was entirely unknown.

To my knowledge, one of the first people to state this concept most clearly was Bob Michell in his seminal review on inositide turnover in 1975. To most people that review is perceived as the first hint of a link between inositides and Ca^{2+}, as indeed it was; but one could view this suggestion as being the less important aspect. What the review predominantly did was to dissect out from a morass of confused data a single signaling system, now known as the phosphoinositidase C pathway, entirely separate from adenylyl cyclase, and consider it as being activated by a separate group of receptors.

In the following sections I shall summarize briefly how our knowledge advanced from 1960 to 1990 in each of the individual major signaling mechanisms, treating them in isolation (for simplicity). But each story is of course not in isolation, and even when we have clarified the major players, the full extent of intracellular cross-talk, from receptors right down to final cell responses, is going to be a major challenge for the next century.

G-Proteins

Trimeric G-Proteins

After the identification of cyclic AMP as an intracellular second messenger (*vide infra*) in the late 1950s, it was a natural progression to investigate the way in which adenylyl cyclase was activated by hormones *in vitro* by using broken cell preparations. Early studies on fat cell membranes (e.g. see Birnbaumer and Rodbell, 1969) showed that more than one hormone can stimulate the same pool of adenylyl cyclase, and so it was likely that the active unit was a multimeric protein complex in which the receptor(s) and adenylyl cyclase were interacting. The physical separation of receptor and cyclase was most directly and elegantly shown by Orly and Schramm (1976) using cell fusion experiments. They inactivated adenylyl cyclase in avian erythrocytes by treatments that do not harm the cells' adrenergic receptors and then fused the erythrocytes with Friend cells that had no receptors but did have a cyclase; the result was a restoration of an adrenaline-sensitive cyclase.

The crucial observations that brought GTP into the picture had already been made in two classic papers from Marty Rodbell's laboratory (Rodbell et al. 1971a,b) which showed that guanyl nucleotides at a low concentration and with a high specificity (i.e. ATP, CTP, and UTP would not mimic GTP's effects) had two profound effects on liver membranes. In the presence of GTP the affinity of the binding of radiolabeled glucagon was decreased and the glucagon-mediated stimulation of adenylyl cyclase was greatly enhanced.

Similar observations in the following years in other tissues and with other receptors (including some not coupled to cyclic AMP) pointed to a commonality in this requirement for GTP. (I should note here that a comprehensive and very readable account of the G-protein story has been written by Lutz Birnbaumer in 1990, and enables me to make this a shorter section than it need otherwise have been.) The crucial demonstration that the GTP-binding part of this coupling was on a separate protein entity, separable by affinity chromatography from both receptor *and* adenylyl cyclase, was first made by Tomas Pfeuffer and colleagues (Pfeuffer and Helmreich, 1975; Pfeuffer, 1977). During the next 10 years the nature and function of the proteins involved in this GTP-requirement were elucidated. As with many examples above, a key component to some of the breakthroughs was the availability of suitable tissues for study: the *cyc*-lymphoma cells introduced by Bourne et al. (1975), turkey erythrocytes, and bovine retinae all contributed to essential parts of the story. The two toxins, cholera and pertussis, which interfere specifically with the functions of particular G-protein α-subunits, also played an important role in the elucidation of G-protein function (see Birnbaumer, 1990). Transducin, which couples rhodopsin to cyclic GMP phosphodiesterase, was the first G-protein to be studied extensively at the biochemical level. The G-protein "cycle" ($\alpha\beta\gamma$ dissociates into αGTP and $\beta\gamma$, and then the three subunits reassociate following GTP hydrolysis) was a central observation that stemmed largely from the work of Lubert Stryers' group on transducin, and that of Al Gilman's group on G_s.

Transducin was also the first trimeric G-protein to have its α-subunits cloned (Harris et al., 1985), and the multiplicity of α-subunits (e.g. Wilkie et al., 1992), plus a large selection of β- and γ-subunits that have now been cloned, is testament to the flexibility of and importance given to G-proteins in signal transduction. As far as we know, all of the serpentine receptors couple obligatorily through G-proteins, and they have been found to regulate a wide variety of cellular processes. Such is the diversity and ubiquity of the G-proteins that a complete description of their discovery would require a whole chapter, but again I refer the reader to the review by Birnbaumer (1990) for a fuller picture.

Monomeric G-Proteins

The binding of GTP to the α-subunit of trimeric G-proteins involves a binding site that is homologous with the site that binds GTP in monomeric G-proteins; so it appears that binding GTP and using its hydrolysis as a molecular "switch" is something that evolved a long time ago and has been put to extensive use. If trimeric G-proteins are widely used because they couple to serpentine receptors, this almost fades into insignificance in comparison with the diversity of monomeric G-proteins and the number of cellular processes in which they are involved. It is virtually impossible to find a complex cellular process that does not use one of these proteins (also known as small G-proteins, or smgs) somewhere in its control mechanism.

In the history of their discovery, as in their evolution, the monomeric G-proteins actually preceded the trimeric variety. In the 1960s the role of GTP in the initiation of protein synthesis and in the elongation of the polypeptide chain was first detailed (for reviews of, see Kaziro, 1978; Siekevitz[2], 1996). We now class the proteins involved as members of the same superfamily as those involved in signal transduction (for reviews of the state of play in 1990 see Bourne et al., 1990; Grand and Owen, 1991). It is the modern term often used for them, the *ras* superfamily, that reveals the link between signal transduction and monomeric G-proteins. The key observation was made by Ed Scolnick and his colleagues in 1979: in characterizing the *ras* oncogenes that they had isolated from sarcoma viruses, they showed that *ras* gene products interacted specifically with guanine nucleotides which could modify their structure (and presumably function).

At this time the involvement of GTP in other cellular processes (e.g. secretion, microtubule function) was also being demonstrated, and so it was during the 1980s that an enormous number of groups, studying the whole spectrum of intracellular functions, purified (by function or GTP binding) and then cloned (by homology with *ras*) the ever-increasing members of this superfamily (see e.g. Zerial and Huber 1995).

Discovery and cloning is certainly way ahead of function at the moment. We are still in the 1990s in the midst of finding out what the *ras* proteins actually do, and we have no clear idea why there is more than one *ras*. But already roles are assigned for many members of the small G-protein superfamily, and these are found in every process that occurs as a direct or indirect result of receptor activation.

Cyclic AMP

The crucial papers on cyclic AMP—its discovery and the generation of the "second messenger" concept—belong to the years before the 1960s (for a review, see Robison et al., 1968). Its mode of action, primarily via a family of protein kinases that it controls (the A kinases, now understood to a remarkable degree at a molecular level) is considered elsewhere (see Ord and Stocken, 1995; this volume, Chapter 7), thus this section will be much shorter than the importance of cyclic AMP as a second messenger might suggest. Cyclic AMP was the "original" second messenger and the paradigm for all that followed.

The regulation of adenylyl cyclase by serpentine receptors via trimeric G-proteins has already been discussed, but that is only part of the story. The regulation of adenylyl cyclase by Ca^{2+} was first shown in rat brain preparations by Bradham et al. (1970), but the effects of Ca^{2+} were complex to understand and it was not really until 1979 when Westcott et al. used calmodulin affinity chromatography to separate the Ca^{2+}/calmodulin-regulated isoform from a nonregulated form of the enzyme that the true complexity of the cyclase family began to be revealed.

The first of these cyclases was characterized and cloned in Al Gilman's laboratory (Krupinski et al., 1989). Since then the family has grown (as usual!) and currently there are at least nine adenylyl cyclases, regulated positively and negatively by Ca^{2+}, α-subunits, and $\beta\gamma$-subunits of G-proteins in different ways (for a recent review, see Surahara et al., 1996). The number of members of the family of proteins that remove cyclic AMP (cyAMP phosphodiesterases) has also grown to a great degree in the 1980s and 1990s since its initial discovery in 1962 by Butcher and Sutherland[15]. Perhaps the most noteworthy observation with regard to its regulation was that by Cheung (1970) who first showed the Ca^{2+} regulation of the activity, and thus not only started the long history of cyclic AMP and Ca^{2+} talking to each other (something they do extensively in many tissues; e.g. see Rasmussen and Barratt, 1984), but also that of calmodulin (*vide infra*).

One particularly interesting development in more recent times has been the elucidation of the effects of cyclic AMP on transcription in the nucleus via protein kinase A which enters the nucleus to phosphorylate negative and positive controllers of the cyclic AMP-responsive element (CRE) (Comb et al., 1986; for a recent review, see Sassone-Cossi, 1994). Because the initial "message", cyclic AMP, is so well-defined, this has provided an excellent model system for studying the regulation of nuclear events by cytosolic influences. Not the least of my reasons for mentioning it here is to introduce another waiver clause—that this review stops at the nuclear membrane. The transfer of information from cytoplasm to the nucleus is currently being unraveled at a remarkable rate. Much of its background, along with huge advances in our understanding of transcriptional regulation (bear in mind that transcription as an idea stems largely from the 1960s), lies in the years covered by this review, but if I don't confine myself to the plasma membrane and cytoplasm this chapter will never be finished!

Calcium

Anyone who has attended one of Jim Putney's lectures will have seen the amusing slide he has of a (notional and entirely biased!) measure of "relative importance of second messengers" in which Ca^{2+} is enormously more important than cyclic AMP, cyclic GMP, etc. But there is a serious side to the flippancy since there is some truth in the statement that Ca^{2+} as a second messenger is so universal and controls acutely and chronically so many aspects of cell function that it probably is the most single "important".

Ever since Ringer and others found that tissues become very unhappy if they are deprived of Ca^{2+} we have known that Ca^{2+} is important for biology, but it is remarkable that it wasn't until the 1980s that this was fully realized. In excitable tissues (neurones and particularly striated muscle) which have voltage-gated Ca^{2+} channels, and where Ca^{2+} controls rapid events in spectacular fashion, the timescale of discovery and elucidation of Ca^{2+} action has been very different from that in other systems. Appreciation of the role of Ca^{2+} in these tissues and how it is involved in excitable coupling goes back into the 1960s and 1970s and will be dealt with in Chapter 4 (Perry). Here I want to trace the acceptance of Ca^{2+} as a second messenger of universal importance which in this content means in nonexcitable tissues.

The earliest measurements and estimates of intracellular Ca^{2+} levels using microinjected aequorin (a calcium-sensitive luminescent protein) were on frog skeletal muscle (Ashley and Ridgeway 1970), followed by investigations on the giant squid axon (Baker et al., 1971). Similar measurements on other giant cells, and the later use of arzenazo III, Ca^{2+} microelectrodes, or null-point determinations during the 1970s, led to the general acceptance that levels of intracellular free Ca^{2+} were submicromolar in the cytoplasm (as opposed to millimolar outside the cell).

However, these measurements were restricted to a few large and mostly excitable cells, while the role of Ca^{2+} in other tissues remained difficult to define. Thus with the exception of the laboratories of a few individuals, such as those of Bill Douglas, Howard Rasmussen, and Peter Baker, Ca^{2+} as a second messenger in other tissues took a back seat compared with cyclic AMP. This was despite the discovery and characterization during the 1970s of calmodulin (Cheung, 1970; Kakiuchi et al., 1970; for a comprehensive review, see Means and Deadman, 1980), and the demonstration that it is a nearly ubiquitous protein, which by being responsive to Ca^{2+} at submicromolar ranges of concentration, regulates many intracellular functions.

During the 1970s, studies on the removal of Ca^{2+} from cells and the efflux of $^{45}Ca^{2+}$ during stimulation led to the concept of a "trigger pool" of Ca^{2+} released from within the cell (again, an idea with its precedent in muscle). For a while the mitochondrion was a favored source, but it was mostly the work of Dick Denton and his colleagues (e.g. Denton and McCormack, 1980) which established that this organelle is a target for, and is indirectly controlled by, cytosolic Ca^{2+}, rather than vice versa. The more likely source was the cellular homologue of the muscle sarcoplasmic reticulum, the endoplasmic reticulum.

The huge increase in interest in and knowledge of Ca^{2+} in all tissues, especially nonexcitable ones, that has happened in the 1980s stems to a significant degree from two giant steps—one methodological and one experimental. The first was the introduction by Roger Tsien of fluorescent dyes for the estimation of intracellular Ca^{2+} and the method of loading them into intact cell populations by supplying them to the cells as cell-permeable esters. The dyes were thus trapped inside the cell by the action of esterases (Tsien, 1980, 1981). The first of these dyes was Quin-2, but later dyes such as Fura-2 or Indo-1 (Grynkiewicz et al.,1985) were great improvements because they required lower loading levels (hence fewer Ca^{2+}-buffering artifacts). Perhaps more importantly, the shift in fluorescent profile on Ca^{2+} binding of Fura-2 and Indo-1 enabled Tsien and his co-workers to introduce the Ca^{2+}-imaging techniques that today so dominate our knowledge of spatio–temporal Ca^{2+} signaling. Oscillations in intracellular Ca^{2+} concentrations have become a major focus of attention. Although their original discovery dates back to the 1970s (for a review, see Berridge and Rapp, 1979), it was the observation of Woods et al. (1986), who ironically used aequorin as a monitor of intracellular Ca^{2+}, and showed that hepatocytes exhibited pulsatile Ca^{2+} signals, thus rekindling interest in them in the "post-InsP_3" days.

The second major advance in our understanding of intracellular Ca^{2+} regulation, which led to a conceptual change, was the identification of inositol 1,4,5-trisphosphate (InsP_3) as the principal second messenger mobilizing Ca^{2+} from the intracellular stores (Streb et al., 1983) of which more will be told below. The discovery that cells use a freely diffusible second messenger to control intracellular Ca^{2+}, together with the invention of techniques to measure Ca^{2+} in real time and space, has enabled us to understand that Ca^{2+} is controlled and used in a very complicated way by cells of many different types. This complexity itself helps us to understand why it took so long for Ca^{2+} to be fully appreciated as a universal controller of cell function.

$Ca^{2+}$is not the only regulatory molecule to oscillate and in fact biological oscillations and how they are generated and interpreted by cells are amongst one of the most exciting areas of signal transduction at the moment (see for example, Goldbeter, 1996 for a full survey). At around the same time as those in electronics have found that frequencies are easier to quantify than amplitudes (which is why FM reception on your radio is much better than AM), biologists are beginning to realize that evolution made this particular discovery a very long time ago.

Inositides and Diacylglycerol

The unraveling of what is now known as the phosphoinositidase C (PIC) pathway was a very long and complex process. It is also the story that I know best, having worked in the field since the 1970s. But despite both these factors, this will not be a long section. This is because three accounts of the events of the 1960s and 1970s have been produced by two of the principle contributors of that period, Michell (1986, 1995) and Hokin (1987). The last two reviews are very much personal accounts, but the first (Michell, 1986) is very full and comprehensive, so I need relate only the highlights here.

The original observation was that acetylcholine stimulated ^{32}P incorporation into phospholipids in pancreatic slices (Hokin[13] and Hokin[14], 1953); the original idea was to investigate RNA synthesis, but as the RNA fraction was cleaned up the radioactivity largely disappeared. Lowell and Mabel Hokin looked for the counts, purified the fraction, and found that it contained labeled phospholipids. The use of Dawson's (1954) deacylation and separation techniques enabled the Hokins to show that phosphatidylinositol (PtdIns) and phosphatidic acid were the lipids involved (Hokin and Hokin, 1953), and they and others extended those observations to a number of other tissues and agonists in a remarkable and extended series of papers (see Michell, 1986). One major conclusion of the Hokins' elegant experiments was that the primary reaction was probably phospholipase C-like action on PtdIns to form diacylglycerol and inositol phosphate (Ins$P$). The recycling of the diacylglycerol to phosphatidate and then, via CMP-diacylglycerol, back to PtdIns formed

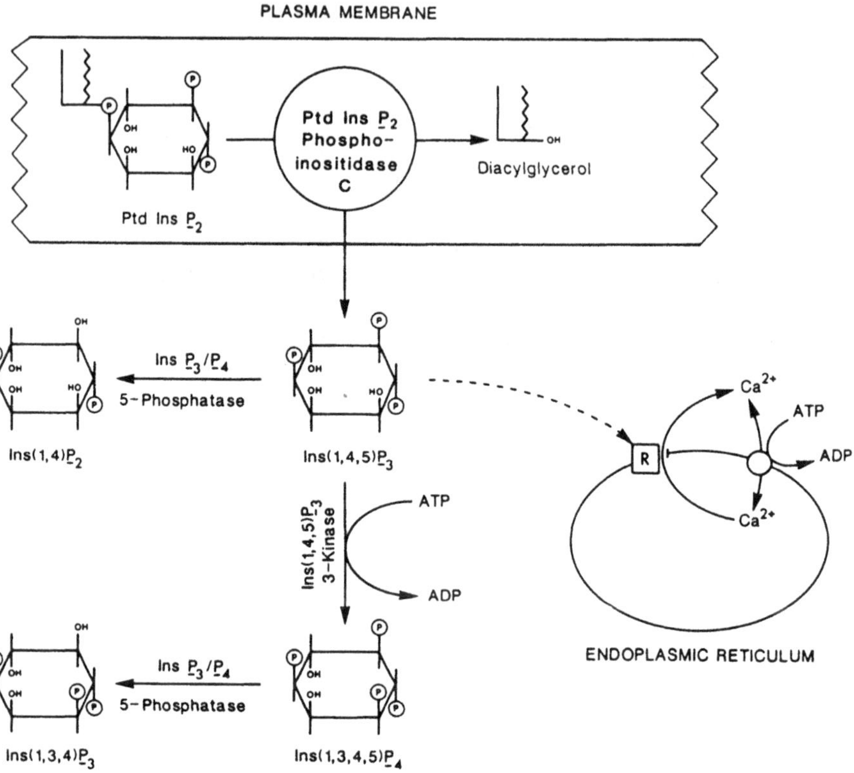

Figure 2. Basic scheme of metabolism and function of Ins(1,4,5)P_3. (Reproduced with permission from Irvine, 1989).

their "classic" PIC cycle (albeit we now know it occurs principally by hydrolysis of the phosphorylated form of PtdIns, PtdInsP_2) (Figure 2). The role of this effect eluded those studying it in the 1960s, although Durell et al. (1969) came very close when they proposed that it might be a facet of receptor activation. They also raised for the first time the possibility that the polyphosphoinositol lipids (PtdIns$P$ and PtdInsP_2) might be acting as substrates for the activated enzyme. Michell et al., (1981) resurrected this idea later, and in 1983 Bernie Agranoff's group working with platelets, Mike Berridge using blowfly salivary glands, and the groups of Marvin Gershengorn, Tom Martin, and Alan Drummond all working independently on rat GH_3 cells (for refs., see Berridge and Irvine, 1984) produced the first unequivocal experimental evidence by demonstrating primary production of InsP_3 from [3H]-inositol-labeled cells.

As I have mentioned, an important influence on the expansion of the study of inositides was Michell's (1975) review. He drew together a huge body of disparate evidence that had accumulated in the previous 20 years and suggested that it represented a signal transduction system akin to, but distinct from, that of cyclic AMP. Moreover, the fact that PtdIns turnover occurred under circumstances where Ca^{2+} was possibly involved led him to suggest that the former was controlling the latter, a prediction that was remarkably accurate. The most convincing experimental evidence to support this hypothesis (until the discovery of InsP_3 as a Ca^{2+} mobilizer) was that of Berridge and Fain (1979) who neatly showed that a "run-down" blowfly salivary gland preparation could be restored to normal Ca^{2+} control only by the specific addition of inositol. For those directing science from on high, obscure tissues can imply obscure science: There is an apocryphal tale that a poster advertizing a talk on the blowfly salivary gland—the tissue that taught us so much about inositides in the 1970s and 1980s—in a well-known American medical school was removed by a senior staff member who thought it was a student prank!

Ca^{2+} was not the first "second messenger" to be directly associated with inositides, however, because Yasukomi Nishizuka and his colleagues (Takai et al., 1979) published a seminal paper in which they showed that a Ca^{2+} and phospholipid-dependent protein kinase from brain, protein kinase C (PKC), was extraordinarily sensitive to stimulation by diacylglycerol. As with many discoveries, impurities in reagents helped Takai et al. to uncover this effect: a stimulation of phosphatidyl serine largely disappeared with a newly purchased batch of phosphatidyl serine, and they traced the difference to diacylglycerol contamination in the earlier batches. Legend has it that GTP contamination of ATP first led Rodbell's group to their discovery of G-proteins (see above) in a similar fashion. Having found this diacylglycerol effect, Takai et al. (1979), with remarkable insight, suggested that this enzyme might be controlled by diacylglycerol generated from the phosphoinositide cycle. Their later identification of PKC as a major target for tumor-promoting phorbol esters (Castagna et al., 1982) gave a further boost to this important family of enzymes. Since the first members of the family were both Ca^{2+}- and diacylglycerol-dependent, the idea that both branches of the inositide signals (diacylglycerol and Ca^{2+}) acted synergistically was very attractive (Berridge, 1984) and still holds true. However, the discovery that some other members of the PKC family are Ca^{2+}-independent or even controlled by lipids other than diacylglycerol (for review, see Nishizuka, 1995) has meant that PKC activation and stimulated inositide turnover are not always associated with each other, and those scientists who study the PKC family must also be interested in lipid signaling pathways additional to the PIC pathway.

Returning to Ca^{2+} and inositides, the demonstration in the early 1980s that the primary target for a stimulated PIC might be PtdInsP_2 (see above) led to the demonstration that InsP_3 itself can be a second messenger that mobilizes

Ca^{2+} (Streb et al., 1983). Sometimes ideas take a while to bear fruit, and sometimes they take off like wildfire; it took 30 years to get from Hokin and Hokin (1953) to Streb et al. (1983), but only 1 year and 1 week later sufficient evidence had accumulated from such a wide range of tissues and preparations, that the basic principles of InsP_3 as a second messenger were established (Berridge and Irvine, 1984).

Much of the history of InsP_3 and Ca^{2+} has been described above under Ca^{2+}, and how cells use this system to generate a remarkably complex spatio-temporal signaling paradigm is still under extensive investigation. Perhaps the single most important additional advance was the purification (Supattapone et al., 1987) and cloning (Furuicha et al., 1989; Mignery et al., 1989) of the InsP_3 receptor. There is one nice historical touch here—Mikoshiba's group had actually been working on the InsP_3 receptor for many years without knowing what it was. Katsuhiko Mikoshoba began working on it in Jean-Pierre Changeux's laboratory where it was first described, and the first report from that group (Mallet et al., 1975) documenting the existence of the protein was published in 1975, the same year as Bob Michell's crucial review. A decade and a half later these very disparate threads finally came together. The PIC enzyme family has proliferated enormously in recent years, and of particular interest has been the discovery of two different subfamilies, the β-enzymes, controlled by G-proteins, and the γ-enzymes controlled by tyrosine kinases. That story comprises the work of many people, though Sue-Goo Rhee's group has made the largest individual and original contributions (see Rhee and Choi, 1992 for a review).

The subsequent story of inositol phosphates is incomplete at the time of writing and, although many of the original discoveries were made in the 1980s, these compounds probably belong to another book for another time. The "mushrooming" of inositol phosphates began with the discovery of inositol 1,3,4-trisphosphate and inositol 1,3,4,5-tetrakisphosphate (InsP_4) (Irvine et al., 1984; Batty et al., 1985). Now there are more than 30 of them; for example, it is not just plants that are making phytic acid (InsP_6); all animal cells (probably) contain it. Sorting out which inositol phosphate is doing what and in which tissue is a major task for now and for the future. The proposed second messenger role of InsP_4 in controlling Ca^{2+} entry (Morris et al. 1987), and how, if at all, this relates to the Ca^{2+} entry mechanism controlled by intracellular stores (Putney, 1986), remains controversial in the 1990s, and there are surely likely to be several other as yet undiscovered functions for these ubiquitous compounds.

The "other" inositide story, the 3-phosphorylated inositol lipid, was similarly started in the 1980s by Lew Cantley and his co-workers (Sugimoto et al., 1984; Whitman et al.. 1988), but its physiological functions are so far undetermined. It is not easy to write a history of seminal observations when you don't yet know who the offspring are!

Cyclic GMP

Although only a single compound is at the heart of cyclic GMP's involvement in cellular signaling (compare this with the multitude of inositides!), there are elements of similarity with inositides insofar as the initial observations and interest were followed by a time of comparative quietness and confusion before considerable clarification in the 1980s.

After the identification of cyclic AMP it was not surprising that other cyclic nucleotides were sought, and cyclic GMP emerged definitively from Earl Sutherland's and from Nelson Goldberg's laboratories in 1969 (Goldberg et al., 1969; Ishikawa et al., 1969). Over the next decade much important groundwork was covered by many groups, but there was considerable confusion as to whether the requisite guanylyl cyclase was membrane bound, soluble, or both. Also, which agonists would and would not stimulate cyclic GMP formation was complex and confused (for a review from this time, see Goldberg and Haddox, 1977). As with inositides, we can now understand why, with the benefit of hindsight, the subsequent clarification came from entirely different routes.

The question of whether cyclic GMP really was a physiological second messenger was answered most definitively in vertebrate photoreceptors (the retina). Above, in the trimeric G-protein section, I noted that transducin, the G-protein activated as a consequence of light falling on rhodopsin, was the first such protein to have its biochemistry mapped out in detail. The effector (= the target) for transducin is a cyclic GMP phosphodiesterase, and thus the predicted consequence for light activation of rhodopsin is a massive activation of this enzyme and a concomitant decrease in cyclic GMP levels. This was first clearly shown experimentally by Miki et al. in 1973. It was already known that retinal rods were depolarized in the dark and that light induced a hyperpolarization; in 1979 Miller and Nicol found that cyclic GMP could mimic this hyperpolarization effect. The mechanism involved cyclic GMP activating a Na^{2+} channel (Fesanko et al., 1985) so that when the cyclic GMP level dropped (because cyclic GMP phosphodiesterase was activated), the channel closed, causing the hyperpolarization. These crucial observations gave a new impetus to the credibility of cyclic GMP as a second messenger, and also of course inspired in parallel the classical studies of the mechanism of transducin action.

Cyclic GMP-controlled channels do exist in other tissues (Yau,1994) but the principle mechanism of action of cyclic GMP in most tissues is via one or more protein kinases, which were first shown to be distinct from cyclic AMP-controlled protein kinases by Kuo and Greengard (1970) and are outside the scope of this review. Where the vertebrate photoreceptor is most unusual is that it is, as far as we know, the only tissue in which acute, receptor-regulated control of cyclic GMP levels is by activation of cyclic GMP phosphodiesterase;

in most tissues it is guanylyl cyclase that is the rate-limiting reaction. Here again the benefit of hindsight makes the telling easier.

As mentioned above, there was controversy in the 1970s as to whether guanylyl cyclase was soluble or membrane bound, and how either form was controlled; we now know that the activities arise from distinct entities subject to different control mechanisms. The breakthrough in our understanding of the membrane-bound form contained what should by now be the familiar components of an "obscure" tissue (of little obvious relevance to those who see science as a user-driven pursuit) that just happened to be "put there" to solve this particular problem. When sea urchin sperm are released into the sea they are attracted to the eggs of the same species by species-specific proteins that are released by the eggs. The classic studies of Dave Garbers and his associates led to the purification (from kilograms of sea urchin eggs!) of these attractants (resact and speract being the two most central to this story). They also purified the particulate guanylyl cyclase activity from the sea urchin sperm (which was already known to be a rich source of that activity), and demonstrated: (1) resact and speract increase cyclic GMP levels in sperm (e.g. Ramarao and Garbers 1985), and (2) the resact "receptor" and guanylyl cyclase activity reside in the same polypeptide (Shimomora et al., 1986). The cloning of the receptor led to subsequent confirmation of this and set the paradigm for the single polypeptide receptor/transducer (Singh et al., 1988).

The wider implications for mammalian studies were soon realized. One of the most pursued receptors around this time was that for atrial naturietic peptide (ANP), which was known from the studies of Ferad Murad's and Pavel Hamet's groups to be probably a membrane-bound guanylyl cyclase. Chinkers et al. (1989) cloned this receptor by guessing it would be homologous to the resact receptor.

Meanwhile, what of the soluble guanylyl cyclase? Kimura and Murad (1974) and Chrisman et al. (1975) showed this to be a distinct entity; it contained heme (Gerzer et al., 1981), the heme being in a regulatory subunit of a heterodimer (Kamisaki et al., 1986). At this time and for some years afterwards there was some confusion about what activated the enzyme—essentially, any agent that could oxidize the heme (i.e. generated an oxidizing free radical) would do it, but what was the physiological activator? Among the free radicals (or generators of such) that were particularly effective was nitric oxide (Arnold et al., 1977). Although at that time NO was of interest only to toxicologists, it became of much wider significance when Furchgott et al. (1987) and Palmer et al. (1987) showed that it is a naturally occurring intracellular/intercellular regulator.

Here, once again, we find two long-running sagas becoming united. For years a number of groups had investigated the phenomenon whereby removal of the endothelium from a piece of artery would prevent acetylcholine-stimulated relaxation (Furchgott and Zawadzlei, 1980), the reason being that the endothelium actually produced a substance that passed into the smooth muscle and caused it to relax. This endothelium-dependent relaxing factor (EDRF; Cherry et al., 1982) was known to exert its action by raising the level of cyclic GMP in the muscle, and it was Furchgott et al. (1987) and Palmer et al. (1987) who put these pieces of the jigsaw together to identify EDRF with NO. The production and action of NO has since been shown to be of enormous physiological significance (e.g. for a review, see Nathan and Xie, 1994) in many diverse tissues (not least in brain, e.g. Bredt and Snyder, 1992), and this expansion, currently still in full force, is something to be historically reviewed in the future. It is obvious, however, that the discovery of NO as a physiological regulator has given its second messenger, cyclic GMP, a whole new lease of life.

Ion Channels and Other Messengers

The nicotinic acetylcholine receptor is a ligand-gated ion channel (see above) as is the γ-aminobutyric acid (GABA) receptor, but most ion channels are not, and so I could leave them out of this chapter altogether. But they play a very central role in signaling and are frequent targets (effectors) for second messengers. The latter can be the soluble messengers themselves, e.g. cyclic AMP or cyclic GMP, protein kinases or, G-proteins (α- or $\beta\gamma$-subunits) (see Figure 1). Perhaps one reason I should mention them is to draw attention to the enormous impact that patch-clamping has had on cellular signaling. This technique, invented and expanded by Erwin Neher and Bert Sakmann (Neher and Sakmann, 1975; Hamill et al., 1981) has enabled researchers to study channels as individual entities (it remains the only readily applicable technique by which biologists can study the behavior of a single molecule). Hand-in-hand with the molecular characterization of ion channels, which continues apace and is not yet ripe for historical review, patch-clamping has proved one of the most powerful and informative ways of understanding what happens at the plasma membrane. No review of cellular signaling for the years 1960–1990 would be complete without citing this Nobel Prize-winning breakthrough.

There are other aspects of membrane receptors and their actions that I have also omitted because they are, like the 3-phosphorylated inositol lipids (see above), too new to be assessed with certainty. For example, lipid signaling now involves very much more than just inositides (e.g. for a review, see Divecha and Irvine, 1995) with phospholipases D, C, and A_2 all being major areas of investigation because of their role in cell activation. Mention of phospholipase A_2 leads to the intercellular mediators, the eicosanoids, which obviously have enormous impact on the regulation of cell function by cell surface receptors, but since they are extracellular messengers I will not deal with their history. The sphingolipids, once thought to be comparatively inactive structural components of membranes, are now also coming into focus as sources of intracellular second messengers (see Divecha and Irvine, 1995).

I feel these stories belong to a future review; it is impossible to draw a clear line because so much is still happening, and, besides, clear lines are so much easier to draw with hindsight—the more hindsight the better. This is the golden age of signal transduction, and I hope it will be clear from the above that the great majority of what we now know and study has been uncovered between the years 1960 and 1990. There is plenty more where that came from, and I for one can't wait to see what happens.

ACKNOWLEDGMENTS

I am very grateful to Ole Petersen and Bob Burgoyne for prior information about their book (Burgoyne and Petersen, 1997), to Sir Arnold Burgen and Derek Lindsay for helpful information and comments, to Sandi for making sense of it all, and to Jennifer Maddock and her assistants at Babraham Library, where this chapter was written.

REFERENCES

Arnold, W.P., Mittal, C.K., Katsaki, S., & Murad, F. (1977). Nitric oxide activates guanylate cyclase and increases guanosine 3':5'-cyclic monophosphate levels in various tissue preparations. Proc. Nat. Acad. Sci. USA 74, 3203-3207.

Ashley, C.C. & Ridgway, E.B. (1970). On the relationships between membrane potential, calcium transient and tension in single barnacle muscle fibers. J. Physiol. 209, 105-130.

Baker, P.F., Hodgkin, A.L., & Ridgway, E.B. (1971). Depolarization and calcium entry in squid giant axons. J. Physiol. 218, 709-755.

Batty, I.R., Nahorski, S.R., & Irvine, R.F. (1985). Rapid formation of inositol 1,3,4,5-tetrakisphosphate following muscarnic receptor stimulation of rat cerebral cortical slices. Biochem. J. 232, 211-215.

Berridge, M.J. (1983). Rapid accumulation of inositol trisphosphate reveals that agonists hydrolyse polyphosphoinositides instead of phosphatidylinositol. Biochem. J. 212, 849-858.

Berridge, M.J. (1984). Inositol trisphosphate and diacylglycerol as second messengers. Biochem. J. 220, 345-360.

Berridge, M.J. & Fain, J.N. (1979). Inhibition of phosphatidylinositol synthesis and the inactivation of calcium entry after prolonged exposure of the blowfly salivary gland to J-hydroxytrypatimine. Biochem. J. 178, 59-69.

Berridge, M.J. & Irvine, R.F. (1984). Inositol trisphosphate, a novel second messenger in cellular signal transduction. Nature 312, 315-321.

Berridge, M.J. & Rapp, P.E. (1979). A comparative survey of the function, mechanism and control of cellular oscillations. J. Exp. Biol. 81, 217-279.

Bird, A.P. (1995). Gene number, noise reduction and biological complexity. Trends in Genetics 11, 94-99.

Birnbaumer,L. (1990). Transduction of receptor signal into modulation of effector activity by G-proteins: the first 20 years or so. FASEB J. 4, 3068-3078.

Birnbaumer, L. & Rodbell, M. (1969). Adenyl cyclase in fat cells. II. Hormone receptors. J. Biol. Chem. 244, 3477-3482.

Blaurock, A.E. & Stoeckenuis, W. (1971). Structure of the purple membrane. Nature New Biol. 233, 152-155.

Bourne, H.R., Coffino, P., & Tomkins, G.M. (1975). Selection of a variant lymphoma cell deficient in adenylate cyclase. Science 187, 750-752.

Bourne, H.R., Sanders, D.A., & McCormick, F. (1990). The GTPase superfamily: a conserved switch for diverse cell functions. Nature 348, 125-132.

Bradham, L.S., Holt, D.A., & Sims, M. (1970). The effect of Ca^{2+} on the adenyl cyclase of calf brain. Biochim. Biophys. Acta 201, 250-260.

Bredt, D.S. & Snyder, S.H. (1992). Nitric oxide, a novel neuronal messenger. Neuron 8, 3-11.

Burgoyne, R. & Peterson, O.H. (1997). Landmarks in Intracellular Signalling. Portland Press, London.

Butcher, R.W. & Sutherland, E.W. (1962). Adenosine 3′,5′-phosphate in biological materials. J. Biol. Chem. 137, 1244-1250.

Castagna, M., Takai, Y., Kaibuchi, K., Sano, K., Kikkawa, U., & Nishizukka, Y. (1982). Direct activation of calcium-activated, phospholipid-dependent protein kinase by tumour-promoting phorbol esters. J. Biol. Chem. 257, 7847-7851.

Cherry, P.D., Furchgott, R.F., Zawadzki, J.V., & Jothianandan, D. (1982). The role of endothelial clels in the relaxation of isolated arteries by bradykinin. Proc. Nat. Acad. Sci. USA 79, 2106-2110.

Cheung, W.Y. (1970). Cyclic 3′,5′-nucleotide phosphodiesterase. Biochem. Biophys. Res. Commun. 38, 533-538.

Chinkers, M., Garbers, D.L., Chang, M.-S., Lowe, D.G., Chin, H., Goeddel, D.V., & Schulz, S. (1989). A membrane form of guanylate cyclase is an atrial natriuretic peptide receptor. Nature 338, 78-83.

Chrisman, T.D., Garbers, D.L., Parks, M.A., & Hardman, J.G. (1975). Characterisation of particulate and soluble guanylate cyclases from rat lung. J. Biol. Chem. 250, 374-381.

Comb, M., Birnberg, N.C., Seasholtz, A., Herbert, E., & Goodman, H.M. (1986). A cyclic AMP- and phorbol ester-inducible DNA element. Nature 323, 353-356.

Cuatrecasas, P.C. (1969). Interaction of insulin with the cell membrane: the primary action of insulin. Proc. Nat. Acad. Sci. USA, 63, 450-457.

Cuatrecasas, P. (1972). Affinity chromatography and purification of the insulin receptor of liver cell membranes. Proc. Nat. Acad. Sci. USA 69, 1277-1281.

Cuatrecasas,P. (1974). Membrane receptors. Ann. Rev. Biochem. 43, 169-214.

Dawson, R.M.C. (1954). The measurement of ^{32}P labelling of individual kephalins and lecithin in a small sample of tissue. Biochim. Biophys. Acta 14, 374-379.

Denton, R.M. & McCormack, J.G. (1980). On the role of the calcium transport cycle in heart and other mammalian mitochondria. FEBS Lett. 119, 1-8.

Divecha, N. & Irvine, R.F. (1995). Phospholipid signalling. Cell 80, 269-278.

Dixon, R.A.F., Kobilka, B.K., Strader, D.J., Benovic, J.L., Dohlman, H.G., Frielle, T., Bolanowski, M.A., Bennett, C.D., Rands, E., Diche, R.E., Mumford, R.A., Slater, E.E., Sigal, I.S., Caron, M.G., Lefkowitz, R.J., & Strader, C.D. (1986). Cloning of the gene and cDNA for mammalian β-adrenergic receptor and homology with rhodopsin. Nature 321, 75-79.

Durell, J., Garland, J.T., & Friedel, R.O. (1969). Acetylcholine action: biochemical aspects. Science 165, 862-866.

Ebina, Y., Ellis, L., Jarnayia, K., Edery, M., Graf, L., Clauses, E., Ou, J., Masiarz, F., Kan, T.W., Goldfine, I.D., Roth, R.A., & Lutter, W.J. (1985) The human insulin receptor cDNA: the structural basis for hormone-activated transmembrane signalling. Cell 40, 747-758.

Ehrenfreis, S. (1962). Immunohistochemical localization of drug-binding protein in tissues of the electric eel. Nature 194, 586-587.

Ek, B., Westermarker, B., Warteson, A., & Heldin, C-H. (1982). Stimulation of tyrosine-specific phosphorylation by platelet-derived growth factor. Nature 295, 419-420.

Evans, R.M. (1988) The steroid and thyroid hormone receptor superfamily. Science 240, 889-895.

Fesenko, E.F., Kolesnikov, S.S., & Lyubarski, A.L. (1985). Induction by cyclic GMP of cationic conductance in plasma membrane of retina rod outer segment. Nature 313, 310-313.

Furchgott, R.F. & Zawadzki, J.N. (1980). The obligatory role of endothelial cells in the relaxation of arterial smooth muscle by acetylcholine. Nature 288, 373-376.

Furchgott, R.F., Khan, M.T., & Jothianandan, D. (1987). Comparison of endothelium-dependent relaxation and nitric oxide-induced relaxation in rabbit aorta. Fed. Proc. 46, 385.

Furuichi, T., Yoshikawa, S., Miyawaki, A., Wada, K., Maeda, N., & Mikoshiba, K. (1989). Primary structure and functional expression of the inositol 1,4,5-trisphosphate-binding protein P_{400}. Nature 342, 32-38.

Gerzer, R., Böhme, E., Hofmann, F., & Schulz, G. (1981). Soluble guanylate cyclase purified from bovine lung contains heme and copper. FEBS Lett. 132, 71-74.

Goldberg, N.D., Dietz, S.B., & O'Toole, A.G. (1969). Cyclic guanosine 3',5'-monophosphate in mammalian tissues and urine. J. Biol. Chem. 244, 4458-4466.

Goldberg, N.D. & Haddox, M.K. (1977). Cyclic GMP metabolism and involvement in biological regulation. Ann. Rev. Biochem. 46, 823-896.

Goldbeter, A. (1996). Biochemical oscillations and rhythms: the molecular basis of periodic and chaotic behavior. Cambridge Uinversity Press, Cambridge, UK.

Grand, R.J.A. & Owen, D. (1991). The biochemistry of *ras* p21. Biochem. J. 279, 609-631.

Grynkiewicz, G., Poenie, M., & Tsien, R.Y. (1985) A new generation of Ca^{++} indicators with greatly improved fluorescence properties. J. Biol. Chem. 260, 3440-3450.

Hamill, O.P., Marty, A., Neher, E., Sakmann, B., & Sigworth, F. (1981). Improved patch-clamp techniques for high-resolution current recording from cells and cell-free membrane patches. Pflügers Archiv. 391, 85-100.

Harris, B.A., Robishaw, J.D., Mumby, S.M., & Gilman, A.G. (1985). Molecular cloning of complementary DNA for the alpha subunit of the G protein that stimulates adenylate cyclase. Science 229, 1274-1277.

Henderson, R. & Unwin, P.N.T. (1975). Three-dimensional model of purple membrane obtained by electron microscopy. Nature 257, 28-32.

Hokin, M.R. & Hokin, L.E. (1953). Enzyme secretion and the incorporation of P^{32} into phospholipids of pancreas slices. J. Biol. Chem. 203, 967-977.

Hokin, L.E. (1987). The road to the phosphoinositide-generated second messengers. Trends. Pharmacol. Sci. 8, 53-56.

Hunter, T. & Sefton, B.M. (1980). Transforming gene product of Rous sarcoma virus phosphorylates tyrosine. Proc. Nat. Acad. Sci. USA 77, 1311-1315.

Irvine, R.F. (1989). Functions of inositol phosphates. In: Inositol Lipids in Cell Signalling, pp. 135-161. Academic Press Ltd., London.

Irvine, R.F., Letcher, A.J., Lander, D.J., & Downes, C.P. (1984). Inositol trisphosphates in carbachol-stimulated rat parotid glands. Biochem. J. 223, 237-243.

Irvine, R.F., Michell, R.H., & Marshall, C.J. (1996) Current understanding of intracellular signalling pathways. Phil. Trans. Roy. Soc. Lond. B, 351, 127-241.

Ishikawa, E., Ishikawa, S., Davis, J.W., & Sutherland, E.W. (1969). Determination of guanosine 3',5'-monophosphate in tissues and of guanyl cyclase in rat intestine. J. Biol. Chem. 244, 6371-6376.

Kakiuchi, S., Yamazaki, R., & Nakajima, H. (1970). Proc. Jap. Acad. Sci. 46, 587-591.

Kamisaki, Y., Sahaki, S., Nakare, M., Palmieri, J.A., Kuno, T., Chang, B.Y., Waldman, S.A., & Murad, F. (1986). Soluble guanyl cyclase from rat lung exists as a heterodimer. J. Biol. Chem. 261, 7236-7241.

Kasugua, M., Fujita-Yamaguchi, Y., Blithe, D.L., & Kahn, C.R. (1983). Tyrosine-specific protein kinase activity is associated with the purified insulin receptor. Proc. Nat. Acad. Sci. USA 80, 2137-2141.

Kaziro, Y. (1978). The role of guanosine 5'-triphosphate in polypeptide chain elongation. Biochim. Biophys. Acta. 505, 95-127.

Kimura, H. & Murad, F. (1974). Evidence for two different forms of guyanylate cyclase in rat heart. J. Biol. Chem. 249, 6910-6916.

Kuo, J.F. & Greengard, P. (1970). Cyclic nucleotide-dependent protein kinases. VI. Isolation and partial purification of a protein kinase activated by guanosine 3',5'-monophosphate. J. Biol. Chem. 245, 2493-2498.

Krupinski, J., Coussen, F., Bakalyar, H.A., Tang, W.-J., Feinstein, P.G., Orth, K., Slaughter, C., Reed, R.R., & Gilman, A.G. (1989). Adenylyl cyclase amino acid sequence: possible channel- or transporter-like structure. Science 244, 1558-1564.

McCarthy, M.P., Earnest, J.P., Young, E.F., Choe, S., & Stroud, R.M. (1986). The molecular neurobiology of the acetylcholine receptor. Ann. Rev. Neuro. Sci. 9, 383-413.

Mallet, J., Huchet, M., Pougeois, R., & Changeux, J.-P. (1975). A protein difference associated with defects of the Purkinje cell in staggerer and nervous mutant mice. FEBS Lett. 52, 216-220.

Martin, G.S. (1970). Rous sarcoma virus: a function required for the maintenance of the transformed state. Nature 227, 1021-1025.

Means, A.R. & Deadman, J.R. (1980). Calmodulin—an intracellular calcium receptor. Nature 285, 73-77.

Michell, R.H. (1975). Inositol phospholipids and cell surface receptor function. Biochim. Biophys. Acta 415, 81-147.

Michell, R.H. (1986). Inositol lipids and their role in receptor function: history and general principles. In: Phosphoinositides and Receptor Mechanisms (Putney, J.W., Jr., Ed.), pp. 1-24. Alan R. Liss, New York.

Michell, R.H. (1995). Early steps along the road to inositol-lipid-based signalling. Trends Biochem. Sci. 20, 326-329.

Michell, R.H., Kirk, C.J., Jones, L.M., Downes, C.P., & Creba, J.A. (1981). The stimulation of inositol lipid metabolism that accompanies calcium mobilization in stimulated cells: defined chracteristics and unanswered questions. Phil. Trans. Roy. Soc. Lond. [B] 296, 123-137.

Mignery, G.A., Südhof, T.C., Takei, K., & DeCamilli, P. (1989). Putative receptor for inositol 1,4,5-trisphosphate similar to synanodine receptor. Nature 242, 192-195.

Miki, N., Keirns, J.-J., Marcus, F.R., Freeman, J., & Bitensky, M.W. (1973). Regulation of cyclic nucleotide concentrations in photoreceptors: an ATP-dependent stimulation of cyclic nucleotide phosphodiesterases by light. Proc. Nat. Acad. Sci. USA 70, 3820-3824.

Miller, W.-H. & Nicol, G.D. (1979). Evidence that cyclic GMP regulates membrane potential in rod photoreceptors. Nature 280, 64-66.

Morris, A.P., Gallacher, D.V., Irvine, R.F., & Petersen, O.H. (1987). Synergism of inositol trisphosphate and tetrakisphosphate in activating Ca^{2+}-dependent K^+ channels. Nature 330, 653-655.

Nathan, C. & Xie, Q. (1994). Nitric oxide synthases: roles, tolls and controls. Cell 78, 915-918.

Nathans, J. & Hogness, D.S. (1983). Isolation, sequence analysis, and intron-exon arrangement of the gene encoding bovine rhodopsin. Cell 34, 807-814.

Neher, E. & Sakmann, B. (1975). Single-channel currents recorded from membrane of denervated frog muscle fibres. Nature 260, 799-802.

Nishizuka, T. (1995). Protein kinase C and lipid signalling for destained cellular responses. FASEB J. 9, 484-496.

Ord, M.G. & Stocken, L.A. (1995). In: Foundations of Modern Biochemistry, vol. 1 (Ord, M.G. & Stocken, L.A., Eds.), pp. 57-60. JAI Press, Greenwich, CT.

Orly, J. & Schramm, M. (1976). Coupling of catecholamine receptor from one cell with adenylate cyclase from another cell by cell fusion. Proc. Nat. Acad. Sci. USA 73, 4410-4414.

Palmer, R.M.J., Ferrige, A.G., & Moncada, S. (1987). Nitric oxide release accounts for the biological activity of endothelium-derived relaxing factor. Nature 327, 524-526.

Pfeuffer, T. (1977). GTP-binding proteins in membranes and the control of adenylate cyclase activity. J. Biol. Chem. 252, 7224-7234.

Pfeuffer, T. & Helmreich, E.J.M. (1975). Activation of pigeon erythrocyte membrane adenylate cyclase by guanyl nucleotide analogs and separation of nucleotide binding protein. J. Biol. Chem. 250, 867-876.

Popot, J.-L. & Changeux, J.-P. (1984). Nicotinic receptor of acetylcholine: structure of an oligomeric integral membrane protein. Physiol. Rev. 64, 1162-1239.

Putney, J.W., Jr. (1986). A model for receptor-regulated calcium entry. Cell Calcium 7, 1-12.

Ramarao, C.S. & Garbers, D.L. (1985). Receptor-mediated regulation of guanylate cyclase activity in spermatozoa. J. Biol. Chem. 260, 8390-8396.

Rasmussen, H. & Barratt, P.Q. (1984). Calcium messenger system: an integrated view. Physiol. Rev. 64, 978-984.

Rhee, S.-G. & Choi, K.-D. (1992). Multiple forms of phospholipase C isozymes and their activation mechanisms. In: Advances in Second Messenger and Phosphoprotein Research, vol. 26 (Putney, J.W. Jr., Ed.), pp. 35-61. Raven Press, New York.

Robison, G.A., Butcher, R.W., & Sutherland, E.W. (1968). Cyclic AMP. Ann. Rev. Biochem. 37, 149-174.

Rodbell, M., Krans, H.M.J., Pohl, S.L., & Birnbaumer, L. (1971a). The glucagon-sensitive adenyl cyclase system in plasma membranes of rat liver. IV. Binding of glucagon: effect of guanyl nucleotides. J. Biol. Chem. 246, 1872-1876.

Rodbell, M., Birnbaumer, L., Pohl, S.L., & Krans, H.M.J. (1971b). The glucagon-sensitive adenyl cyclase system in plasma membranes of rat liver. V. An obligatory role for guanyl nucleotides in glucagon action. J. Biol. Chem. 246, 1877-1882.

Ross, E.M. (1996). Pharmacodynamics. In: The Pharmacological Basis of Therapeutics (Hardman, J.G., Limbird, L.E., Molinoff, P.B., Ruddon, R.W., & Gilman, A.G., Eds.), 9th ed., pp. 29-41. McGraw Hill, New York.

Rowse, A.L. (1950). The England of Elizabeth. MacMillan, London.

Sadowski, I., Stone, J.C., & Pawson, T. (1986). A non-catalytic domain conserved among cytoplasmic protein-tyrosine kinases modifies the kinase function and transforming activity of Fujinami sarcoma virus $P130^{gag-fps}$. Mol. Cell Biol. 6, 4391-4408.

Sassone-Corsi, P. (1994). Goals for signal transduction pathways: linking up with transcriptional regulation. EMBO J. 13, 4717-4728.

Scolnick, E.M., Papageorge, A.G., & Shih, T.Y. (1979). Guanine nucleotide-binding activity as an assay for *src* protein of rat-derived murine sarcoma viruses. Proc. Nat. Acad. Sci. USA 76, 5355-5359.

Shimomura, H., Dangott, L.J., & Garbers, D.L. (1986). Covalent coupling of a resact analogue to guanylate cyclase. J. Biol. Chem. 261, 15778-15782.

Siekevitz, P. (1996). Protein synthesis and the ribosome. In: Foundations of Modern Biochemistry (Ord, M.G. & Stocken, L.A., Eds.), Vol. 2, pp. 109-131. JAI Press, Greenwich, CT.

Singh, S., Cowe, D.G., Thorpe, D.S., Rodriguez, H., Kuang, W.-J., Dangott, L.J., Chinkers, M., Goedde, D.V., & Garbers, D.L. (1988). Membrane guanylate cyclase is a cell-surface receptor with homology to protein kinases. Nature 334, 708-712.

Sivilotti, L. & Colquhoun, D. (1995) Acetylcholine receptors: too many channels, too few functions. Science 269, 1681-1682.

Steitz, T.A., Goldman, A., & Engelman, D.M. (1982). Quantitative application of the helical hairpin hypothesis to membrane proteins. Biophys. J. 37, 121-125.

Stiles, G.L., Caron, M.G., & Lefkowitz, R.J. (1984). β-Adrenergic receptors: biochemical mechanisms of physiological regulation. Physiol. Revs. 64, 661-743.

Stoeckenius, W. & Kunau, W.H. (1968) Further characterization of particulate fractions from lysed cell envelopes of *Halobacterium halobumin* and isolation of gas vacuole membranes. J. Cell Biol. 38, 337-357.

Streb, H., Irvine, R.F., Berridge, M.J., & Schulz, J. (1983). Release of Ca^{2+} from a non-mitochondrial intracellular store in pancreatic acimar cells by inositol 1,4,5-trisphosphate. Nature 306, 67-69.

Sugimoto, Y., Whitman, M., Cantley, L.C., & Erikson, R.L. (1984). Evidence that the Rous sarcoma virus transforming gene product phosphorylates phosphatidylinositol and diacylglycerol. Proc. Nat. Acad. Sci. USA 82, 2117-2121.

Sunahara, R.K., Dessauer, C.W., & Gilman, A.G. (1996). Complexity and diversity of mammalian adenylyl cyclases. Ann. Res. Pharmacol. Toxicol. 36, 461-480.

Supattapone, S., Worley, P.F., Baraban, J.M., & Snyder, S.H. (1988). Solubilization, purification, and characterization of an inositol trisphosphate receptor. J. Biol. Chem. 263, 1530-1534.

Sutherland, W., Rall, T.W., & Menon, T. (1962). Adenyl cyclase. I. Distribution, preparation and properties. J. Biol. Chem. 237, 1220-1227.

Takai, Y., Kishimoto, A., Kikkawa, U., Mori, T., & Nishizuka, Y. (1979). Unsaturated diacylglycerol as a possible messenger for the activation of calcium-activated, phospholipid-dependent protein kinase system. Biochem. Biophys. Res. Comm. 91, 1218-1224.

Tsien, R.Y. (1980). New calcium indicators and buffers with high selectivity agonist magnesium and proteons: design, synthesis, and properties of prototype structures. Biochemistry 19, 2396-2404.

Tsien, R.Y. (1981). A non-disruptive technique for loading calcium buffers and indicators into cells. Nature 290, 527-528.

Ullrich, A., Bell, J.B., Chen, E.Y., Herrera, R., Setruzzeli, C.M., Null, T-J., Gray, A., Coussens, L., Liao, Y-C., Tsubokawa, M., Mason, A., Seeburg, P.H., Graunfell, C., Rosen, O.M., & Ramadandran, J. (1987). Human insulin receptor and its relationship to the tyrosine kinase family of oncogenes. Nature 313, 756-761.

Unwin, N. (1995). Acetylcholine receptor channel imaged in the open state. Nature 373, 37-43.

Ushio, H. & Cohen, S. (1980). Identification of phosphotyrosine as a product of epidermal growth factor-activated protein kinase in A-431 cell membranes. J. Biol. Chem. 255, 8363-8365.

van der Geer, P., Hunter, T., & Lindberg, R.A. (1994). Receptor protein-tyrosine kinases and their signal transduction pathways. Ann. Rev. Cell Biol. 10, 251-337.

Westcott, K.R., La Porte, D.C., & Storm, D.R. (1979). Resolution of adenylate cyclase sensitive and insensitive to Ca^{2+} and calcium-dependent regulatory protein (CDF) by CDR-Sepharose affinity chromatography. Proc. Nat. Acad. Sci. USA 76, 204-208.

Wilkie, T.M., Gibert, D.J., Olsen, A.S., Owen, X-N., Amatuda, T.T., Korenberg, J-R., Trask, B.J., de Jong, S., Reed, R.R., Simon, M.I., Jenkins, N.A., & Copeland, N.G. (1992). Evolution of the mammalian G protein α subunit multigene family. Nature Genetics 1, 85-91.

White, M.F., Maron, R., & Kahn, C.R. (1985). Insulin rapidly stimulates tyrosine phosphorylation of a M_r-185,000 protein in intact cells. Nature 318, 183-186.

Whitman, M., Downes, C.P., Keeler, M., Keller, T., & Cantley, L. (1988). Type I phosphophatidylinositol kinase makes a novel inositol phospholipid, phosphatidylinositol-3-phosphate. Nature 332, 644-646.

Woods, N.M., Cuthbertson, K.S.R., & Cobbold, P.H. (1986). Repetitive transient rises in cytoplasmic free calcium in hormone stimulated hepatocytes. Nature 319, 600-602.

Yau, K-W. (1994). Cyclic nucleotide-gated channels: an expanding family of ion channels. Proc. Nat. Acad. Sci. USA 91, 3481-3483.

Zerial, M. & Huber, L.A. (Eds.) (1995). Guidebook to the Small GTPases. OUP, Oxford, New York, Tokyo.

Chapter 7

MECHANISMS IN REGULATION:
PROTEIN PHOSPHORYLATION

Philip J. Randle

INTRODUCTION

Protein phosphorylation is widespread in nature as a regulatory device.* It is present in animals, plants, and microorganisms, and within individual organisms operates in most major metabolic pathways. Cohen (1993) estimates that about one in three proteins in mammalian cells contain covalently bound phosphate and that protein kinases and phosphatases may account for about 5% of all human genes. Protein phosphorylation was originally discovered through work on the regulation of glycogen metabolism (glycogen phosphorylase and later glycogen synthase). The discovery of cyclic AMP-dependent protein kinase (Walsh et al., 1968) as an enzyme with substrates outside the pathways of glycogen synthesis and degradation; and of reversible phosphorylation in the pyruvate dehydrogenase complex (Linn et al., 1969a) led to more widespread interest in this regulatory device (see Krebs, 1983).

These studies have provided a vast number of examples of serine and less frequently of threonine or histidine phosphorylation. Further and novel impetus to interest in protein phosphorylation was added by the later discovery of protein tyrosine phosphorylation (Eckhart et al., 1979) and that Rous sarcoma virus gene product is a tyrosine protein kinase (Collet et al., 1980; Hunter and Sefton, 1980a,b).

The explosive growth in knowledge of protein phosphorylation and its role in diverse biochemical functions which began in the 1960s was made possible by the commercial availability of enzymes, biochemicals, isotopically labeled biochemicals, and the discovery of a battery of specific enzyme inhibitors. Also important were advances to the level of routine in techniques for protein fractionation, amino acid sequencing of proteins (including physical techniques such as mass spectrometry), gene cloning and DNA sequencing, tissue culture, and genetic manipulation of cells.

THE DISCOVERY OF PROTEIN PHOSPHORYLATION/DEPHOSPHORYLATION CYCLES AND THEIR ROLE AS A REGULATORY MECHANISM

Early Work

The earliest phosphoproteins to be described were the caseins and the vitellins—major proteins of milk and egg yolk, respectively. In casein phosphate, orthophosphoric acid is esterified to the hydroxyamino acid serine. Posternak (1927) was apparently the first to suggest such an ester linkage; Rimington (1927) isolated from casein a phosphopeptone containing much of the organically bonded phosphate and also showed that bone and kidney phosphatases could liberate phosphate from casein; and Levene and Hill (1933) and Schmidt (1934) described the isolation from casein of the phosphodipeptide glutamyl serine phosphate. Lipmann[1] (1933) showed that the -OH group of serine in casein is esterified with phosphoric acid by isolation of the serine phosphate. Thus evidence prescribing the existence of biochemical reactions resulting in phosphorylation and dephosphorylation of serine residue(s) in a protein was already available in 1927.

The work in the laboratories of Carl[1] and Gerti Cori[1] in Buffalo and then in St. Louis which led through to the discovery of (serine) protein phosphorylation/dephosphorylation cycles and their role in enzyme regulation had already begun in 1927. It was to come to fruition some 30 years later. In what follows, protein phosphorylation will refer to formation of phosphoserine residues unless stated otherwise; the two other phospho amino acids of note in respect of enzyme regulation by reversible phosphorylation are phosphothreonine and phosphotyrosine. Phosphohistidine, labeled *in vivo*, has been detected in a number of proteins (Chen et al., 1974, 1977). It is unstable at low pH and hence is lost following acid precipitation of proteins. Three of the serine/threonine protein phosphatases (1, 2A, and 2C) are protein histidine phosphatases (Kim et al., 1993). Phosphohistidine has also been identified as an intermediate in enzyme mediated phosphate transfer reactions, e.g. in the bifunctional 6-phosphofructo-2-kinase-fructose-2,6-bisphosphate-2-phosphatase (reviewed in Pilkis et al., 1987) (see also, Chapter 5).

Discovery of Glycogen Phosphorylase and Phosphorylation/Dephosphorylation Cycles

By 1929 it was known that adrenaline stimulates breakdown of liver glycogen to blood glucose, and of muscle glycogen to blood lactate, and that blood lactate can give rise to liver glycogen and to blood glucose (the Cori cycle; for bibliography see Cori, 1931). Later studies in muscle led the Coris to demonstrate with R.E. Fisher and A.H. Hegnauer that breakdown of glycogen induced by adrenaline, electrical stimulation, anaerobiosis, or 2:4-dinitrophenol results in accumulation of hexose monophosphates. Subsequently Baranowski and Parnas showed that glycogen is broken down, with consumption of P_i, by muscle extracts devoid of ATP and creatine phosphate. Parnas coined the term "phosphorolysis" to describe this reaction.

The Coris proceeded to study the phosphorolysis reaction in minced frog muscle extracted with water—a procedure that removed acid soluble phosphates but not glycogen (and fortuitously phosphoglucomutase but not phosphorylase). Incubation in phosphate buffer gave rise to hexose monophosphates but only in small amounts. They then discovered that a boiled extract of muscle stimulates phosphorolysis and traced the effect to 5′-AMP. They went on (with S.P. Colowick) to identify as the product of phosphorolysis, glucose-1-phosphate (Cori ester)—a hitherto unknown hexosemonophosphate (Colowick et al., 1937). They named the enzyme phosphorylase (Cori et al., 1939) and with A.A. Green purified and crystallized it (Cori et al., 1942) [this was phosphorylase *a*; phosphorylase *b* was not crystallized until later (Cori and Cori, 1945)].

In 1943 with A.A. Green the Coris described two forms of phosphorylase in muscle; the *b* form requiring 5′-AMP for activity, and the *a* form active in its absence (Cori and Green, 1943), both of which they crystallized (see above)**. The conversion of phosphorylase *a* to *b* was shown to be catalyzed by an enzyme which they named PR (prosthetic removing) enzyme in the belief that it catalyzed removal of an organic phosphate prosthetic group assumed to be 5-AMP***. They were unable to achieve conversion of phosphorylase *b* to *a*. Further work failed to show 5′-AMP in phosphorylase *a* but it did show that the phosphate content of the *a* form is four times that of the *b* form (Cori and Cori, 1945, 1947). The Coris postulated that interconversion of the *a* and *b* forms is a physiologically significant mechanism and C.F. Cori with E.W. Sutherland[15] went on to show that the glycogenolytic effects of adrenaline and glucagon in liver were associated with conversion of phosphorylase *b* to *a* (Sutherland and Cori, 1951). Subsequently two former associates of the Coris, E.W. Sutherland working with liver, and E.G. Krebs (with E.H. Fischer) working with muscle, discovered that phosphorylase *a* is a phosphoenzyme and that the interconversion reactions involved enzymes catalyzing phosphorylation and dephosphorylation, respectively. The two enzymes were named phosphorylase kinase and phosphorylase phosphatase (Fischer and Krebs, 1955; Krebs and Fischer, 1956; Rall et al., 1956; Wosilait and Sutherland, 1956). The reactions can now be written (single site of phosphorylation):

$$2 \text{ phosphorylase } b \text{ (dimers)} + 4\text{ATP} \rightarrow \text{phosphorylase } a + 4\text{ADP}$$

$$\text{phosphorylase } a + 4\text{H}_2\text{O} \rightarrow 2\text{-phosphorylase } b \text{ (dimers)} + 4 \text{ P}_i$$

Discovery of cAMP and PKA

In 1957, Sutherland and colleagues (Rall et al., 1957) reported the crucial observation that the actions of adrenaline and of glucagon to stimulate the conversion of phosphorylase *b* to phosphorylase *a* in cat and dog liver homogenates could be separated into two distinct phases. They discovered that a particulate fraction forms a heat stable factor in the presence of ATP and glucagon or adrenaline, and that this heat stable factor stimulates conversion of phosphorylase *b* to phosphorylase *a* in a supernatant fraction in which the hormones themselves have no effect. They concluded that the hormones stimulate the formation from ATP of an activator of phosphorylase *b* to *a* conversion. Subsequently Sutherland and Rall (1958a,b) were able to to utilize the particulate system to fractionate, crystallize, and characterize the heat stable factor as an adenine ribonucleotide which however lacked monoesterified phosphate but which was converted into 5′AMP on incubation with an enzyme from heart muscle.

The author was priveleged to participate in the Eli Lilly Insulin Symposium in Indianapolis in May 1957, at which Earl Sutherland reported on the current state of this research at a crucial phase. At this point in time Sutherland was uncertain as to the structure of the unknown compound. He reported that it contained stoichiometrically equivalent amounts of adenine, ribose, and phosphate but considered that it might be a cyclic dinucleotide of AMP. Unbeknown to Sutherland and his colleagues, Roy Markham, the Cambridge nucleic acid biochemist and virologist, had discovered a new adenine nucleotide formed by the action of $Ba(OH)_2$ on ATP and which was available in quantity. Markham went to the Chemistry Department at Washington University in St. Louis to characterize this nucleotide with Lipkin and Cook (Sutherland had by then moved from St. Louis to Cleveland). By happenstance Sutherland and Rall, and Cook, Lipkin, and Markham wrote more or less simultaneously to Dr. Lern Heppel of the National Institutes of Health for a sample of purified spleen phosphodiesterase to assist in characterization of their unknown adenine nucleotide(s). Dr. Heppel wrote to the two groups and suggested that they were possibly working on the same nucleotide and the upshot was that their compounds were identified as adenosine 3′,5′ (cyclic) monophosphate (cAMP) (Cook et al., 1957; Sutherland and Rall, 1957). Sutherland and Rall also discovered in tissues a phosphodiesterase which in the presence of Mg^{2+} inactivated cAMP by hydrolysis to 5′AMP and showed that this enzyme is inhibited by caffeine (Sutherland and Rall, 1958a,b).

Subsequent studies showed that adenylate cyclase, the enzyme which forms cAMP from ATP, is found in all animal tissues studied except erythrocytes of man and dog (nucleated erythrocytes contain the enzyme). With the aid of a specially developed pressure homogenizer it was possible to show that most or perhaps all adenylate cyclase in avian erythrocytes and rat liver is in the plasma membrane (Davoren and Sutherland, 1963a,b). It was thus established that the binding of glucagon or adrenaline to a membrane receptor leads to activation of adenylate cyclase in the plasma membrane with the formation of cAMP + PP_i from cytosolic ATP, and that the action of cAMP is terminated by hydrolysis to 5′AMP by cAMP phosphodiesterase. cAMP was shown to activate phosphorylase kinase by phosphorylation (by what mechanism was unknown; Krebs et al., 1959, 1964) and hence stimulate the conversion of phosphorylase *b* to *a*. It was not until 1968 that Walsh et al. isolated cAMP-dependent protein kinase, establishing that this is the enzyme that phosphorylates and activates phosphorylase kinase, and catalyzes other cAMP-mediated protein phosphorylations. This was the first example of a linear signal transduction pathway from cell surface receptor to intracellular protein phosphorylation. It was also an example involving more than one protein kinase in the sequence.

SERINE AND THREONINE PROTEIN KINASES AND PHOSPHATASES AND THEIR ROLE IN THE REGULATION OF METABOLISM

General Comments

There is a complex interplay between substrates, products, and allosteric effectors of enzymes and their regulation by reversible phosphorylation of serine or threonine residues. In many instances the *b* form of the enzyme can be activated by an allosteric effector as well as by phosphorylation or dephosphorylation. Phosphorylase *b* is activated by 5′-AMP (Cori and Green,

1943) as well as by phosphorylation by phosphorylase kinase. Other allosteric effectors of the *b* form include ATP and glucose 6-phosphate which can reverse activation by 5′-AMP (Parmeggiani and Morgan, 1962; Morgan and Parmeggiani, 1964). Glycogen synthase *b* is activated by glucose 6-phosphate as well by the *b* to *a* conversions effected by protein phosphatases 1 and 2A (PP1 and 2A) (Friedman and Larner, 1963; Cohen, 1989). The *b* forms (phosphorylated forms) of mammalian PDH and branched chain ketoacid dehydrogenase can only be reactivated by dephosphorylation; no allosteric activator of the *b* form has yet been established. While it is dangerous to formulate general rules there are examples which appear to show that substrates and activators of the *b* form may facilitate $b \rightarrow a$ conversion of enzymes whereas products and inhibitors of the *a* form may facilitate $a \rightarrow b$ conversion. A striking example is provided by the PDH complex (see later section).

Reversible phosphorylation of enzymes is not always accompanied by any marked change in activity . One of the earliest examples of these so-called *silent phosphorylations* (Cohen, 1982) was ATP citrate lyase (Linn and Srere, 1979) and others have included protein kinases which are known to undergo autophosphorylation without any obvious change in activity. This does not mean necessarily that phosphorylation has no functional role because other possibilities remain to be investigated. These could include for example effects on binding to other proteins or macromolecules or on rate of turnover (i.e. degradation). Artifacts are possible as denaturation or unfolding has been reported to change otherwise inert proteins into substrates for protein kinases (Bylund and Krebs, 1975).

The criteria to be satisfied in establishing that reversible phosphorylation of a protein is of physiologically regulatory significance were defined by Krebs and Beavo (1979), namely that a protein substrate for PKA should bear a functional relationship and a rate relationship to the cAMP-mediated biochemical change; that the functional change in the protein should be demonstrated *in vitro* to accompany phosphorylation catalyzed by PKA and dephosphorylation by a protein phosphatase; and that phosphorylation *in vivo* in response to a hormone should be at the same site(s) phosphorylated by PKA *in vitro*.

There are many examples of multiple sites of phosphorylation on individual proteins and not all are necessarily involved in regulation of enzyme activity. In glycogen synthase, phosphorylation of five of nine sites decrease activity (Cohen, 1993). In the PDH complex, occupancy of one or other of two of three phosphorylation sites is inactivating (Teague et al., 1979; Tonks et al., 1982) though *in vivo*; inactivation is for all practical purposes effected by one phosphorylation of one site (Sale and Randle, 1982a,b). Phosphorylation at multiple sites affecting enzyme activity, as in glycogen synthase, affords opportunities for signal integration, while phosphorylation at multiple sites not affecting enzyme activity, as in the PDH complex, can slow reactivation by dephosphorylation (Sugden et al., 1978; Kerbey et al., 1981; Sale and Randle, 1982b). Sometimes phosphorylation of a second site by a different kinase can modify or ablate a change in activity induced by phosphorylation of the first site. Thus the increase in activity of hormone sensitive lipase effected by phosphorylation of site 1 with PKA is nullified by phosphorylation of site 2 by AMP kinase (AMPK) (Yeaman et al., 1994). The phosphorylation of site 1 of the glycogen binding subunit of PP1 by ISPK activates dephosphorylation of glycogen synthase and phosphorylase kinase whereas phosphorylation of site 2 by PKA leads to release of PP1 from the glycogen particle and cancellation of the insulin effect (e.g. by adrenaline) (Hiraga and Cohen[16], 1986; Dent et al., 1990).

In what follows, more detailed consideration is given to cyclic AMP-dependent protein kinase (PKA), and to the metabolic role of its protein phosphorylation/ dephosphorylation cycles; to the regulation of glycogens synthase; to tyrosine phosphorylation, the mechanism of action of insulin and other growth factors; to the regulation of the mitochondrial protein kinases—PDH kinases and BCDH kinase; to Ca^{2+} signaling; and to AMP-kinase.

Serine/Threonine/Histidine Protein Phosphatases

Classification

Understanding of non-mitochondrial serine/threonine/histidine specific protein phosphatases was revolutionized by the realization that they could be divided simply into two groups (PP1 and PP2) (for reviews see Ingebritsen and Cohen, 1983; Cohen, 1989, 1994). By definition PP1 group isoenzymes dephosphorylate specifically the β-subunit of phosphorylase kinase and are inhibited by nM concentrations of the protein inhibitors-1 and -2. PP2 isozymes dephosphorylate the α-subunit of phosphorylase kinase preferentially and are insensitive to inhibitor-1 and inhibitor-2; they have been subclassified into three distinct families (2A, 2B, 2C) based on their dependence on divalent cations (2A none; 2B Ca^{2+}; 2C Mg^{2+}) and other criteria. All dephosphorylate phosphoserine and phosphothreonine residues and all but 2B dephosphorylate phosphohistidine (Kim et al,. 1993). Structurally the isoforms that exist show substantial homology. cDNA cloning has shown that there are two distinct gene families: one comprising PP1, PP2A, and PP2B, and the second PP2C (Cohen and Cohen, 1989; Cohen, 1994).

PP1 is predominantly particulate because of specific binding subunits resulting in its targeting and attachment, for example,to the glycogen particle (PP1G), to myosin (PP1M-reversal of myosin light chain kinase phosphorylation), and to the sarcoplasmic reticulum (likely substrate phospholamban). In effect the targeting subunits determine the substrate specificity (or priority) of individual PP1 molecules and provide one example of the targeting subunit hypothesis (Hubbard and Cohen, 1993). Free (i.e. cytosolic) PP1 is rendered largely inactive by two cytosolic inhibitor proteins. Targeting is regulated by reversible phosphorylation. Thus phosphorylation of the glycogen binding subunit by PKA causes dissociation of PP1 from glycogen, whereas phosphorylation by ISPK enhances it (PKA is overriding). Other examples include protein kinase C which is targeted by cytosolic and nuclear membrane proteins, and PKA through proteins of the endoplasmic reticulum, mitochondria, peroxisomes, Golgi and microtubules (for recent review see Faux and Scott, 1996).

Inhibitor I is activated by PKA phosphorylation thus allowing cAMP to regulate the phosphorylation of proteins which are not substrates for PKA but which are dephosphorylated by PP1. Inhibitor 1 can be dephosphorylated by PP2B; the latter potentially allows Ca^{2+} signaling to modulate other signaling pathways, e.g. to attenuate signals acting via cAMP. The regulation of PP1 distribution by phosphorylation of its binding proteins provides a further mechanism by which one signaling pathway can influence another through protein phosphatase activity. Details of these mechanisms follow and are also reviewed in Cohen (1989, 1994).

Cyclic AMP-Dependent Protein Kinase (PKA)

PKA is composed of regulatory (R) and catalytic (C) subunits and activation by cAMP and subsequent reversal proceeds according to Reactions 1 (PKA) and 2 (cAMP phosphodiesterase):

$$\text{RC} + \text{cAMP} \longleftrightarrow \text{RcAMP} + \text{C} \qquad (1)$$

$$\text{cAMP} + \text{H}_2\text{O} \rightarrow 5'\text{AMP} \qquad (2)$$

This mechanism for PKA was first suspected from activation by dilution and by aging and by the observation that cAMP increases V_{max} without changing Km (Brostrom et al., 1970). The mechanism was established unequivocally following purification and characterization of enzymes from heart and skeletal muscles, liver and brain, from M_r measurements, and from demonstration of cAMP-dependent dissociations. The stoichiometry of R and C units in the bovine heart enzyme is generally quoted as R_2C_2 though it has also been deduced to be RC_2 after corrections for subunit axial asymmetry (see Rubin et al., 1972; Ehrlichman et al., 1973). The K_m for ATP was 13 μM (it is commonly of this order for protein kinases); the $K_{0.5}$ for cAMP was 60 nM (Rubin et al., 1972). Isozymes of PKA are known (Krebs, 1983).

Table 1. Substrates for Cyclic AMP-Dependent Protein Kinase

Protein	*Activity Status*	*Organ*[1]	*References*
Phosphorylase kinases	activated	L, M	
Protein phosphatase inhibitor I	activated		Cohen (1989)
Fructose 2,6-bisphosphatase[2]	activated	L	Van-Schaftingen et al. (1981); El-Maghrabi et al. (1982); Murray et al. (1984); Pilkis et al. (1987)
Hormone sensitive lipase/ cholesterol esterase	activated	Ad, Ac, Cl	Stralfors et al. (1987); Yeaman et al. (1994)
Phenylalanine hydroxylase	activated	L	Kaufman (1987)
Tyrosine hydroxylase	activated	L	
Glycogen synthase	inhibited	L, M	Cohen (1985, 1993)
Protein phosphatase 1 glycogen binding subunit	inhibited		Cohen (1993)
Acetyl CoA carboxylase	inhibited	L, Ad, M	Brownsey & Denton (1987)
6-Phosphofructo-2-kinase[1]	inhibited	L	Van-Schaftingen et al. (1981); El-Maghrabi et al. (1982); Murray et al. (1984); Pilkis et al. (1987)
Pyruvate kinase	inhibited	L	Engström et al. (1987)

Notes: [1] L, liver; M, muscle; Ad, adipocyte; Ac, adrenal cortex; Cl, corpus luteum.
[2] The bifunctional hepatic 6-phosphofructo-2-kinase-fructose 2,6-bisphosphatase which is switched from the 6-phosphofructo-2-kinase to the fructose 2,6-bisphosphatase mode by PKA phosphorylation.

Substrates for PKA

Physiological substrates for PKA have shown two or more basic amino acid residues (arginine, lysine) N-terminal to the phosphate acceptor serine (threonine in inhibitor 1) (reviewed in Cohen, 1985). Major metabolic enzyme substrates in mammalian cell types are given in Table 1. PKA catalyzed protein phosphorylations are entrained by stimulatory effects on adenylate cyclase of β-adrenergic agonists in skeletal and cardiac muscles, liver, and adipose tissue; by effects of glucagon in liver, cardiac muscle, and adipose tissue; and by effects of ACTH in the adrenal cortex and of LH in the corpus luteum.

The discovery that cAMP-dependent protein phosphorylations mediate metabolic effects of β-adrenergic agonists and glucagon led to consideration that they might mediate other effects of β-adrenergic agonists and glucagon on aspects of myocardial contraction, and of secretory responses of endocrine and exocrine glands to these and other agonists. These considerations led to the discovery of PKA phosphorylation of cardiac troponin I, which is phosphorylated on two serine residues within a 26 amino acid N-terminal extension absent from its skeletal muscle counterpart (Solaro et al., 1976; England, 1977; Moir et al., 1977; Sulakhe and Vo, 1995; Zhang et al., 1995) and of its contribution to the increased rate of cardiac relaxation induced by adrenaline. Likewise, PKA phosphorylation of cardiac phospholamban may also contribute to faster relaxation through accelerated Ca^{2+} uptake by sarcoplasmic reticulum (Hollinworth and England, 1978; Zhang et al., 1995). The inotropic effects of β-adrenergic agonists and of glucagon in the heart may be mediated by PKA phosphorylation of L-type Ca^{2+} channels (Flockerzi et al., 1986; Hartzell et al., 1991; Haase et al., 1993). This may also play a part in cAMP-mediated secretory responses although there is evidence that other mechanisms may be of major significance in PKA-mediated enhancement of insulin secretion (reviewed in Ashcroft et al., 1994).

Studies of the role of PKA in the regulation of cellular metabolism have been greatly facilitated by the use of bioactive derivatives which more readily enter cells. These have included analogues of cAMP (e.g. dibutyryl cAMP; 8-bromo cAMP) and inhibitors of cAMP phosphodiesterase (e.g. isobutylmethylxanthine).

cAMP and the Cell Nucleus

cAMP has effects on specific protein synthesis in a number of cells. One important example is provided by its longer term effects to induce enzymes of gluconeogenesis in liver and repress those of glycolysis in animals deprived of dietary carbohydrate. In rat liver cAMP increases the expression of genes for PEP carboxykinase and fructose 1,6 bisphosphatase, and decreases the expression of genes for glucokinase, 6-phosphofructo-1-kinase, pyruvate kinase, and the bifunctional 6-phosphofructo-2-kinase/fructose 2,6-bisphosphatase (see Granner and Pilkis, 1990). The effect of cAMP on gene transcription is mediated by translocation of the catalytic subunit of PKA to the nucleus where it phosphorylates cAMP response elements (CRE's), CREB's (CRE binding proteins), and CREM (CRE modulator) (reviewed in Karin and Hunter, 1995). This transfer of an activated protein kinase to the nucleus is one of two more general mechanisms for signal transmission from the cell surface to the nucleus. The other is translocation of transcription factors from the cytosol following activation by protein phosphorylation (for general review see Karin and Hunter, 1995).

The development of knowledge of PKA phosphorylations over the past 28 years has provided mechanisms for the overall effects of glucagon and of β-adrenergic agonists on energy metabolism. These are to mobilize the principle metabolic fuels (glucose from glycogen and gluconeogenic precursors; fatty acids from adipocyte and muscle triglycerides) and to shut off opposing biosynthetic pathways (i.e. glycogen synthesis and fatty acid synthesis from acetylCoA).

cAMP Phosphodiesterase

Much evidence accrued to show that the antilipolytic effect of insulin in adipose tissue, which is only readily apparent when cAMP is increased by a lipolytic hormone, is brought about through a decrease in cAMP and hence of cAMP-dependent protein kinase activity (Wong et al., 1978; Wong and Loten, 1981; Londos et al., 1985; Eriksson et al., 1995). Insulin was found to activate a plasma membrane-associated, cGMP-inhibited, low K_m cAMP phosphodiesterase [a 64-kDa protein (Degerman et al., 1987)] in adipocytes (Senft et al., 1968; Loten and Sneyd, 1970; Manganiello and Vaughan, 1973; Loten et al., 1978) and in 3T3-L1 cells (Elks et al., 1983). This activation is apparently mediated through phosphorylation by a serine protein kinase (Smith et al., 1991) later identified as the wortmannin-sensitive PI-3-kinase (Rahn et al., 1994). This cAMP phosphodiesterase may also be activated by glucagon and isoproterenol through phosphorylation of a different serine residue by PKA (Rascon et al., 1994; Robles-Flores et al., 1995). Evidence for an additional effect of insulin to inhibit adenylate cyclase is inconclusive (reviewed in Stralfors et al., 1987).

Details of the cAMP/PKA regulatory pathway are summarized in diagrammatic form in Figure 1. This is a linear regulatory pathway. Networks are also known. These arise when a protein kinase or phosphatase in one pathway modifies the activity of a protein kinase or phosphatase in another pathway by reversible phosphorylation (see later section on mechanism of insulin action).

Regulation of Glycogen Synthesis by Reversible Phosphorylation

The activity of glycogen synthase (GS) which forms the 1:4 α-glucosyl links in glycogen is rate-limiting for glycogen synthesis in muscle and liver, and in both tissues the enzyme is activated by insulin and inactivated by hormones which increase cAMP or Ca^{2+} (cf. contraction in muscle). It provides a notable example of multisite phosphorylation and of regulation by several protein kinases. Rabbit muscle GS is phosphorylated on nine serines by at least seven protein kinases. Five of these are inactivating, namely N7, N10, and C30, C34, and C38. Adrenaline inactivates GS directly by increasing phosphorylation of all five by PKA; and indirectly (and possibly more importantly) by dissociating PP1 from the glycogen particle through PKA phosphorylation of PP1 glycogen

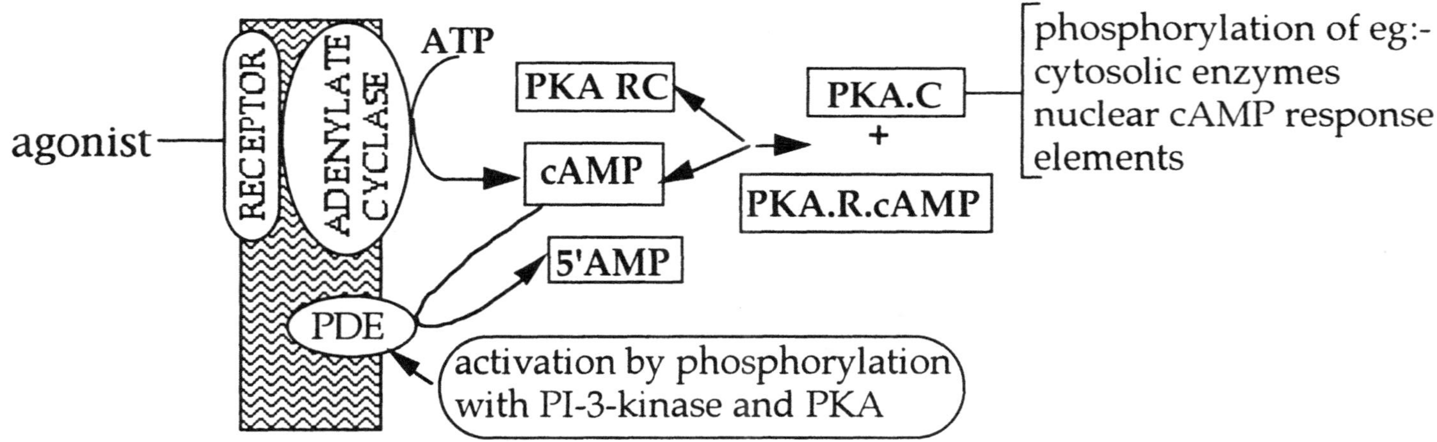

Figure 1. The cyclic AMP/protein kinase A/cyclic AMP phosphodiesterase pathway. PKA, cAMP-dependent protein kinase; R, regulatory subunit; C, catalytic subunit; PDE, phosphodiesterase; PI, phosphatidyl inositol.

binding protein (Cohen and Hardie, 1991). Insulin decreases the phosphorylation of C30, C34, and C38—serines that are phosphorylated by glycogen synthase kinase 3 (GSK3) (reviewed in Cohen. 1993). GSK3 is inhibited 40–50% by insulin in L6 myotubes and was known to be inactivated by phosphorylation with MAPKAP kinase-1 or $p70^{s6k}$. However the recent study of Cross et al., (1995) has shown that inhibitors of insulin activation of these kinases in myotubes (PD 98059, rapamycin) do not block the phosphorylation and inhibition of GSK3. Under these conditions the effect of insulin is mediated by protein kinase B (PKB).

The other mechanism by which insulin activates GS is through activation of PP1 with resultant dephosphorylation. This is brought about by ISPK (insulin sensitive protein kinase) catalyzed phosphorylation of site 1 on the glycogen binding subunit of PP1, an effect which also brings about dephosphorylation and inactivation of phosphorylase kinase (which phosphorylates glycogen synthase on site N7 (see Cohen, 1993) and of phosphorylase (the action of which is opposed to that of glycogen synthase).

Cyclic GMP-Dependent Protein Kinase (PKG)

The discovery of cAMP and of PKA prompted a search for other cyclic nucleotodes and their dependent kinases. PKG was originally discovered in invertebrate tissues (Kuo and Greengard, 1970) but is now known to be present in mammalian tissues, especially brain (cerebellum highest), lung, kidney and heart. The purified enzyme is a dimer (subunit M_r 74,000); each subunit has two binding sites for cGMP and a catalytic domain (Hofmann and Flockerzi, 1983). PKG can be activated in tissues by 8-bromocyclic-GMP and by phosphodiesterase inhibitors.

A substantial number of proteins have been shown to be phosphorylated by purified PKG in extracts of rat brain (Wang and Robinson, 1995) but their physiological significance is unknown. Hormone sensitive lipase is phosphorylated and activated by PKG *in vitro*, but there is no evidence that this occurs in adipose tissue. There is substantial evidence that cGMP is produced and PKG activity increased by stimulation of receptors that induce phospholipid turnover and Ca^{2+} entry (e.g. muscarinic cholinergic and α-adrenergic receptors). A role of cGMP and of PKG would appear to be to limit the extent and duration of $[Ca^{2+}]$ increase. There is substantial evidence that cGMP mediated phosphorylation may limit phospholipid breakdown and that it inhibits L-type Ca^{2+} channels, e.g. in brain and in heart cells (Chik et al., 1995; Tohse et al., 1995). There is also evidence that nitric oxide stimulates cGMP production in brain and mediates stimulation by *N*-methyl-D-aspartate receptors (Southam and Garthwaite, 1993).

Mitochondrial 2-Oxoacid Dehydrogenase Complexes

Pyruvate Dehydrogenase Complex (PDH Complex)

The mitochondrial PDH complex catalyzes the irreversible reaction:

$$\text{pyruvate} + \text{CoA} + \text{NAD}^+ \rightarrow \text{acetylCoA} + \text{NADH} + \text{H}^+$$

It subserves two functions, namely the oxidation of glucose for bioenergetics and the generation of acetylCoA for fat synthesis from carbohydrate. In animals this reaction results in irreversible loss of glucose with its vital role in cerebral metabolism, and the PDH complex reaction is subject to rather stringent regulation. The complex is inactivated by phosphorylation, and reversal can only be effected by dephosphorylation (Linn et al., 1969). The discovery was of considerable general importance to the development of the protein phosphorylation field because it was the first example of a reversible phosphorylation in mitochondria and because it was unrelated to the then known cAMP-mediated protein phosphorylations. It is important physiologically in adaptations to carbohydrate deprivation in animals.

Popov et al. (1993, 1994) showed by cDNA cloning there are two PDH kinases in the rat. The catalytic subunit of a PDH phosphatase has been cloned which shows sequence homology with protein phosphatase 2C (Lawson et al., 1993); a flavoprotein regulatory subunit (Pratt et al., 1982); dependence on Mg^{2+}; and activation by Ca^{2+} at concentrations within the physiological range (Randle et al., 1974).

The PDH complex contains four proteins, pyruvate dehydrogenase (E1); dihydrolipoyl acetyltransferase (E2); dihydrolipoyl dehydrogenase (E3); and protein X. The core is composed of 60 copies of E2 and attached to it are 30–60 copies of E1($\alpha_2\alpha_2$), 12 copies of E3 dimer, and 5 copies of PDH kinase. PDH phosphatase is not integral to the complex. The elementary reactions catalyzed by E1, E2, and E3 and the roles of coenzymes [CoA, NAD^+, TPP (thiamin pyrophosphate), and lip (lipoate)] are:

$$\text{pyruvate} + \text{TPP}\cdot\text{E1} \rightarrow [\text{hydroxyethyl}\cdot\text{TPP}\cdot\text{E1}] + \text{CO}_2 + \text{H}^+ \quad (3)$$

$$[\text{hydroxyethyl}\cdot\text{TPP}\cdot\text{E1}] + \text{E2}[\text{lipS}\cdot\text{S}] \longleftrightarrow \text{E2}[\text{lipSH}\cdot\text{Sacetyl}] + \text{TPP}\cdot\text{E1} \quad (4)$$

$$\text{E2}[\text{lipSH}\cdot\text{Sacetyl}] + \text{CoASH} \longleftrightarrow \text{E2}[\text{lipSH}\cdot\text{SH}] + \text{acetylCoA} \quad (5)$$

$$\text{E2}[\text{lipSH}\cdot\text{SH}] + \text{NAD}^+ \longleftrightarrow \text{E2}[\text{lipS}\cdot\text{S}] + \text{NADH} + \text{H}^+ \quad (6)$$

$$2[\text{hydroxyethylTPP}'\cdot\text{E1}] \longleftrightarrow 2\text{TPP}\cdot\text{E1} + \text{acetoin} \quad (7)$$

as shown in Reactions 3 to 6. Reaction 7 only proceeds in the absence of CoA and NAD^+ and is facilitated by acetaldehyde.

PDH kinases catalyze phosphorylation of the E1 component and more specifically one α-chain in the E1 tetramer (a_2b_2)—an example of half-site phosphorylation. There are three sites of phosphorylation on serines, only two of which (sites 1 and 2) are inactivating. With purified complexes and in rat heart mitochondria relative rates of phosphorylation are site 1>> site 2> site 3. PDH phosphatase catalyzes all three dephosphorylations; relative rates are sites 2 > 3 > or = 1 (Davis et al., 1977; Sugden and Randle, 1978; Teague et al., 1979; Sale and Randle, 1982a). *In vivo*, inactivation is due almost wholly to phosphorylation of site 1; phosphorylation of sites 2 and 3 inhibits dephosphorylation of site 1 and reactivation (Sale and Randle, 1982b). Measurement of site occupancies in rat heart *in vivo* have shown that phosphorylation of site 1 is related linearly with inactivation, whereas phosphorylation of sites 2 and 3 occurs mainly over the range of 80–100% inactivation of the complex. It is significant in this regard that the percentage of complex in the inactive phosphorylated form in the heart is increased from approximately 70% (resting normal rats fed high carbohydrate diet) to 95–99% (in rats starved for 48 h or fed a low carbohydrate diet or rendered diabetic).

Comparable changes are to be expected in other tissues (liver, skeletal muscle, kidney, gut) in which the percentage of complex in the inactive form is likewise increased by diet and diabetes. The effect of this is to lock the complex in an inactive form and to put a brake on dephosphorylation and reactivation in conditions associated with restricted availability of carbohydrate (i.e. a hysteresis mechanism).

In perfused or isolated tissues *in vitro*, phosphorylation and inactivation of PDH complex are enhanced by oxidation of fatty acids, and (in muscles) of ketone bodies . Studies with purified complexes have shown that the PDH kinase reactions are enhanced by products of the PDH complex reaction and inhibited by substrates for it (Cooper et al., 1974, 1975; Pettit et al., 1975) and also by ADP. The result *in vivo* is that inactivation of the complex by phosphorylation is enhanced by increased mitochondrial concentration ratios of [acetyl CoA]/[CoA], [NADH]/[NAD^+], and [ATP]/[ADP] and inhibited by pyruvate (Hansford, 1976, 1977; Kerbey et al., 1976, 1977). The effects of [acetyl CoA]/[CoA] and [NADH]/[NAD^+] on PDH kinase activity are achieved through reductive acetylation or reduction of particular lipoate residues in the complex through reactions shown in Reactions 3–7 as first proposed by Cooper et al., (1974,1975), Roche and Cate (1976), and Randle et al., (1978), and established in detail by Ravindran et al. (1996). The effects of oxidation of fatty acids and ketone bodies are mediated largely by increases in [acetyl CoA]/[CoA] and of [NADH]/[NAD^+] ratios.

The effect of diabetes and of starvation to enhance phosphorylation and inactivation of PDH complex in muscles and in liver involves a stable and longer term increase in PDH kinase activity (several hours) which differs from the reversible activation induced by altered [acetyl CoA]/[CoA] and [NADH]/[NAD^+] ratios (seconds). The effects of starvation/diabetes are reversed by several hours of refeeding/insulin treatment. Studies in tissue culture (hepatocytes, cardiac myocytes, soleus muscle) have implicated longer term and persistent effects of cAMP and fatty acids (Hutson and Randle, 1978; Fatania et al., 1986; Marchington et al., 1990; Stace et al., 1992). *In vivo*, this effect of starvation is to increase the specific activity of PDH kinase; its concentration is unchanged (Priestman et al., 1992). In tissue culture cAMP produces a stable increase in PDH kinase actiivity which is blocked by inhibitors of protein synthesis, whereas a comparable effect of fatty acid is blocked by inhibitors of mitochondrial uptake of fatty acids but not by inhibitors of protein synthesis. ELISA assays indicated that these effects in culture were also on specific activity of PDH kinase and not its concentration (Priestman et al., 1994).

Current views of mechanisms that mediate effects of diabetes and of dietary deprivation of carbohydrate to inactivate PDH complex by phosphorylation are summarized in Figure 2. The pathway can be interrupted at three points by inhibitors of lipolysis (e.g. acipimox), fatty acid oxidation (e.g. methyl tetradecylglycidate) and PDH kinase (e.g. dichloroacetate, Whitehouse et al., 1974). These compounds lower blood glucose and promote glucose oxidation in diabetic animals including man (e.g. see Wells et al., 1980; Tutweiler, 1989; Piatti et al., 1996).

Insulin *in vitro* rapidly effects conversion of PDHb to PDHa in rat adipose tissue and to a lesser extent in rat liver as part of its action to stimulate fatty acid biosynthesis. The effect is apparently mediated by activation of PDH phosphatase, but the mechanism is as yet unknown (Denton et al., 1989; Denton and Tavare, 1995).

Branched Chain 2-Oxoacid Dehydrogenase Complex (BCDH Complex)

The mitochondrial BCDH complex catalyzes reactions strictly analagous to those of the PDH complex but the substrates are the 2-oxoacids formed by the deamination of L-valine, L-leucine and L-isoleucine The enzyme is inactivated by phosphorylation and reactivated by dephosphorylation. As with the PDH complex, no activator of the *b* form of the enzyme is known (for review see Randle et al., 1987). The BCDH kinase cDNA has been cloned and sequenced.

2-Oxoacid Dehydrogenase Kinases

cDNAs for three mitochondrial 2-oxoacid dehydrogenase kinases have been cloned, two for the PDH complex and one for the BCDH complex (for review

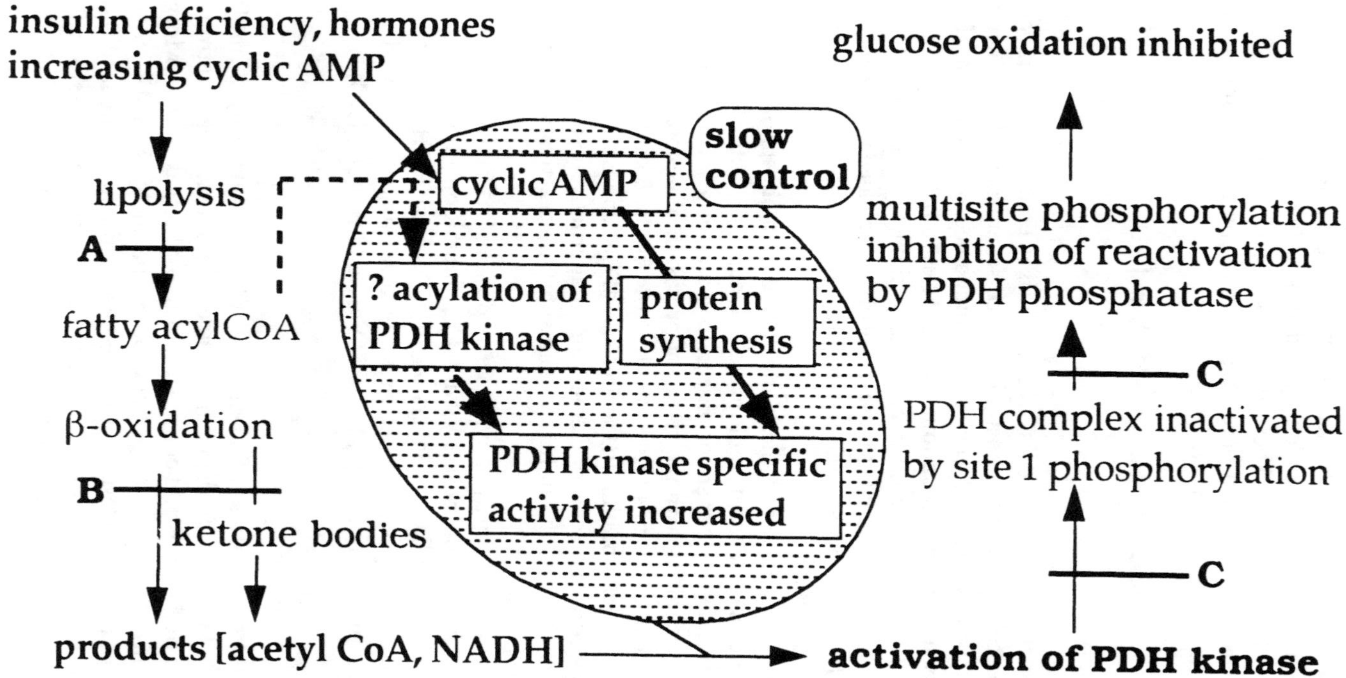

Figure 2. Mechanisms mediating effects of dietary carbohydrate deprivation and diabetes to inhibit glucose oxidation via phosphorylation/inactivation of the pyruvate dehydrogenase complex. **A, B,** and **C** are respectively inhibitors of lipolysis (e.g. acipimox), β-oxidation of fatty acids (e.g. tetradecylglycidate), and PDH kinase (e.g. dichloracetate).

see Harris and Popov, 1996). Detailed sequence analysis indicates that these serine kinases are only distantly related to other serine protein kinases though exhibiting considerable sequence identity and similarity with the prokaryotic histidine protein kinase family. These protein kinases, which phosphorylate aspartate residues, are so named because their catalytic mechanism involves autophosphorylation of a histidyl residue.

Signaling by Ca^{2+}, 5′-AMP, and ADP

Calcium

Muscle contraction and the process of exocytosis in endocrine and exocrine glands and other cells are energy-consuming processes entrained by an increase in cytosolic Ca^{2+} concentration. They are accompanied by activation of key reactions in fuel supply and oxidation. Inadequacies in fuel or oxygen supply especially in skeletal muscles result in an increase in the ratio of [ADP]/[ATP] which in the presence of adenylate kinase (cytosol) is translated into an increase in the [5′-AMP]/[ATP] ratio, and accelerated provision of substrate for enhanced glycolytic flux.

In the pathways of glycogen synthesis and degradation,phosphorylase kinase and the calmodulin-dependent multiprotein kinase are activated by Ca^{2+} resulting in conversion of phosphorylase *b* to *a* and of glycogen synthase *a* to *b* and to glycogenolysis with the resultant provision of hexose phosphates. Phosphorylase kinase contains four subunits ($\alpha\beta\gamma\delta$), the δ subunit being the Ca^{2+} binding protein calmodulin. In the presence of Ca^{2+}, phosphorylase kinase can also bind a second molecule of calmodulin or alternatively troponin-C can be bound to this site resulting in further and substantial activation (reviewed in Cohen, 1983, 1986).

In the mitochondrial pathway of pyruvate oxidation, Ca^{2+} activation of PDH phosphatase (in association possibly with Ca^{2+} inhibition of PDH kinase–Cooper et al., 1974) mediates conversion of PDH*b* to PDH*a* effected by contraction in skeletal muscles and by pressure development (increased work load) in cardiac muscle (Hennig et al., 1975; Vary and Randle, 1984).

5′-AMP and ADP

Anoxia (and uncouplers of oxidative phosphorylation) stimulate glycogenolysis. Two mechanisms have been implicated: activation of phosphorylase *b* by 5′-AMP and conversion of phosphorylase *b* to *a* (Morgan and Parmeggiani, 1964). The latter may be mediated by 5′-AMP-induced transformation of phosphorylase *b* into the conformational R state facilitating phosphorylation by phosphorylase kinase. A further possibility is phosphorylation by the 5′-AMP-activated kinase (AMPK) (Carling et al., 1989) as AICAR (see below) activates glycogenolysis (Young et al., 1996).

AMPK is believed to assist in protection of cells against environmental stresses which deplete ATP (e.g. anoxia, heat stress, fructose-induced ATP depletion in liver), the key signal being 5′-AMP (Hardie, 1994). 5-Aminoimidazole-4-carboxamide ribonucleoside (AICAR) is a useful agent which enters cells and is converted to a monophosphorylated derivative (ZMP) which can activate AMPK directly. AMPK is also activated by upstream protein kinases (AMPK kinase, AMPKK, and, less prominently, Ca^{2+} calmodulin protein kinase) and indirectly through inhibition of dephosphorylation of AMPKK by 5′-AMP.

It is of interest to note that cell membrane damage associated with environmental stresses that deplete ATP also lead to Ca^{2+} entry. AMPK inhibits lipogenic enzymes (acetylCoA carboxylase; HMGCoA reductase) and blocks activation of hormone-sensitive lipase by cAMP kinase (Corton et al., 1995). Recognition motifs similar to those of mammalian AMPK have been identified in higher plant HMG CoA reductase kinase A; and yeast SNF1, the enzyme required for the stress response to glucose starvation in yeast (Dale et al., 1995).

TYROSINE PROTEIN KINASES (PTKs) AND PHOSPHATASES (PTPs) AND THEIR ROLE IN THE REGULATION OF METABOLISM

Discovery and General

The product of the *src* gene of avian sarcoma virus, which causes sarcomata in birds and transforms fibroblasts in tissue culture, is a 60-kDa phosphoprotein ($pp60^{src}$). The normal homologue is the *sarc* gene which gives rise to the functionally and structurally similar $pp60^{sarc}$ (Collett et al., 1978; Wang et al., 1978). These phosphoproteins were identified as cAMP independent protein kinases (Collett and Erikson, 1978; Erikson et al., 1978, 1979). It was then shown that $pp60^{src}$ phosphorylates IgG heavy chain, and the phosphoamino acid was initially identified as phosphothreonine (Collett and Erikson, 1978).

Hunter and colleagues (Eckhart et al., 1979) in parallel studies of phosphorylating activity in immunoprecipitates containing polyoma virus T-antigens identified phosphotyrosine as the phosphorylated amino acid. They showed moreover that the conventional paper electrophoretic technique (pH 1.9) fails to separate phosphothreonine and phosphotyrosine. Subsequent studies (Hunter and Sefton, 1980a,b; Hunter et al.,1990) showed that IgG is phosphorylated on tyrosine, not threonine, by $pp60^{src}$ and $pp60^{sarc}$, and that $pp60^{src}$ is autophosphorylated on tyrosine. This led to widespread interest in tyrosine phosphorylation and its role.

Phylogenetically PTKs and PTPs have not been found in prokaryotes (although serine/threonine protein kinases are well established) and PTKs are scarce in unicellular eukaryotes such as yeast. According to Hunter et al. (1990), PTKs and PTPs appear to have emerged with metazoa and to have increased in frequency as the complexity of multicellular organisms has increased. Thus tyrosine phosphorylation may have emerged as part of cell-to-cell signaling mechanisms initiated either by cell-to-cell contact or humoral factors. Many PTKs were first identified as oncogenes derived from cellular genes (protooncogenes) encoding normal cellular PTKs. Some 50% of known oncogenes were held to be PTKs by the end of the 1980s (Hunter et al., 1990).

By 1990 some 40 distinct receptor phosphotyrosine kinases and phosphatases had been described. The kinases included seven main families of growth factor (GF) receptors: EGF (epidermal), insulin, PDGF (platelet derived), FGF (fibroblast), NGF (nerve) and the ECK and Ark receptor-like proteins. They are characterized by an amino-terminal glycosylated and ligand binding domain, a single membrane spanning region, a spacer region, a PTK catalytic domain, and a carboxy-terminal region. Ligand binding (e.g. of insulin) to the external domain activates the cytoplasmic catalytic domain leading to autophosphorylation followed by phosphorylation of cytoplasmic protein(s) (reviewed in Hunter et al., 1990, 1992).

The profile of substrates phosphorylated by any one PTK depends on the number and nature of the PTK autophosphorylation sites, and on spectrum of substrates (SH2-containing proteins and others). The insulin receptor is discussed in more detail as an example below. Many of the other PTKs may be anchored to the inner surface of the plasma membrane by N-terminal myristylation and include those in the *src*, *fps*, and *abl* groups. These PTKs have no external domain, but one possibility is that they may function as subunits of cell surface receptors lacking a cytoplasmic domain (Hunter et al., 1990).

The first protein tyrosine phosphatase to be characterized was purified from human placenta by Tonks et al. (1988). As with the PTKs, the PTPs may be divided into two main groups: a membrane spanning group (i.e. with extracellular, membrane, and intracellular domains) and a cytoplasmic group (Hunter et al., 1992).

The Insulin Receptor and Aspects of the Network of Protein Kinases Involved in the Mechanism of Action of Insulin

Insulin has a large number of effects in different metabolic pathways. These include regulation of enzymes or carriers in: (1) glucose metabolism, namely glucose transport, glycogen synthesis/glycogenolysis, glycolysis/gluconeogenesis, and pyruvate dehydrogenase; (2) fatty acid and triacylglycerol synthesis/lipolysis, namely acetylCoA carboxylase and triacylglycerol lipase;

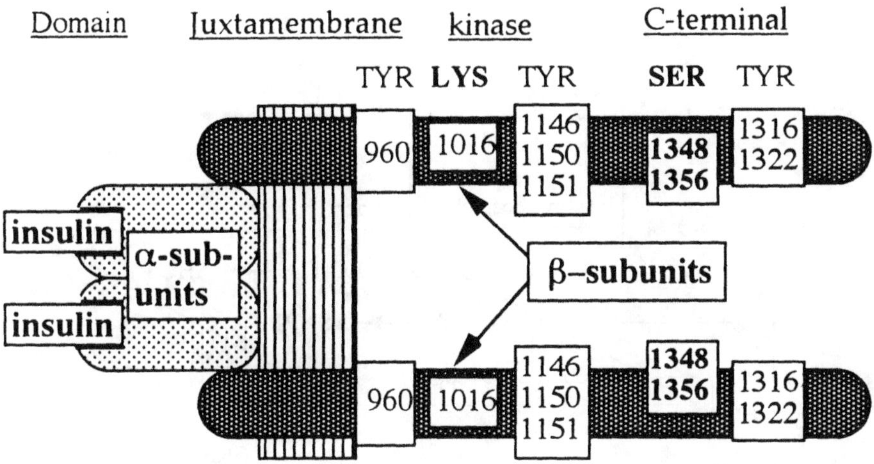

Figure 3. Topography of the insulin receptor and its phosphorylation sites. Receptor signal transduction is initiated by tyrosine phosphorylation and inhibited by serine phosphorylation.

(3) protein synthesis/ribosomal protein S6 kinase; and (4) enzyme induction/repression (transcription). In all of these pathways protein phosphorylation/dephosphorylation is involved and in no instance is the complete signaling pathway or network known. Three stages are to be distinguished in these processes: (1) binding of insulin to its receptor with activation of the receptor tyrosine kinase and phosphorylation of receptor tyrosine residues; (2) phosphorylation by the receptor tyrosine kinase of tyrosine residues in intracellular proteins which act as *onward transmitters*; and (3) further reactions of *onward transmitters* resulting in activation or inhibition of intracellular protein kinases. Knowledge of steps 2 and 3 is rather incomplete.

The insulin receptor (see Figure 3) is a heterotetrameric glycoprotein consisting of two α- and two β-subunits (M_r α, 135,000; β, 95,000) linked by disulfide bonds (Pilch and Czech, 1979; Massague et al., 1981; Kahn, 1994). The α-subunits are extracellular and provide the insulin binding site. The β-subunits span the membrane, with a single transmembrane domain, and exhibit tyrosine kinase activity. They contain the ATP binding site (which includes Lys1016 shown in Figure 3) and three clusters of tyrosine autophosphorylation sites (Kasuga et al., 1982; White et al., 1985; Tavaré et al., 1988; see Figure 3).

Phosphorylation of the three tyrosines in the kinase domain (Figure 3) further increases the tyrosine kinase activity towards exogenous protein substrates (see Kahn, 1994; Herrera and Rosen, 1986), the first stage in onward transmission of the signal. These exogenous protein substrates include: IRS-1 (insulin

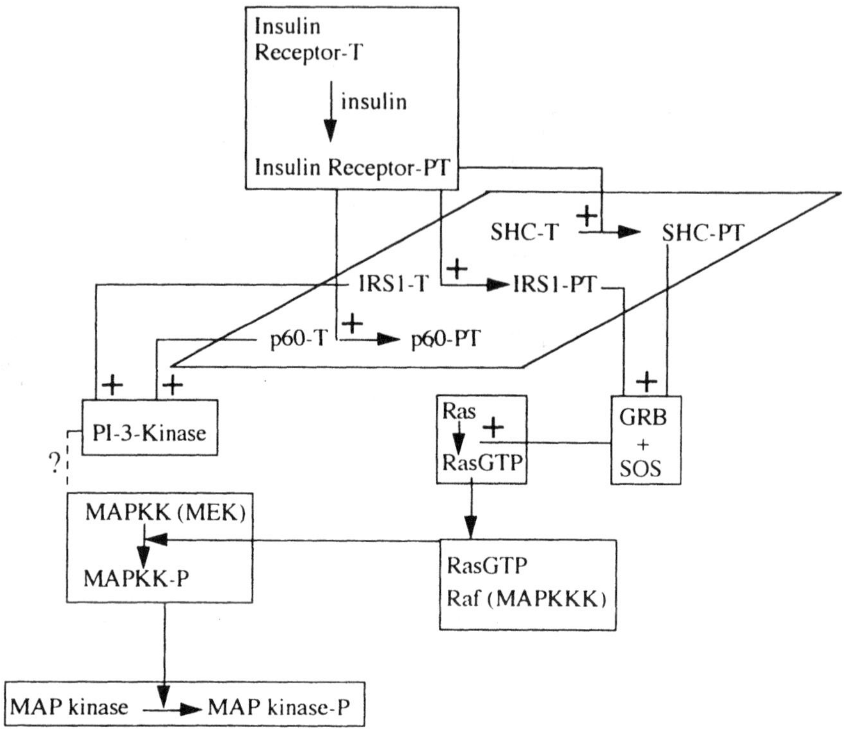

Figure 4. Possible links between the insulin-insulin receptor interaction and activation of the MAP kinase cascade. PT, phosphotyrosine. For other definitions see the appendix. For more detailed discussion of this pathway see Kahn (1994) and Denton and Tavare (1995).

receptor substrate 1); Shc, which can bind to GRB-2/SOS and activate the Ras complex (a family of GTP binding proteins); and p60 (and IRS-1) which activate PI-3-kinase. For example, IRS-1 tyrosines are rapidly phosphorylated following insulin stimulation of the receptor and this multisite phosphorylation leads directly to binding of IRS-1 to SH2 (*src* homology 2) domains in GRB2. This binding leads in some instances to stimulation of enzymic activity (e.g. PI-3 kinase) and in others (e.g. GRB2) to coupling of IRS to other signaling pathways such as Ras. Activation of GRB2 by IRS-1 may lead to interaction through SH3 (*src* homology 3) domains resulting in formation of GTP-Ras.

Activation of Ras leads directly to activation of a serine kinase cacade—the MAP kinase cascade (see Figure 4). For reviews see Cohen (1993), Kahn (1994), Denton and Tavaré (1995), and Karin and Hunter (1995) (also Sun et al., 1991; Skolnik et al., 1993; Sung et al., 1994).

Cohen has concluded from the work of his group that the key to activation of glycogen synthase by insulin is twofold: inhibition of GSK3 by serine phosphorylation, and activation of the binding of PP1 to glycogen through the glycogen binding (G) subunit. He has presented evidence that the latter is effected by phosphorylation of site 1 in the G-subunit by an insulin-sensitive protein kinase (ISPK). He and his colleagues purified an ISPK to homogeneity (Lavoinne et al., 1991; Sutherland et al., 1993; Sutherland and Cohen, 1994) and established it as a mammalian homologue of frog S6 kinase II (insulin stimulates ribosomal protein S6 phosphorylation—part of the mechanism by which it stimulates protein synthesis; Cohen, 1993). This ISPK is inactivated by dephosphorylation (PP2A). It is reactivated by phosphorylation with MAP kinase and was named MAP kinase activator protein kinase-1 (MAPKAPK1). This led Cohen to consider that insulin activation of insulin receptor tyrosine kinase might initiate activation of glycogen synthase through the MAP kinase cascade leading to phosphorylation and activation of MAPKAPK1 and thence of the G-subunit—the binding of PP1 to glycogen and dephosphorylation of glycogen synthase. This pathway is thus of potential importance to the regulation of glycogen synthase dephosphorylation by insulin.

Inhibition of GSK3 by phosphorylation is also involved in the stimulation of glycogen synthesis by insulin. However studies by Cross et al. (1995) with compounds that prevent activation of MAPKAPkinase-1 (8-bromo-cAMP, PD98059) or of $p70^{s6k}$(rapamycin) indicated that neither kinase was essential for this action of insulin. Under these conditions the inhibition of GSK3 by insulin appeared to be mediated by phosphorylation catalyzed by another insulin-stimulated protein kinase [protein kinase-B (PK-B)]. The inhibition of GSK3 by PK-B phosphorylation, like the inhibition of GSK3 by insulin (Cross et al., 1994; Welsh et al., 1994), is inhibited by the PI-3 kinase inhibitors, wortmannin, and LY 294002 (Cross et al., 1995).

In a recent critical review Denton and Tavaré (1995) have summarized evidence which led them to conclude that: activation of the MAP kinases Erk-1 and Erk-2 may mediate growth-promoting effects of insulin; the regulation of *Glut-1* and *c-fos* expression and AP-1 transcriptional activity; and possibly control of mRNA translation by insulin. They also conclude that MAP kinase activation does not explain activation by insulin of glucose transport, glycogen synthesis, acetylCoA carboxylase, or (in rat adipose tissue) pyruvate dehydrogenase. The evidence included, for example, experiments with EGF which, like insulin, activated MAP kinases Erk-1 and Erk-2, but failed to stimulate glucose transport or glycogen synthase. There the matter currently rests.

Other Topics

There are many other important aspects of regulation by protein kinases and phosphatases which have not been considered because of limitations of space and time. It may be helpful to draw attention to some of these and provide references to recent reviews. Protein kinase C (PKC), discovered some 20 years ago by Nishizuka and his colleagues and which exists in multiple isoforms has been reviewed recently by Dekker and Parker (1994), Tanaka and Nishizuka (1994), Dekker et al., (1995), and Nishizuka (1995). Other topics include the regulation of the cell cycle and the role of cyclins and cyclin-dependent kinases (Lew and Wang, 1995; Poon and Hunter, 1995); detailed consideration of transcriptional control by protein phosphorylation (Karin and Hunter, 1995); domain structure and classification of protein kinases (Lindberg et al., 1992; Hanks and Hunter, 1995).

It is apparent from what has been reviewed that protein kinases and phosphatases are of immense importance in pathology and hence of considerable potential importance in pharmacology. Philip Cohen in 1994 drew attention to their importance in toxicology and the economic importance of this toxicology (see Cohen, 1994). One example which he cites is provided by okadaic acid, a potent inhibitor of PP1 and PP2A (Bialojan andTakai, 1988). Okadaic acid is a major toxin associated with diarrhoeal seafood poisoning and has resulted in substantial lost revenues in the shellfish industry. Other examples of the importance of protein phosphorylation in relation to toxicology and therapeutics are to be found in Cohen's review.

NOTES

* In this field there is a good deal of jargon which is not readily understood by the uninitiated and a definition of these terms and of the author's abbreviations is given in the appendix.

** It is conventional, with interconvertible enzyme forms, to refer to the more active as the *a* form and the less active as the *b* form.

*** When conversion of phosphorylase a to b was shown to be associated with halving of molecular weight the Coris suggested that PR might now refer to "phosphorylase rupturing." The discovery that conversion of phosphorylase *b* to *a* involves dephosphorylation led to the further appreciation that PR could be translated to "phosphate removing".

APPENDIX

Protein Kinases, Protein Phosphatases, Substrates, and Other Related Proteins

Kinase	*Abbreviation*	*Comments*
AMP-kinase	AMP-K	5'-AMP activated kinase Also activated by Ca/CAM PK
AMP-K kinase	AMP-KK	Activates AMP-K
Branched chain ketoacid dehydrogenase kinase	BCDH-K	Inhibited by branched chain ketoacids

(continued)

Appendix. *(continued)*

Kinase	*Abbreviation*	*Comments*
Calmodulin-dependent multiprotein kinase	Ca/CAM PK	Ca/Calmodulin activated kinase
Extracellular signal related kinase	Erk-1, Erk-2	MAP kinases
Glycogen synthase kinase 3	GSK-3	Inactivates glycogen synthase
Insulin-stimulated protein kinase	ISPK	also known as MAPKAPK1
Mitogen-activated protein kinase	MAP-K	Activated by phosphorylation cascade from PTKs (MAP-KK)
MAP-K kinase	MAP-KK	Activated phosphorylation (MAP-KKK)
MAP-KK kinase	MAP-KKK	Activated by phosphorylation
MAP kinase activated protein kinase	MAPKAPK	Also known as $p90^{rsk}$
Myosin light-chain kinase	MLCK	Ca^{2+}-regulated
Phosphatidyl inositol-3-kinase	PI-3K	Wortmannin sensitive
Phosphorylase kinase		Phosphorylates phosphorylase b and glycogen synthase
Protein kinase A	PKA	cAMP dependent
Protein kinase B (*a,b*)	PKB(a,b)	serine and threonine specific homologues of the viral oncogene v-akt; also known as Akt/RAC
Protein kinase C	PKC	Activated by diacylglycerol
Protein kinase G	PKG	cGMP dependent
Pyruvate dehydrogenase kinases	PDH-K	Activated by products and inhibited by substrates of PDH complex reaction
Ribosomal protein S6 kinase	p70S6K	Activated by phosphorylation
Protein Tyrosine Kinase Receptors (PTKs)		
Epidermal growth factor-R Plateket derived GF-R Fibroblast GF-R Nerve GF-R Insulin-R	EGF-R PDGF-R FGF-R NGF-R	Tyrosines in cytoplasmic domains autophosphorylated. All activated substrates contain SH2 (*src* homology region 2) domains which bind to receptor phosphorylated tyrosines
Protein Phosphatases		
Class 1*	PP1	Dephosphorylate β-subunit of phosphorylase kinase. Inhibited by nM concentrations of protein inhibitors 1 and 2
Class 2	PP2A*, PP2B PP2C*	Dephosphorylate α-subunit phosphorylase kinase. Not inhibited by inhibitors 1 or 2
	PP2A	No cation dependency
	PP2B	Ca^{2+}-dependent
	PP2C	Mg^{2+}-dependent

(continued)

Appendix. *(continued)*

Kinase	*Abbreviation*	*Comments*
Pyruvate dehydrogenase phosphatase		Mg^{2+}-dependent; Ca^{2+}-dependent, sequence homology with PP2C
Protein tyrosine phosphatases	PTPs	
Other Proteins		
	Shc	Substrate of insulin receptor kinase; alternative to IRS-1 as activator of Ras
cAMP response elements	CRE	are nuclear; include CRE binding proteins (CREBs) and CRE modulator (CREM)
Growth factor receptor binding	GRB2	Couples tyrosine phosphorylated proteins
Protein 2		Ras through its SH2 domains
Son of Sevenless	SOS	G binding protein exchanging GTP for GDP on RAS thus activating
src Homology 2	SH2	Region of protein that binds to tyrosine phosphorylation sites

Note: * Also dephosphorylate proteins containing histidine phosphate.

REFERENCES

Ashcroft, F.M., Proks, P., Smith, P.A., Ämmälä, C., & Bokvist, K. (1994). Stimulus-secretion coupling in pancreatic b cells. J. Cell. Biochem. 55, Suppl., 54-65.

Bialojan, C. & Takai, A. (1988). Inhibitory effect of a marine-sponge toxin, okadaic acid, on protein phosphatases. Biochem. J., 256, 283-290.

Brostrom, M.A., Reimann, E.M., Walsh, D.A., & Krebs,. E.G. (1970). A cyclic 3′5′-AMP-stimulated protein kinase from cardiac muscle. Adv. in Enzyme Reg. 8, 191-203.

Brownsey, R. & Denton, R.M. (1987). AcetylCoA carboxylase. In: The Enzymes (Boyer, P.D. & Krebs, E.G., Eds.), Vol. XVIII, pp. 123-143. Academic Press, Orlando.

Bylund, D.B. & Krebs, E.G. (1975). Effect of denaturation on the susceptibility of proteins to enzymic phosphorylation. J. Biol. Chem. 250, 6355-6361.

Carling, D. Clarke, P.R., Zammit, V.A., & Hardie, D.G. (1989). Purification and characterization of the AMP-activated protein kinase. Copurification of acetyl-CoA carboxylase kinase and 3-hydroxy-3-methylglutaryl-CoA reductase kinase activities. Eur. J. Biochem. 186, 129-136.

Chen, C.C., Smith, D.L., Bruegger, B.B., Halpern, R.M., & Smith, R.A. (1974). Occurrence and distribution of acid-labile histone phosphates in regenerating rat liver. Biochemistry 13, 3785-3789.

Chen, C.C., Bruegger, B.B., Kern, C.W., Lin, Y.C., Halpern, R.M.m & Smith, R.A. (1977). Phosphorylation of nuclear proteins in rat regenerating liver. Biochemistry 16, 4852-4855.

Chik, C.L., Liu, Q.Y., Li, B., Karpinski, E., & Ho, A.K. (1995). cGMP inhibits L-type Ca^{2+} channel currents through protein phosphorylation in rat pinealocytes. J. Neurosci. 15, 3104-3109.

Cohen, P. (1982). The role of protein phosphorylation in neural and hormonal control of cellular activity. Nature (London) 296, 613-620.

Cohen, P. (1983). Protein phosphorylation and the control of glycogen metabolism in skeletal muscle. Phil. Trans. R. Soc. Lond. B. 302, 13-25.

Cohen, P. (1985). The role of protein phosphorylation in the hormonal control of enzyme activity. Eur. J. Biochem. 121, 439-448.

Cohen, P. (1986). Muscle glycogen synthase. The Enzymes XV11, 461-497.

Cohen, P (1989). The structure and regulation of protein phosphatases. Ann. Rev. Biochem. 58, 453-508.

Cohen, P. (1993). Dissection of the protein phosphorylation cascades involved in insulin and growth factor action. Biochem. Soc. Trans. 21, 555-567.

Cohen, P. (1994). The discovery of protein phosphatases: from chaos and confusion to an understanding of their role in cell regulation and human disease. Bioessays 16, 583-588.

Cohen, P. & Cohen, P.T.W. (1989). Protein phosphatases come of age. J. Biol. Chem. 264, 21435-21438.

Cohen, P. & Hardie, D.G. (1991). The actions of cyclic AMP on biosynthetic processes are mediated indirectly by cyclic AMP dependent protein kinase. Biochim. Biophys. Acta 1094, 292-299.

Collett, M.S. & Erikson, R.L. (1978). Protein kinase activity associated with the avian sarcoma virus *src* gene product. Proc. Natl. Acad. Sci. USA 75, 2021-2024.

Collett, M.S., Brugge, J.S., & Erikson, R.L. (1978). Characterization of a normal avian cell protein related to avian sarcoma virus transforming gene product. Cell 15, 363-370.

Collett, M.S., Erikson, E., & Erikson, R.L. (1979). Structural analysis of the ASV transforming protein: sites of phosphorylation. J. Virol. 20, 770-781.

Collett, M.S., Erikson, E., & Erikson, R.L. (1980). Avian sarcoma virus-transforming protein, pp60src shows protein kinase activity specific for tyrosine. Nature (London) 285, 167-169.

Colowick, S.P., Cori, G.T., & Cori, C.F. (1937). The isolation and synthesis of glucose-1-phosphate. J. Biol. Chem. 121, 465-471.

Cook, W.H., Lipkin, D., & Markham, R. (1957). The formation of a cyclic dianhydrodiadenylic acid by the alkaline degradation of adenosine 5′-triphosphoric acid. J. Am. Chem. Soc. 79, 3607.

Cooper, R.H., Randle, P.J., & Denton, R.M. (1974). Regulation of heart muscle pyruvate dehydrogenase kinase. Biochem. J. 143, 625-641.

Cooper, R.H., Randle, P.J., & Denton, R.M. (1975). Stimulation of phosphorylation and inactivation of pyruvate dehydrogenase by physiological inhibitors of the pyruvate dehydrogenase reaction. Nature 257, 808-809.

Cori, C.F. (1931). Mammalian carbohydrate metabolism. Physiol. Rev. 11, 143-275.

Cori, C.F., Green, A.A., & Cori, G.T. (1942). Crystalline muscle phosphorylase. J. Biol. Chem., 447-448.

Cori, C.F. & Cori, G.T. (1947). Polysaccharide phosphorylase. In: Les Prix Nobel en 1947, 1949, pp. 216-223. Stockholm Imprimiere Royale.

Cori, G.T. & Cori, C.F. (1945). The enzymatic conversion of phosphorylase *a* to *b*. J. Biol. Chem. 158, 321-352.

Cori, G.T., Colowick, S.P., & Cori, C.F. (1939). The activity of the phosphorylating enzyme in muscle extracts. J. Biol. Chem. 127, 771-782.

Cori, G.T. & Green, A.A. (1943). Crystalline muscle phosphorylase. II Prosthetic group. J. Biol. Chem. 151, 31-38.

Corton, J.M., Gillespie, J.G., Hawley, S.A., & Hardie, D.G. (1995). 5-aminoimidazole-4-carboxamide ribonucleoside. A specific method for activating AMP-activated protein kinase in intact cells. Eur. J. Biochem. 229, 558-565.

Cross, D.A., Alessi, D.R., Vandenheede, J.R., McDowell, H.E., Hundal, H.S., & Cohen, P. (1994). The inhibition of glycogen synthase kinase-3 by insulin or insulin-like growth factor 1 in the rat skeletal muscle cell line L6 is blocked by wortmannin, but not by rapamycin: evidence that wortmannin blocks activation of the mitogen-activated protein kinase pathway in L6 cells between Ras and Raf. Biochem. J. 303, 21-26.

Cross, D.A.E., Alessi, D.R., Cohen, P., Andjelkovich, M., & Hemmings, B.A. (1995). Inhibition of glycogen synthase kinase-3 by insulin mediated by protein kinase B. Nature (London) 378, 785-789.

Dale, S., Wilson, W.A., Edelman, A.M., & Hardie, D.G. (1995). Similar substrate recognition motifs for mammalian AMP-activated protein kinase, higher plant HMG-CoA reductase kinase-A, yeast SNF1, and mammalian calmodulin-dependent protein kinase I. FEBS Lett. 361, 191-195.

Davis, P.F., Pettit, F.H., & Reed, L.J. (1977). Peptides derived from pyruvate dehydrogenase as substrates for pyruvate dehydrogenase kinase and phosphatase. Biochem. Biophys. Res. Commun. 75, 541-549.

Davoren, P.R. & Sutherland, E.W. (1963a). The effect of epinephrine and other agents on the synthesis and release of adenosine 3′5′ monophosphate by whole pigeon erythrocytes. J. Biol. Chem. 238, 3009-3015.

Davoren, P.R. & Sutherland, E.W. (1963b). The cellular location of adenyl cyclase in the pigeon erythrocyte. J. Biol. Chem. 238, 3016-3023.

Degerman, E., Belfrage, P., Newman, A.H. Rice, K.C., & Manganiello, V.C. (1987). Purification of the putative hormone-sensitive cyclic AMP phosphodiesterase from rat adipose tissue using a derivative of cilostamide as a novel affinity ligand. J. Biol. Chem. 262, 5797-5807.

Dekker, L. & Parker, P.J. (1994). Protein kinase C—a question of specificity. TIBS 19, 73-77.

Dekker, L.V., Palmer, R.H., & Parker, P.J. (1995). The protein kinase C and protein kinase C related gene families. J. Curr. Opin. Struct. Biol. 5, 396-402.

Dent, P., Lavoinne, A., Nakielny, S., Caudwell, F.B., Watt, P., & Cohen, P. (1990). The molecular mechanism by which insulin stimulates glycogen synthesis in mammalian skeletal muscle. Nature 348, 286-287.

Denton, R.M., Midgely, P.J.W., Rutter, G.A., Thomas, A.P., & McCormack, J.G. (1989). Studies into the mechanism whereby insulin activates pyruvate dehydrogenase in adipose tissue. Ann. N.Y. Acad. Sci. 573, 285-296.

Denton, R.M. & Tavaré, J.M. (1995). Does mitogen-activated-protein kinase have a role in insulin action? The cases for and against. Eur. J. Biochem. 227, 597-611.

Eckhart, W., Hutchinson, M.A., & Hunter, T. (1979). An activity phosphorylating tyrosine in polyoma T antigen immunoprecipitates. Cell 18, 925-933.

Ehrlichman, J., Rubin, C.S., & Rosen, O.M. (1973). Physical properties of a purified cyclic adenosine 3′,5′-monophosphate dependent protein kinase from bovine heart muscle. J. Biol. Chem. 248, 7607-7609.

Elks, M.L., Manganiello, V.C., & Vaughan, M. (1983). Hormone-sensitive particulate cAMP phosphodiesterase activity in 3T3-L1 adipocytes. Regulation of responsiveness by dexamethasone. J. Biol. Chem. 258, 8582-8587.

El-Maghrabi, M.R., Fox, E., Pilkis, J., & Pilkis, S.J. (1982). Cyclic AMP-dependent phosphorylation of rat liver 6-phosphofructo 2-kinase, fructose 2,6-bisphosphatase. Biochem. Biophys. Res. Commun. 106, 794-802.

England, P.J. (1977). Phosphorylation of the inhibitory subunit of troponin in perfused hearts of mice deficient in phosphorylase kinase. Evidence for the phosphorylation of troponin by adenosine 3′:5′-phosphate-dependent protein kinase *in vivo*. Biochem. J. 168, 307-310.

Engström, L., Ekman, P., Humble, E., & Zetterqvist, Ö. (1987). Pyruvate kinase. In: The Enzymes (Boyer, P.D. & Krebs, E.G., Eds.), Vol. XVIII, pp. 47-72. Academic Press, Orlando.

Erikson, E., Collett, M.S., & Erikson, R.L. (1978). *In vitro* synthesis of of a functional avian sarcoma virus-transforming gene product. Nature 274, 919-921.

Erikson, R.L., Collett, M.S., Erikson, E., & Purchio, A.P. (1979). Evidence that the avian sarcoma virus-transforming gene product is a cAMP-independent protein kinase. Proc. Natl Acad. Sci. USA 78, 6260-6264.

Eriksson, H., Ridderstrale, M., Degerman, E., Ekholm, D., Smith, C.J., Manganiello, V.C., Belfrage, P., & Tornqvist, H. (1995). Evidence for the key role of the adipocyte cGMP-inhibited cAMP phosphodiesterase in the antilipolytic action of insulin. Biochim. Biophys. Acta 1266, 101-107.

Fatania, H.R., Vary, T.C., & Randle, P.J. (1986). Modulation of pyruvate dehydrogenase kinase activity in cultured hepatocytes by glucagon and *n*-octanoate. Biochem. J. 234, 233-236.

Faux, M.C. & Scott, J.D. (1996). More on target with protein phosphorylation conferring specificity by location. TIBS 21, 312-315.

Fischer, E.H. & Krebs, E.G. (1955). Conversion of phosphorylase b to phosphorylase a in muscle extracts. J. Biol. Chem. 216, 121-132.

Flockerzi, V., Oeken, H.J., Hofmann, F., Pelzer, D., Cavalie, A., & Trautwein, W. (1986). Purified dihydropyridine-binding site from skeletal muscle t-tubules is a functional calcium channel. Nature (London) 323, 66-68.

Friedman, D.L. & Larner, J. (1963). Studies on UDPG-α-glucan transglucosylase. III. Interconversion of two forms of muscle UDPG-α-glucan transglucosylase by a phosphorylation dephosphorylation reaction sequence. Biochemistry 2, 669-675.

Granner, D. & Pilkis, S. (1990). The genes of hepatic glucose metabolism. J. Biol. Chem. 265, 10173-10176.

Haase, H., Karczewski, P., Beckert, R., & Krause, E.G. (1993). Phosphorylation of the L-type calcium channel beta subunit is involved in beta-adrenergic signal transduction in canine myocardium. FEBS Lett. 335, 217-222.

Hanks, S.E. & Hunter, T. (1995). The eukaryotic protein kinase superfamily: kinase (catalytic) domain structure and classification. FASEB J. 9, 576-596.

Hansford, R.G. (1976). Studies on the effects of coenzyme A-SH: acetyl coenzyme A, nicotinamide adenine dinucleotide: reduced nicotinamide adenine dinucleotide, and adenosine diphosphate: adenosine triphosphate ratios on the interconversion of active and inactive pyruvate dehydrogenase in isolated rat heart mitochondria. J. Biol. Chem. 251, 5483-5489.

Hansford, R.G. (1977). Studies on inactivation of pyruvate dehydrogenase by palmitoylcarnitine oxidation in isolated rat heart mitochondria. J. Biol. Chem. 252, 1552-1560.

Hardie, D.G. (1994). An emerging role for protein kinases: the response to nutritional and environmental stress. Cell Signal 6, 813-821.

Harris, R.A. & Popov, K.M. (1996). The mitochondrial α-ketoacid dehydrogenase kinases: molecular cloning, tissue specific expression and primary structure analysis. In: Alpha Keto Acid Dehydrogenase Complexes (Patel, M.S., Roche, T.E., & Harris, R.A., Eds.), pp. 139-149. Birkhâuser Verlag, Basel.

Hartzell, H.C., Mery, P.F., Fischmeister, R., & Szabo, G. (1991). Sympathetic regulation of cardiac calcium current is due exclusively to cAMP-dependent phosphorylation. Nature (London) 351, 573-576.

Hennig, G., Löffler, G., & Wieland, O.H. (1975). Active and inactive forms of pyruvate dehydrogenase in skeletal muscle. FEBS Lett. 59, 142-145.

Herrera, R. & Rosen, O.M. (1986). Autophosphorylation of the insulin receptor *in vitro*: designation of phosphorylation sites and correlation with receptor kinase activation. J. Biol. Chem. 261, 11980-11985.

Hiraga, A. & Cohen, P. (1986). Phosphorylation of the glycogen-binding subunit of protein phosphatase-1G by cyclic-AMP-dependent protein kinase promotes translocation of the phosphatase from glycogen to cytosol in rabbit skeletal muscle. Eur. J. Biochem. 161, 763-769.

Hofmann, F. & Flockerzi, V. (1983). Characterization of phosphorylated and native cGMP-dependent protein kinase. Eur. J. Biochem. 130, 599-603.

Hollinworth, D.N. & England, P.J. (1978). The stimulation of calcium uptake into sarcoplasmic-reticulum vesicles from rat heart by adenosine 3′,5′-phosphate-dependent protein kinase. Biochem. Soc. Trans. 6, 573-576.

Hubbard, M.J. & Cohen, P. (1993). On target with a new mechanism for the regulation of protein phosphorylation. Trends Biochem. Sci. 18, 172-177.

Hunter, T. & Sefton, B.M. (1980a). The transforming gene product of Rous sarcoma virus phosphorylates tyrosine. Proc. Natl. Acad. Sci. USA 77, 1311-1315.

Hunter, T. & Sefton, B.M. (1980b). The Rous sarcoma virus *src* gene product is a protein kinase that phosphorylates tyrosine. In: Protein Phosphorylation and Bio-regulation, pp. 193-208. FMI-EMBO Workshop, Basel, 1979. Karger, Basel.

Hunter, T., Lindberg, R.A., & Middlemas, D.S. (1990). Novel receptor tyrosine kinases. In: The Biology and Medicine of Signal Transduction (Nishizuka, Y. et al., Eds.), pp. 260-266. Raven Press, New York.

Hunter, T., Lindberg, R.A., Middlemas, D.S., Tracy, S., & van der Greer, P. (1992). Receptor tyrosine kinases and phosphatases. Cold Spring Harbor Symposia on Quant. Biol. 57, 25-40.

Hutson, N.J. & Randle, P.J. (1978). Enhanced activity of pyruvate dehydrogenase kinase in rat heart mitochondria in alloxan-diabetes or starvation. FEBS Lett. 92, 73-76.

Ingebritsen, T.S. & Cohen, P. (1983). Protein phosphatases: properties and role in cellular regulation. Science 221, 331-338.

Kahn, C.R. (1994). Insulin action, diabetogenes, and the cause of Type II diabetes. Diabetes 43, 1066-1084.

Karin, M. & Hunter, T. (1995). Transcriptional control by protein phosphorylation; signal transmission from the cell surface to the nucleus. Curr. Biol. 5, 747-757.

Kasuga, M., Zick, Y., Blith, D.L., Karlsson, F.A., Haring, H.U., & Kahn, C.R. (1982). Insulin stimulation of phosphorylation of the beta subunit of the insulin receptor. Formation of both phosphoserine and phosphotyrosine. J. Biol. Chem. 257, 9891-9894.

Kaufman, S. (1987). Aromatic amino acid hydroxylases. In: The Enzymes (Boyer, P.D. & Krebs, E.G., Eds.), Vol. XVIII, pp. 218-277. Academic Press, Orlando.

Kerbey, A.L., Randle, P.J., Cooper, R.H., Whitehouse, S., Pask, H.T., & Denton, R.M. (1976). Regulation of pyruvate dehydrogenase in rat heart. Mechanism of regulation of proportions of dephosphorylated and phosphorylated enzyme by oxidation of fatty acids and ketone bodies and of effects of diabetes; role of coenzyme A, acetyl coenzyme A and reduced and oxidized nicotinamide adenine dinucleotide. Biochem. J. 154, 327-348.

Kerbey, A.L., Radcliffe, P.M., & Randle, P.J. (1977). Diabetes and the control of pyruvate dehydrogenase in rat heart mitochondria by concentration ratios of ATP/ADP, NADH/NAD^+ and acetylCoA/CoA. Biochem. J. 164, 509-519.

Kerbey, A.L., Radcliffe, P.M., Randle, P.J., & Sugden, P.H. (1979). Regulation of kinase reactions in pig heart pyruvate dehydrogenase complex. Biochem. J. 181, 427-433.

Kerbey, A.L., Randle, P.J., & Kearns, A. (1981). Dephosphorylation of pig heart pyruvate dehydrogenase phosphate complexes by pig heart pyruvate dehydrogenase phosphate phosphatase. Biochem. J. 195, 51-59.

Kim, Y., Huang, J., Cohen, P., & Matthews, H.R. (1993). Protein phosphatases 1, 2A and 2C are protein histidine phosphatases. J. Biol. Chem. 268, 18513-18518.

Krebs, E.G. (1983). Historical perspectives on protein phosphorylation and a classification system for protein kinases. Phil. Trans. R. Soc. Lond. B 302, 3-11.

Krebs, E.G. & Beavo, J.A. (1979). Phosphorylation-dephosphorylation of enzymes Ann. Rev. Biochem. 48, 923-959.

Krebs, E.G. & Fischer, E.H. (1956). The phosphorylase b to a converting enzyme of rabbit skeletal muscle. Biochim. Biophys. Acta 20, 150-157.

Krebs, E.G., Graves, D.J., & Fischer, E.H. (1959). Factors affecting the activity of phosphorylase b kinase. J. Biol. Chem. 234, 2867-2873.

Krebs, E.G., Gonzalez, C., Posner, J.B., Love, D.S., Bravold, G.E., & Fischer, E.H. (1964). Interconversion of muscle phosphorylases *b* and *a*. Ciba Foundation Symposium on Control of Glycogen Metabolism, pp. 200-210. J & A Churchill, London.

Kuo, J.F. & Greengard, P. (1970). Cyclic nucleotide-dependent protein kinases. VI. Isolation and partial purification of a protein kinase activated by guanosine 3',5'-monophosphate. J. Biol. Chem. 245, 2493-2498.

Lavoinne, A., Erikson, E., Maller, J.L., Price, D.J., Avruch, J., & Cohen, P. (1991). Purification and characterisation of the insulin-stimulated protein kinase from rabbit skeletal muscle; close similarity to S6 kinase II. Eur. J. Biochem. 199, 723-728.

Lawson, J.E., Niu, X.D., Browning, K.S., LeTrong, H., Yan, J., & Reed, L.J. (1993). Molecular cloning and expression of the catalytic subunit of bovine pyruvate dehydrogenase phosphatase and sequence similarity with protein phosphatase 2C. Biochemistry 32, 8987-8993.

Levene, P.A. & Hill, D.W. (1933). On a dipeptide phosphoric acid isolated from casein. J. Biol. Chem. 101, 711-718.

Lew, J. & Wang, J.H. (1995). Neuronal cdc-2-like kinase. TIBS 20, 33-37.

Lindberg, R.A., Quinn, A.M., & Hunter, T. (1992). Dual-specificity protein kinases: will any hydroxyl do? TIBS 17, 114-119.

Linn, T.C., Pettit, F.H., & Reed, L.J. (1969a). Regulation of the activity of pyruvate dehydrogenase complex from beef kidney mitochondria by phosphorylation and dephosphorylation. Proc. Natl. Acad. Sci. USA 62, 234-241.

Linn, T.C., Pettit, F.H., & Reed, L.J. (1969b). Alpha-keto acid dehydrogenase complexes. X. Regulation of the activity of the pyruvate dehydrogenase complex from beef kidney mitochondria by phosphorylation and dephosphorylation. Proc. Natl. Acad. Sci. USA 62, 234-241.

Linn, T.C. & Srere, F.A. (1979). Identification of ATP citrate lyase as a phosphoprotein. J. Biol. Chem. 254, 1691-1698.

Lipmann, F. (1933). Über die Bindung der phosphorsäure in phosphoproteinen. I. Isolierung einer phosphorhaltigen aminosaüre (serinephosphorsaüre). aus casein. Biochem. Z. 262, 3-8.

Londos, C., Honnor, R.C., & Dhillon, G.S. (1985). cAMP-dependent protein kinase and lipolysis in rat adipocytes. III. Multiple modes of insulin regulation of lipolysis and regulation of insulin responses by adenylate cyclase regulators. J. Biol. Chem. 260, 15139-15145.

Loten, E.G. & Sneyd, J.G.T. (1970). An effect of insulin on adipose-tissue adenosine 3':5'-cyclic monophosphate phosphodiesterase. Biochem. J. 120, 187-193.

Loten, E.G., Assimacopoulos-Jeannet, F.D., Exton, J.H., & Park, C.R. (1978). Stimulation of a low Km phosphodiesterase from liver by insulin and glucagon. J. Biol. Chem. 253, 746-757.

Manganiello, V.C. & Vaughan, M. (1973). An effect of insulin on cyclic adenosine 3':5'-monophosphate phosphodiesterase activity in fat cells. J. Biol. Chem. 248, 7164-7170.

Marchington, D.R., Kerbey, A.L., & Randle, P.J. (1990). Longer term regulation of pyruvate dehydrogenase in cultured rat cardiac myocytes. Biochem. J. 267, 245-247.

Massague, J., Pilch, F., & Czech,M.P. (1981). Electrophoretic resolution of three major insulin receptor structures with unique subunit stoichiometries. Proc. Natl. Acad. Sci. USA 77, 7137-7141.

Moir, A.J.G., Cole, H.A., & Perry, S.V. (1977). The phosphorylation sites of troponin T from white skeletal muscle and the effects of interaction with troponin C on their phosphorylation by phosphorylase kinase. Biochem. J. 161, 371-382.

Morgan, H.E. & Parmeggiani, A. (1964). Regulation of glycogenolysis in muscle. Ciba Foundation Symposium on Control of Glycogen Metabolism, pp. 254-270. J & A Churchill, London.

Murray, K.J., El-Maghrabi, M.R., Kountz, P.D., Lukas, T.J., Soderling, T.R., & Pilkis, S.J. (1984). Amino acid sequence of the phosphorylation site of rat liver 6-phosphofructo-2-kinase/fructose-2,6-bisphosphatase. J. Biol. Chem. 259, 7673-7681.

Nishizuka, Y. (1995). Protein kinase C and lipid signaling for sustained cellular responses. FASEB J., Apr., 9(7), 484-496.

Parmeggiani, A. & Morgan, H.E. (1962). Effect of adenine nucleotides and inorganic phosphate on muscle phosphorylase activity. Biochem. Biophys. Res. Commun. 9, 252-257.

Pettit, F.H., Pelley, J., & Reed, L.J. (1975). Regulation of pyruvate dehydrogenase kinase and phosphatase by acetyl-CoA/CoA and NADH/NAD ratios. Biochim. Biophys. Res. Commun. 65, 575-582.

Piatti, P.M., Monti, L.D., Davis, S.N., Conti, M., Brown, M.D., Pozza, G., & Alberti, K.G.G.M. (1996). Effects of an acute decrease in non-esterified fatty acid levels on muscle glucose utilization and forearm indirect calorimetry in lean NIDDM patients. Diabetologia 39, 103-112.

Pilch, P.F. & Czech, M.P. (1979). Interaction of cross-linking agents with the insulin effector system of isolated fat cells. Covalent linkage of ^{125}I-insulin to a plasma membrane receptor of 140,000 daltons. J. Biol. Chem. 254, 3375-3381.

Pilkis, S.J., Claus, T.H., Kountz, M., & El-Maghrabi, R. (1987). Enzymes of the fructose 6-phosphate-fructose 1,6-bisphosphate substrate cycle. In: The Enzymes (Boyer, P.D. & Krebs, E.G., Eds.), Vol. XVIII, pp. 3-46. Academic Press, Orlando.

Poon, Y.C. & Hunter, T. (1995). Innocent bystanders or chosen collaborators. Cur. Biol. 5, 1243-1247.

Popov, K.M., Kedishvili, N.Y., Zhao, Y., Shimomura, Y., Crabb, D.W., & Harris, R.A. (1993). Primary structure of pyruvate dehydrogenase kinase establishes a new family of eukaryotic protein kinases. J. Biol. Chem. 268, 26602-26606.

Popov, K.M., Kedishvili, N.Y., Zhao, Y., Gudi, R., & Harris, R.A. (1994). Molecular cloning of the p45 subunit of pyruvate dehydrogenase kinase. J. Biol. Chem. 269, 29720-29724.

Posternak, S. (1927). Sur le noyau phosphoré de la caséine. Compt. Rend. 184, 217-220.

Pratt, M.L., Maher, J.F., & Roche, T.E. (1982). Purification of bovine kidney and heart pyruvate dehydrogenase phosphatase on sepharose derivatized with the pyruvate dehydrogenase complex. Eur. J. Biochem. 125, 349-355.

Priestman, D.A., Mistry, S.C., Kerbey, A.L., & Randle, P.J. (1992). Purification and partial characterization of rat liver pyruvate dehydrogenase kinase activator protein (free pyruvate dehydrogenase kinase). FEBS Lett. 308, 83-86.

Priestman, D.A., Mistry, S.C., Halsall, A., & Randle, P.J. (1994). Role of Protein synthesis and of fatty acid metabolism in the longer-term regulation of pyruvate dehydrogenase kinase. Biochem. J. 300, 659-664.

Rahn, T., Ridderstrale, M., Tornqvist, H., Manganiello, V., Fredrikson, G., Belfrage, P., & Degerman, E. (1994). Essential role of phosphatidylinositol 3-kinase in insulin-induced activation and phosphorylation of the cGMP-inhibited cAMP phosphodiesterase in rat adipocytes. Studies using the selective inhibitor wortmannin. FEBS Lett. 350, 314-318.

Rall, T.W., Sutherland, E.W., & Wosilait, W.D. (1956). The relationship of epinephrine and glucagon to liver phosphorylase. III. Reactivation of liver phosphorylase in slices and in extracts. J. Biol. Chem. 218, 483-495.

Rall, T.W., Sutherland, E.W., & Berthet, J. (1957). The relationship of glucagon and epinephrine to liver phosphorylase. J. Biol. Chem. 224, 463-475.

Randle, P.J., Denton, R.M., Pask, H.T., & Severson, D.L. (1974). Calcium ions and the regulation of pyruvate dehydrogenase. Biochem. Soc. Symp. 39, 75-87.

Randle, P.J., Sugden, P.H., Kerbey, A.L., Radcliffe, P.M., & Hutson, N.J. (1978). Regulation of pyruvate oxidation and conservation of glucose. Biochem. Soc. Symp. 43, 47-67.

Randle, P.J. (1986). Fuel selection in animals. Nineteenth Ciba Medal Lecture. Biochem. Soc. Trans. 14, 799-806.

Randle, P.J., Patston, P.A., & Espinal, J. (1987). Branched chain ketoacid dehydrogenase. In: The Enzymes, 3rd ed. (Boyer, P.D. & Krebs, E.G., Eds.), Vol. XVIII, pp. 97-121. Academic Press, London.

Rascon, A., Degerman, E., Taira, M., Meacci, E., Smith, C.J., Manganiello, V., Belfrage, P., & Tornqvist, H. (1994). Identification of the phosphorylation site *in vitro* for cAMP-dependent protein kinase on the rat adipocyte cGMP-inhibited cAMP phosphodiesterase. J. Biol. Chem. 269, 11962-11966.

Ravindran, S., Radke, G.A., Guest, J.R., & Roche, T.E. (1996). Lipoyl domain-based mechanism for the integrated feedback control of the pyruvate dehydrogenase complex by enhancement of pyruvate dehydrogenase kinase activity. J. Biol. Chem. 271, 653-662

Reed, L.J. & Pettit, F.H. (1981). Phosphorylation and dephosphorylation of pyruvate dehydrogenase. Cold Spring Harb. Conf. Cell Prolif. 8, 701-711.

Rimington, C. (1927). The phosphorus of caseinogen. I. Isolation of a phosphorus containing peptone from tryptic digests of caseinogen. II. Constitution of phosphopeptone. Biochem. J. 21, 1179-1193.

Robles-Flores, M., Allende, G., Pina, E., & Garcia-Sainz, J.A. (1995). Cross-talk between glucagon- and adenosine-mediated signalling systems in rat hepatocytes: effects on cyclic AMP-phosphodiesterase activity. Biochem. J. 312, 763-767.

Roche, T.E. & Cate, R.L. (1976). Evidence for lipoic acid mediated NADH and acetyl-CoA stimulation of liver and kidney pyruvate dehydrogenase kinase. Biochem. Biophys. Res. Commun. 72, 1375-1383.

Rossell-Perez, M. & Larner, J. (1954). UDP glucose-glycogen-α 4 glucosyl transferase. V. Two forms of the enzyme in dog skeletal muscle and their interconversion. Biochemistry 3, 81-88.

Rubin, C.S., Ehrlichman, J., & Rosen, O.M. (1972). Molecular forms and subunit composition of a cyclic adenosine 3′,5′ monophosphate dependent protein kinase purified from bovine heart muscle. J. Biol. Chem. 247, 36-44.

Sale, G.J. & Randle, P.J. (1982a). Role of individual phosphorylations sites in inactivation of pyruvate dehydrogenase complex in rat heart mitochondria. Biochem. J. 203, 99-108.

Sale, G.J. & Randle, P.J. (1982b). Occupancy of phosphorylation sites in pyruvate dehydrogenase phosphate complex in rat heart *in vivo*. Biochem J. 206, 221-229.

Schlender, K.K., Wei, S.H.m & Villar-Palasi, C. (1964). UDPglucose-glycogen-α 4 glucosyl transferase. I.Kinase activity of purified muscle protein kinase. Biochem. Biophys. Acta 191, 272-278.

Schmidt, G. (1934). Über den phosphorsäure haltigen baustein des caseinmoleküls Z. Physiol. Chem. 223, 86-88.

Senft, G., Schultz, G., Munske, K., & Hoffman, M. (1968). Influence of insulin on cyclic 3′,5′-AMP phosphodiesterase activity in liver, skeletal muscle, adipose tissue and kidney. Diabetologia 4, 322-329.

Skolnik, E.Y., Batzer, A., Li, N., Lee, C.H., Lowenstein, E., Mohammadi, M., Margolis, B., & Schlesinger, J. (1993). Role of p85 subunit of phosphatidylinositol-3-kinase as an adaptor molecule linking the insulin receptor, p62, and GTPase-activating protein. Science 260, 1953-1955.

Smith, C.J., Vasta, V., Degerman, E., Belfrage, P., & Manganiello, V.C. (1991). Hormone-sensitive cyclic GMP-inhibited cyclic AMP phosphodiesterase in rat adipocytes. Regulation of insulin- and cAMP-dependent activation by phosphorylation. J. Biol. Chem. 266, 13385-13390.

Solaro, R.J., Moir, A.J.G., & Perry, S.V. (1976). Phosphorylation of troponin I and the inotropic effect of adrenaline in the perfused rabbit heart. Nature (London) 262, 615-617.

Southam, E. & Garthwaite, J. 1993. The nitric oxide-cyclic GMP signalling pathway in rat brain. Neuropharmacology 32, 1267-1277.

Stace, P.B., Fatania, H.R., Jackson, A., Kerbey, A.L., & Randle, P.J. (1992). Cyclic AMP and free fatty acids in the longer-term regulation of pyruvate dehydrogenase kinase in rat soleus muscle. Biochim. Biophys. Acta 1135, 201-206.

Stralfors, P., Olsson, H., & Belfrage, P. (1987). Hormone sensitive lipase. In: The Enzymes (Boyer, P.D. & Krebs, E.G., Eds.), Vol. XVIII, pp. 147-172. Academic Press, Orlando.

Stull, J.T. & Buss, J.E. (1977). Phosphorylation of cardiac troponin by cyclic adenosine 3':5'-monophosphate-dependent protein kinase. J. Biol. Chem. 252, 851-857.

Sugden, P.H. & Randle, P.J. (1978). Regulation of pig heart pyruvate dehydrogenase by phosphorylation. Studies on the subunit and phosphorylation stoichiometries. Biochem. J. 173, 659-668.

Sugden, P.H., Hutson, N.J., Kerbey, A.L., & Randle, P.J. (1978). Phosphorylation of additional sites on pyruvate dehydrogenase inhibits its re-activation by pyruvate dehydrogenase phosphatase. Biochem. J. 169, 433-435.

Sulakhe, P.V. & Vo, X.T. (1995). Regulation of phospholamban and troponin-I phosphorylation in the intact rat cardiomyocytes by adrenergic and cholinergic stimuli: roles of cyclic nucleotides, calcium, protein kinases and phosphatases and depolarization. Mol. Cell. Biochem. 149-150, 103-126.

Sun, X.J., Rothenberg, P., Kahn, C.R., Backer, J.M., Araki, E., Wilden, P., Cahill, D.A., Goldstein, B.J., & White, M.F. (1991). The structure of the insulin receptor substrate IRS-1 defines a unique signal transduction protein. Nature (London) 352, 73-77.

Sung, C.K., Sanchezmargelet, V., & Goldfine, J.D. (1994). The function of GRB2 in linking the insulin receptor to Ras signaling pathways. J. Biol. Chem. 269, 12503-12507.

Sutherland, C., Campbell, D.G., & Cohen, P. (1993). Identification of insulin-stimulated protein kinase-1 as the rabbit equivalent of rskmo-2. Identification of two threonines phosphorylated during activation by mitogen-activated protein kinase. Eur. J. Biochem. 212, 581-588.

Sutherland, C. & Cohen, P. (1994). The alpha-isoform of glycogen synthase kinase-3 from rabbit skeletal muscle is inactivated by p70 S6 kinase or MAP kinase-activated protein kinase-1 *in vitro*. FEBS Lett. 338, 37-42.

Sutherland, E.W. & Cori, C.F. (1951). Effect of hyperglycaemic-glycogenolytic factor and epinephrine on liver phosphorylase. J. Biol. Chem. 188, 531-543.

Sutherland, E.W. & Wosilait, W.D. (1955). Inactivation and activation of liver phosphorylase. Nature (London) 175, 169-171.

Sutherland, E.W. & Rall, T.W. (1957). The properties of an adenine ribonucleotide produced with cellular particles, ATP, Mg^{2+} and epinephrine or glucagon. J. Am. Chem. Soc. 79, 3608.

Sutherland, E.W. & Rall, T.W. (1958a). Formation of a cyclic adenine ribonucleotide by tissue particles. J. Biol. Chem. 232, 1065-1076.

Sutherland, E.W. & Rall, T.W. (1958b). Fractionation and characterization of a cyclic adenine ribonucleotide formed by tissue particles. J. Biol. Chem. 232, 1077-1091.

Tanaka, C. & Nishizuka, Y. (1994). The protein kinase C family for neuronal signaling. Annu. Rev. Neurosci. 17, 551-567.

Tavaré, J.M., O'Brie, R.M., Siddle, K., & Denton, R.M. (1988). Analysis of insulin receptor phosphorylation sites in intact cells by two-dimensional phosphopeptide mapping. Biochem. J. 253, 783-788.

Teague, W.M., Pettit, F.H., Yeaman, S.J., & Reed, L.J. (1979). Function of Phosphorylation Sites on Pyruvate Dehydrogenase. Biochem. Biophys. Res. Commun. 87, 244-252.

Tohse, N., Nakaya, H., Takeda, Y., & Kanno, M. (1995). Cyclic GMP-mediated inhibition of L-type Ca^{2+} channel activity by human natriuretic peptide in rabbit heart cells. Br. J. Pharmacol. 114, 1076-1082.

Tonks, N.K., Kearns, A., & Randle, P.J. (1982). Pig heart [^{35}S]thiophosphoryl pyruvate dehydrogenase complexes. Eur. J. Biochem. 122, 549-551.

Tonks, N.K., Diltz, C.D., & Fischer, E.H. (1988). Purification of the major protein-tyrosine-phosphatases of human placenta. J. Biol. Chem. 263, 6722-6730.

Tutweiler, G.F. (1989). Glucose fatty acid cycle—possible therapeutic implications. In: Diabetes (Larkins, R., Zimmet, P., & Chisholm, D., Eds.), pp. 175-179. Elsevier, Amsterdam.

Van-Schaftingen, E., Davies, D.R., & Hers, H.G. (1981). Inactivation of phosphofructokinase 2 by cyclic AMP dependent protein kinase. Biochim. Biophys. Res. Commun. 103, 362-368.

Vary, T.C. & Randle, P.J. (1984). The effect of ischaemia on the activity of pyruvate dehydrogenase complex in rat heart. J. Mol. Cell. Cardiol. 15, 723-733.

Walsh, D.A., Perkins, J.P., & Krebs, E.G. (1968). An adenosine 3′,5′-monophosphate-dependant protein kinase from rabbit skeletal muscle. J. Biol. Chem. 243, 3763-3765.

Wang, L.-H., Halpern, C.C., Nadel, M., & Hanafusa, H. (1978). Recombination between viral and cellular sequences generates transforming sarcoma virus. Proc. Natl. Acad. Sci. USA 75, 5812-5816.

Wang, X. & Robinson, P.J. (1995). Cyclic GMP-dependent protein kinase substrates in rat brain. J. Neurochem. 65, 595-604.

Wells, P.G., Moore, G.W., Rabin, D., Wilkinson, G,R., Oates, J.A., & Stacpoole, P.W. (1980). Metabolic effects and pharmacokinetics of intravenously administered dichloroacetate in humans. Diabetologia 19, 109-113.

Welsh, G.I., Foulstone, E.J., Young, S.W., Tavare, J.M., & Proud, C.G. (1994). Wortmannin inhibits the effects of insulin and serum on the activities of glycogen synthase kinase-3 and mitogen-activated protein kinase. Biochem. J. 303, 15-20.

White, M.F., Takayama, S., & Kahn, C.R. (1985). Differences in the sites of phosphorylation of the insulin receptor *in vivo* and *in vitro*. J. Biol. Chem. 260, 9470-9478.

Whitehouse, S., Cooper, R.H., & Randle, P.J. (1974). Mechanism of activation of pyruvate dehydrogenase by dichloroacetate and other halogenated carboxylic acids. Biochem. J. 141, 761-767.

Wong, E.H.A., Loten, E.G., & Park, C.R. (1978). The correlation of cyclic AMP and protein kinase activity in adipocytes with lipolysis stimulated by ACTH: the effect of adenosine deaminase and actinomycin D. J. Cyclic Nucleotide Res. 5, 359-374.

Wong, E.H.A. & Loten, E.G. (1981). The antilipolytic action of insulin on adrenocorticotrophin-stimulated rat adipocytes. The roles of adenosine 3′,5′-monophosphate and the protein kinase dependent on adenosine 3′,5′-monophosphate. Eur. J. Biochem. 115, 17-22.

Wosilait, W.D. & Sutherland, E.W. (1956). The relationship of epinephrine and glucagon to liver phosphorylase. II. Enzymatic inactivation of liver phosphorylase. J. Biol. Chem. 218, 469-481.

Yeaman, S.J., Smith, G.M., Jepson, C.A., Wood, S.L., & Emmison, N. (1994). The multifunctional role of hormone-sensitive lipase in lipid metabolism. Adv. Enzyme Regul. 34, 355-370.

Young, M.E., Radda, G.K., & Leighton, B. (1996). Activation of glycogen phosphorylase and glycogenolysis in rat skeletal muscle by AICAR—an activator of AMP-activated protein kinase. FEBS Letts. 382, 43-47.

Zhang, R., Zhao, J., Mandveno, A., & Potter, J.D. (1995). Cardiac troponin I phosphorylation increases the rate of cardiac muscle relaxation. Circ. Res. 76, 1028-1035.

Chapter 8

REGULATION OF EXPRESSION OF MICROBIAL GENES

Patricia H. Clarke

INTRODUCTION

In 1961 François Jacob[19] and Jacques Monod[20] published the review "Genetic Regulatory Mechanisms in the Synthesis of Proteins" in the third volume of the still infant *Journal of Molecular Biology* and opened up an entirely new way of thinking about genes and enzymes. Most of the paper was concerned with experiments with the bacterial enzyme, β-galactosidase, but the authors pointed out that their model of bacterial gene regulation had much wider implications:

> Enzymatic adaptation, as studied in microorganisms, offers a valuable model for the interpretation of biochemical coordination within tissues and between organs in higher organisms.

It was to be many years before the Jacob–Monod model could be tested in higher organisms and many surprises and complexities would appear. But questions about adaptation in microorganisms, particularly in relation to bacterial metabolism, had long been exercising microbial biochemists. During the 1950s a series of papers on the β-galactosidase of *Escherichia coli* had emerged from the Pasteur[1] Institute in Paris and rumors were rife about a new and exciting theory. The definitive review had been eagerly awaited. In essence, the model proposed that the expression of a structural gene for an enzyme was prevented by repression by the product of its regulator gene unless, and until, an appropriate inducing molecule appeared in the growth environment. The two authors had already presented their ideas at several well-attended lectures and François Jacob delighted the audience at one lecture with the following analogy. He suggested that we should think of the enzyme structural gene being in an airplane carrying a bomb. The regulator gene was constantly sending out a message saying "don't drop, don't drop, don't drop." When this message was interrupted by an overriding command the repression would be removed and the enzyme/bomb could be dropped.

Regulation of enzyme synthesis is now at the core of molecular biology, but at the beginning of this story it was of interest mainly to microbiologists. Many ideas, plausible at the time, sank without trace in a few years. Many problems that were difficult to resolve with the methods then available are now amenable to solution by modern techniques of genetic manipulation *in vitro*. The history of the operon has been well documented in articles and reviews and in the proceedings of many conferences. In the following account I have directed attention to those publications that seemed to give the truest impression of the climate of opinion at the time they were written. Of particular interest are the contributions of the main participants. Two publications are especially valuable: *Selected Papers in Molecular Biology by Jacques Monod* and *Origins of Molecular Biology: A Tribute to Jacques Monod*, both edited by André Lwoff[18] and Agnes Ullman (1978, 1979).

ADAPTATION

Adaptive Enzymes

Adaptation of microbial cultures to different growth environments had been recognized for a very long time. One hundred years before Jacob and Monod were involved the great Pasteur himself had asked questions about the metabolic activities of microorganisms. He examined growth and sugar utilization by yeast under aerobic and anaerobic conditions and showed that oxygen repressed the rate of fermentation; this became known as the Pasteur effect (see Senez, 1968). By the end of the nineteenth century it had been recognized that a number of microbial enzymes were produced only in particular growth media. The term enzyme was introduced in 1876 (see Gutfreund, 1976).

The description of this kind of metabolic variation in terms of adaptive and constitutive enzymes was contributed by Karstrom in 1930. He grew various bacteria in media containing a range of different sugars, prepared washed suspensions, and examined their fermentative abilities. He found that the ability to ferment certain sugars could be related to the presence or absence of that sugar in the growth medium. For example, he found that, irrespective of the sugar used as a growth substrate, all the washed suspensions of *Betacoccus arabinosaceus* (sic) were able to ferment glucose. On the other hand, cultures grown on glucose were unable to ferment galactose, arabinose, or lactose but could do so if they had been grown on these sugars. In another experiment, a washed suspension of *Bacterium aerogenes* (sic), grown in the *absence of xylose*, was incubated with xylose and a nitrogen source. Fermentation of xylose was observed after 2 h incubation, whereas a washed suspension of the bacteria grown *with xylose* was able to ferment xylose immediately. These results led Karstrom to suggest that there were two types of metabolic enzymes; constitutive enzymes that were synthesized under all environmental conditions and adaptive enzymes that were synthesized only in the presence of their substrates. At that time this was a helpful distinction and a specific question could now be asked about the regulation of enzyme synthesis in the form of "what are the special characteristics of an adaptive enzyme?"

E. coli mutabile

Several examples were known of another type of adaptation. One of these was a lactose-negative strain of *E. coli* which after an incubation period of several days produced colonies of lactose-positive bacteria. Massini (1907) noticed that a strain of *E. coli* plated on nutrient agar containing lactose and a pH indicator initially produced white colonies. After a few days red lactose-fermenting colonies appeared as knobs on the top of the white colonies. Some of the cultures derived from these colonies remained lactose-positive but others lost the lactose-positive character after subculture on lactose-free medium. These findings were not fully understood but it was thought probable that a mutation was involved and the strain was designated *E. coli mutabile.* It was found some years later that many strains of enterobacteria were "late lactose fermenters". The adaptation of *E. coli mutabile* to lactose fermentation was correctly assumed to be a mutational event, but the nature of the enzyme adaptations described by Karstrom was less clear The two types of adaptation were confused for several decades.

Formic Hydrogenlyase

Microbial biochemistry developed very rapidly in Cambridge during the 1920s and 1930s under the leadership of Marjory Stephenson[17]. In the preface to the second edition of *Bacterial Metabolism* in 1939 she commented that:

> In the problem of bacterial growth, advances have been made along new lines. Happily this subject now attracts mathematicians and statisticians less than formerly but has passed into the hands of biochemists interested in problems of nutrition.

Her studies on metabolic pathways and the enzymes involved introduced many biochemists to a new field of research. The third edition of *Bacterial Metabolism* in 1947 provided evidence of the rapid progress made during the intervening years. One of the main techniques used in those investigations was to examine the activities of bacteria grown in a variety of different media. The bacteria were harvested by centrifugation from cultures grown overnight (or for a few days) and resuspended in an appropriate buffer solution. These washed suspensions of bacteria were used in experiments on enzyme activities. In many cases Warburg[1] manometers (see Volume 1 in this series) were used to measure the uptake or output of gas. Oxidative activities could be measured by oxygen uptake, fermentation by carbon dioxide output, and for some experiments the manometers were filled with special gas mixtures. One of the standard tests used in bacterial classification is gas production by bacteria grown in broth containing sugars. Stephenson and Stickland (1932) showed that *E. coli* grown in media containing fermentable carbohydrates produced formate and converted it into carbon dioxide and hydrogen. They also showed that washed suspensions grown in tryptic broth did not liberate hydrogen until they had been in contact with formate for 24 h. At first it was thought that the bacteria might have been growing and increasing in numbers during that time. Karstrom's results suggested that it would be worth investigating whether a specific adaptive enzyme was appearing in the absence of growth.

Formic Hydrogenlyase
($HCOOH \longleftrightarrow H_2 + CO_2$)

They established that when formate was added to a culture of *E. coli* growing in tryptic broth there was no increase in the overall growth rate. They then set up a large-scale experiment in which the time course of formic hydrogenlyase synthesis was measured in the presence of formate. In this rather heroic experiment a flask containing 3 L of tryptic broth was inoculated with *E. coli* and grown for 10 h. A-500 mL sample was removed as a control and formate was added to the rest of the culture. Further 500-mL samples were removed at intervals and bacterial counts and formic hydrogenlyase activities measured. After 2 h there was a dramatic increase in formic hydrogenlyase activity with no significant increase in the bacterial count. They concluded that mutation and natural selection played no part in the adaptive formation of formic hydrogenlyase (Stephenson and Stickland, 1933).

Later, an experiment with formic hydrogenlyase was devised for a student practical class as an example of enzyme adaptation. A modified form of that experiment is shown in Figure 1. *E. coli* cultures were grown in broth with or without the addition of glucose or formate. Washed suspensions were incubated in Warburg flasks to measure gas output or uptake. Flasks filled with nitrogen were set up to measure hydrogen production from formate with the carbon dioxide being absorbed with potassium hydroxide. One flask had no formate and was filled with a mixture of hydrogen and carbon dioxide to measure gas uptake and the contents of that flask were tested for formate at the end of the experiment. One of the pleasing features of this experiment was that it not only showed that hydrogen was produced from formate by cultures grown with glucose or formate, but also demonstrated the reversibility of the reaction by showing that formic acid was produced from a mixture of hydrogen and carbon dioxide (Clarke, 1981).

The Glucose Effect

Ernest Gale extended the studies on enzyme synthesis and showed that there were many factors that could affect bacterial enzyme production during growth. He pointed out that some of the so-called adaptive enzymes were produced in low amounts in the absence of their substrates. This feature led Yudkin (1938) to suggest that enzyme synthesis might be regulated by mass action. He envisaged that if bacteria normally produced very low levels of adaptive enzymes then, whenever a potential substrate appeared in the growth medium, it would become bound to its enzyme and the equilibrium would be disturbed. This would be expected to result in the synthesis of more of the enzyme. Yudkin's suggestion was one of the first theories to account for adaptive enzyme synthesis in the presence of enzyme substrates. Although it was not correct it had the virtue of being precise and was only refuted when a great deal more experimental evidence had been obtained.

Gale found that many substances that were not enzyme substrates affected enzyme synthesis in various ways. He showed that glucose in the medium reduced the rate of synthesis of a number of enzymes but it was not necessarily a specific effect. In some cases the glucose effect was due to acid production and consequent reduction in the pH of the growth medium. An interesting group of enzymes affected in this way were the six amino acid decarboxylases of *E. coli* acting on arginine, glutamic acid, histidine, lysine, ornithine, and tyrosine. These enzymes are produced only at the end of active cell division and their synthesis is dependent on the presence of the specific substrate and

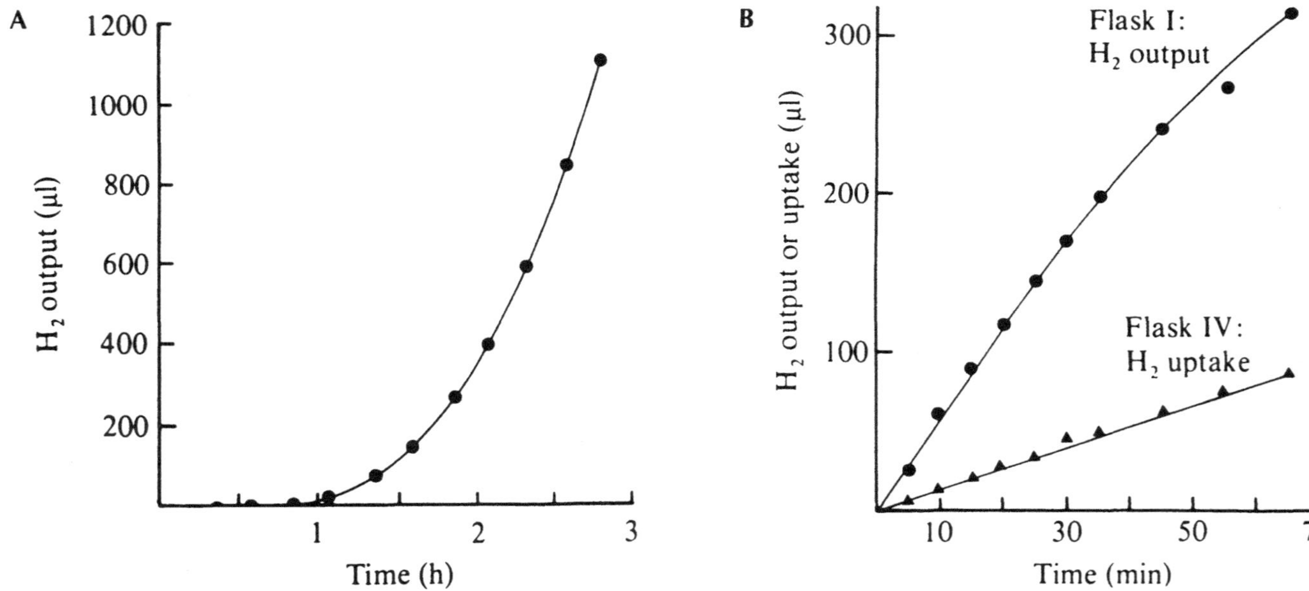

Figure 1. (**A**) Formic hydrogenlyase activity in *E. coli* during adaptation with formate. The bacteria were grown aerobically in the absence of formate and harvested aseptically. A suspension was introduced into a sterile manometer flask containing phosphate buffer, formate, and tryptic broth and gassed with N_2. After a lag period of 1 h hydrogen began to be produced. No significant increase in the viable count occurred during the experimental period (Stephenson and Stickland, 1933). (**B**) Formic hydrogenlyase activity in washed suspensions of *E. coli*. Flask I (●) contained 0.1 mL suspension of bacteria (grown in tryptic broth containing 0.5% sodium formate), 100 μL sodium formate, phosphate buffer, and N_2 as the gas phase. Flask IV (▲) contained a suspension of formate-grown bacteria (but formate as substrate was omitted), bicarbonate buffer, and 5% CO_2 in H_2 as the gas phase. The output of gas from flask I and the uptake of gas from flask IV was due to H_2 alone. Flask III (not shown) was set up as flask I but with a suspension of bacteria grown aerobically in the absence of formate; no gas production was observed. Flask II (not shown) was set up as flask I but without formate as substrate and with H_2 as the gas phase; no gas uptake was observed (see Clarke, 1981).

a low pH. The amines produced by the decarboxylases raise the pH of the culture and have a protective effect by buffering the acid produced during fermentation. Gale (1943) pointed out that, as well as suppressing or promoting the formation of specific enzymes, glucose could alter growth conditions in the following ways:

1. by the production of acid with subsequent alteration of pH;
2. by the production of gas with consequent anaerobiosis;
3. by an increase in growth rate and final cell density; and
4. by the transient formation of polysaccharide.

These data and the information available from experiments with many different bacterial species emphasized the difficulties in deriving a unified theory of regulation of enzyme synthesis. However, Gale provided a useful way of proceeding by suggesting that the metabolic capabilities of bacteria should be thought of as:

1. the total biochemical potential of an organism; and
2. the actual activities of a culture that could be related to the growth environment.

Training Bacteria

By hindsight it might be thought that the emphasis in these early experiments on whether or not growth occurred during enzyme adaptation was somewhat excessive, but at the time it was critical to establish this point. The term "adaptation" was being used to describe at least two different types of change in the properties of bacterial cultures. It seemed that mutation could explain the appearance of the lactose-fermenting strains of *E. coli mutabile,* while the synthesis of formic hydrogenlyase in the absence of growth required some other explanation, such as Yudkin's mass action theory. Another idea, current at that time, was that bacteria could be "trained" by serial subculture to cope in some way with the absence of certain growth requirements. Metabolic studies in Paul Fildes' laboratory had produced some evidence that this could occur. He was a distinguished medical bacteriologist who was interested in bacterial nutrition. During the 1930s the research carried out by Fildes, Gladstone, Knight, and Woods at the Bacterial Chemistry Unit in London was complementary to that being carried out by Marjory Stephenson and her colleagues in Cambridge. The nutritional requirement of various species of bacteria were determined by growing them in chemically defined media supplemented with amino acids and growth factors. It was found that tryptophan was essential for the growth of a strain of *Bacterium typhosum* (sic). The strain was then grown in a minimal medium with decreasing amounts

Table 1. Serial Cultivation of *B. typhosus* to Become Tryptophan-Independent[a]

	Days of incubation[b]								
Tryptophan added (%)	*1*	*10*		*1*	*10*		*1*		*1*
				No Tryptophan			*No Tryptophan*		
0.08	++	+++		·	·		·		·
0.0032	+	++		·	·		·		·
0.0064	+	+	→	0	+	→	+	→	++
None	0	0	→	0	0		·		·

Notes: [a] Data taken from Fildes et al. (1933).
[b] Serial subcultivation at the time indicated by the arrows. Amount of growth observed indicated by +, ++. +++.

of tryptophan and after a series of subcultures they were able to obtain a culture which grew in the absence of trytophan (Fildes et al., 1933). Marjory Stephenson said that this was an interesting experiment "the *modus operandi* of which is at present obscure." Similar results were obtained in experiments in which bacterial cultures were "trained" to do without other hitherto essential amino acids.

An outline protocol of the tryptophan experiment is shown in Table 1.

The method of serial subculture was used later by Hinshelwood and colleagues to "train" bacteria to be resistant to drugs and antibiotics (Hinshelwood, 1946). However, the interpretation of the results of those experiments was so wrapped up in the mathematical analysis disliked by Marjory Stephenson as to be inaccessible to all but the most intrepid microbiologists. This application of the concepts of physical chemistry to a complex biological system was not successful. They commented that bacteria can survive and grow in a very wide range of conditions and "it seems rather unlikely that this would be possible unless they were capable of various internal adjustments on passing from one medium to another." In these experiments changes in populations of bacteria were measured over fairly long periods of time and Hinshelwood rejected much of the evidence for spontaneous mutation and selection in favor of a state of "metastability" in the presence of the "inducing agent". This neo-Lamarckian view was one of the theories that disappeared without trace.

GENES AND ENZYMES

Bacterial Genes?

In *The Bacterial Cell,* Dubos (1945) gave a very perceptive account of the variability that had been observed in bacterial morphology and metabolism.

He noted that the ability to produce certain enzymes to synthesize pigments, amino acids and growth factors could be reversibly gained and lost by a bacterial culture. He said that:

> Granted that the transmission of hereditary characters in bacteria presents at least some analogy with the same process as occurs in higher cells, one may assume that this process takes place through the agency of genes.

His concluding words to the book are:

> The extraordinary plasticity of bacteria, the ease with which they adapt themselves to the environment, either by reversible modification, or by hereditary variation, has not only determined their importance in the economy of nature; it also makes them ideal subjects for the study of that organization and integration of independent characters which define and characterize life.

Most species of bacteria studied by biochemists up to 1945 were those consisting of single cells reproducing by binary fission. Inherited characters had been discussed but there was little or no information about genetic structures. However, by using various. staining methods several workers claimed to have seen "chromosomes" or "nucleoids" and resting and dividing nuclear structures (Dubos, 1945). The techniques and concepts of bacterial genetics had not been developed and there were as yet no experimental methods for identifying hereditary elements controlling bacterial enzymes.

Inborn Errors

Genes and enzymes had already met each other in respect of metabolic diseases of humans. In *Inborn Errors of Metabolism,* Garrod[1] (1923) ascribed the causes of certain inherited diseases to defects in enzymes of metabolic pathways (later confirmed to be due to mutations in genes for particular enzymes). It took many years for ideas on the relationship between genes and enzymes to progress much further. Geneticists studied genes and biochemists studied enzymes and there was little communication between them. But there were some notable exceptions. Beadle[2] and Ephrussi[2] (1936), working on eye-color mutants of *Drosophila,* concluded that there were at least two genes in control of the biochemical reactions leading to the synthesis of the eye pigments they had examined. It was difficult at that time to pursue this relationship in depth using *Drosophila.*

One Gene–One Enzyme

In a fascinating account of his scientific career, Beadle (1974) relates that one day in 1940 he was sitting in on some lectures by Tatum[2] on comparative biochemistry and realized that it would be easier to examine gene–enzyme relationships by starting with known biochemical pathways and selecting mutants rather than by starting with morphological mutants and looking for the enzyme defects. Tatum and Beadle decided to work with mutants of *Neurospora crassa* carrying defects in biosynthetic pathways. The advantage of this organism was that it had a known genetic system and could be grown on a chemically defined medium. They isolated large numbers of mutants able to grow on a complex medium but unable to grow on a chemically defined medium. By supplementing the minimal medium with amino acids and growth factors they were able to identify the pathways in which the mutations had occurred and assign genes to steps in the pathway. This work was in fact very useful in elucidating biosynthetic pathways but, even more important, it led to the very powerful concept of "one gene–one enzyme".

Bacteria Have Sex

The investigations with *Neurospora* had provided valuable information about genes for the enzymes of biosynthetic pathways. However, the methods for mutant isolation and identification of enzyme defects were still relatively tedious. It proved to be much easier to work with bacteria and developments in bacterial genetics were to produce rapid advances within a very few years. Although a few bacteriologists had previously claimed to have seen mating pairs of bacteria under the microscope there was no convincing evidence for genetic recombination. This was to change when Lederberg and Tatum (1946) reported that bacteria with novel genotypes were formed when two auxotrophic strains with multiple gene defects were mixed together. For example, a mixture of a strain unable to synthesize biotin and methionine with a strain unable to synthesize threonine and leucine gave rise to fully prototrophic recombinants.

The first step as before was to isolate auxotrophic mutants that were able to grow on complex media but not on a minimal medium. A method that became widely used was that of replica plating. Bacterial cultures were grown in nutrient broth and irradiated, or treated with mutagenic agents, and plated out on a complex medium at a dilution that would produce individual colonies. The colonies were impressed onto a velvet pad and transferred to minimal agar plates with various nutritional supplements. Comparison with the original plates identified the auxotrophic mutants. In early mating experiments between pairs of mutants carrying different nutritional defects they obtained stable prototrophic colonies at a frequency of one per 10^6 to 10^7 bacteria plated. Although there were some confusing features the results indicated that there was a single linkage group.

It became clear that this was a novel type of recombination when it was found that the contributions of the two partners in the mating pair were unequal and that only one of them needed to be viable. Hayes (1953) examined what happened if one of the parental auxotrophs was streptomycin-resistant and the other streptomycin-sensitive. In a cross between strain A (*str-s*) and strain B (*str-r*) the prototrophic recombinants appeared at the standard rate on minimal agar plates containing streptomycin. If the streptomycin markers were reversed then no recombinants appeared. This indicated that only strain B was essential for the formation of recombinant progeny and led to the designation of strain A as the donor or male strain and strain B as the recipient or female strain. The ability of a strain to act as a donor of genetic material was found to be due to the presence of a fertility factor which became known as the sex factor F. Donor strains carrying the factor were designated as F^+ and recipient strains as F^-.

Genetic analysis of the bacterial chromosome was made much easier by the discovery of donor strains generating 1,000-fold more recombinants and these were designated high-frequency or *Hfr* donors. In *Hfr* strains the sex factor becomes inserted into the bacterial chromosome. These developments in bacterial genetics were of critical importance in the analysis of the control of gene expression. Genetic exchange by conjugation was found to occur in many other species of bacteria (Hayes, 1968) and is also mediated by sex factors or other extrachromosomal plasmids.

Catabolic Pathways

Microbiologists were aware that heterotrophic microorganisms were able to use a variety of different compounds as growth substrates and that the compounds utilized varied greatly from species to species. In a remarkable study, den Dooren de Jong (1926) found that many common laboratory bacteria could use a surprisingly large number of carbon compounds for growth in minimal media with ammonium salts as nitrogen source. Table 2 lists a few of the compounds tested. Some species were very versatile; *Pseudomonas putida,* for example, was able to utilize 77 of the 200 compounds tested. It seemed clear that not only enzymes but also corresponding regulatory systems would be required to fit all these metabolic activities into the overall cell economy. Although Marjory Stephenson reported this work *in extenso* it was not widely known until the publication of the survey of the physiological activities of the *Pseudomonas* species by Stanier and colleagues in 1966. Many different strains of Pseudomonads had been isolated from soil and water by biochemists who were interested in catabolic pathways for the biodegradation of organic compounds. When the pathways had been elucidated the regulation of synthesis of enzymes of these complex pathways could be investigated.

Stanier was intrigued by the problems of enzyme adaptation. From his studies of growth on aromatic compounds he developed the concept of "simultaneous adaptation". He suggested that this could be used as a general

Table 2. Carbon Sources Supporting the Growth of Strains of *Pseudomonas spp.*[a,b]

Acids	*Alcohols & Sugars*	*Amino Acids*	*Amines & Amides*
Acetic	Methyl	Glycine	Acetamide
Propionic	Ethyl	Alanine	Propionamide
Butyric	Propyl	Aspartic	Butyramide
Valeric	Butyl	Glutamic	Succinamide
Caproic	Octyl	Phenylalanine	Asparagine
Oleic	Decyl	Histidine	Butylamine
Lactic	Glycol	Tyrosine	Amylamine
Pyruvic	Glucose	Benzoylalanine	Benzylamine
Malonic	Fructose	Leucine	Histamine
Phenylacetic	Mannose	Asparagine	Cadaverine
Benzoic	Rhamnose	Tryptophan	Allantoin

Notes: [a] Data from den Dooren de Jong (1926), Kluyver (1931), Stanier et al. (1966).
[b] A strain of *Pseudomonas putida* was able to utilize 77 of the 200 compounds tested.

test for the identification of the intermediates of a degradative pathway. The method adopted was to grow the bacteria on the initial substrate and a range of possible intermediates. The oxygen uptake by washed suspensions was measured on each compound in turn. He argued that if the organism grown on compound A was also able to oxidize without lag compounds B and C it could be concluded that the pathway for the catabolism of compound A was A → B → C. He assumed that the synthesis of each enzyme in turn would be controlled by its substrate and that other possible intermediates could be eliminated if there was a significant lag before they were oxidized.

The hypothesis of simultaneous adaption was a powerful one since it was precise and could be tested by experiment. It was very useful for analyzing complex catabolic pathways (Stanier, 1947). However, in its original form it proved to be too simplistic. For example, Karlsson and Barker (1948) applied the test of simultaneous adaptation to their studies on terminal oxidation in *Azotobacter*. Their results led to the erroneous conclusion that bacteria operated a dicarboxylic acid for terminal oxidation (Couldn't count up to three perhaps?). In this, and other cases, the problem was due to the use of whole cells that were impermeable to some of the test compounds unless a specific uptake system had been induced. Later, permeability was to present a significant complication in unravelling the *lac* operon. However the concept of simultaneous adaptation was an important precursor of the operon theory in that it drew attention to the regulation of sequential enzymes of a pathway. Studies on *Pseudomonads* have continued to contribute to ideas on the regulation of enzyme synthesis.

FROM GROWTH TO ENZYME ADAPTATION

At the Institut Pasteur

Jacques Monod was originally a zoologist and his early research in the 1930s was on the growth of ciliates. In 1940 he was introduced to the advantages of bacteria by Andre Lwoff who told him that "ciliates were the worst material to attack the problems of growth" and advised him to use a bacterium able to grow in a synthetic medium, for example *E. coli* (see Lwoff and Ullman, 1979). This advice led to Monod's Ph.D thesis, *Recherches sur la Croissance de Cultures Bactériennes* (Monod, 1941). The main findings of this work appeared later in the *Annual Review of Microbiology* (1949). Monod measured growth rates as a function of the concentration of the limiting carbon source using a number of different sugars. Having measured growth rates with a single sugar as the energy source he then examined growth in the presence of two sugars. With some combinations of sugars, for example glucose plus lactose, he observed two separate growth curves separated by a lag period which he described as "diauxy". Glucose was used in the first growth phase and lactose in the second phase after a lag period. Lwoff suggested that it might have something to do with enzyme adaptation and gave him his copy of Marjory Stephenson's *Bacterial Metabolism* and a reprint of Karstrom's Ph.D. thesis on bacterial enzyme synthesis. This was the moment that initiated the brilliant studies on induced enzyme synthesis at the Institut Pasteur over the next decades.

It is interesting to reflect how the choice of experimental material led to this line of research. First Monod, a zoologist, wanted to know about growth of organisms. Lwoff advised him to abandon protozoa for bacteria because they were simpler organisms and more amenable to detailed physiological studies. Having measured growth yields and growth kinetics of bacterial populations, Monod looked at growth on specific compounds: first single sugars, then pairs of sugars, and discovered diauxic growth curves. One interpretation of growth diauxy was that adaptive enzyme synthesis was involved. The logical next step was to identify the factors controlling the synthesis of a single adaptive enzyme. This did not start straight away. Among the reasons why this did not happen was that Europe was at war. Twenty-five years later, in his Nobel Prize lecture, Jacques Monod referred to this dark period in occupied France.

β-Galactosidase

After 1945 β-galactosidase began to emerge as the requisite single enzyme. A stable lactose-positive strain (L^+) isolated from *E. coli mutabile* strain ML (Mutabile Lwoffi from the digestive tract of M. Lwoff!) was compared with the parental lactose-negative strain (L^-). The L^+ mutant was able to oxidize lactose if it had been grown on lactose but not if it had been grown on glucose. The conclusion was that the utilization of lactose by cultures of *E. coli mutabile* was in part due to a rare mutational event and in part due to adaptive enzyme formation (Monod and Audureau, 1946). The mechanism of enzyme adaptation was still mysterious and in his contribution to the Growth Symposium in 1947 Monod summed up the position at that time in the following words:

1. The formation of most enzymes attacking exogeneous subtrates is specifically increased in the presence of the specific substrate.
2. In many cases, no appreciable enzyme formation occurs in the absence of the specific substrate.
3. Enzymatic adaptation is as highly specific as enzymatic activity.
4. Enzymatic adaptation occurs only in growing cells, or at least, does not occur under conditions preventing protein synthesis.

Two important questions were soon answered. First, they showed that the synthesis of β-galactosidase could be induced by galactosides other than lactose, and second that some inducing galactosides were not good enzyme substrates (Monod and Cohn, 1952). Some of these findings are shown in Table 3.

In the introduction to *The Lactose Operon*, Jacob and Monod (1970) pointed out that the ability to synthesize a large range of galactoside analogues was a key factor in choosing to concentrate on lactose rather than on one of the other systems being studied in the Institut Pasteur. Of particular importance was the introduction of thiogalactosides that were not substrates but were very effective inducers. The distinction between the properties of inducers and substrates can be seen in a simple experiment in which *E. coli* was tested for growth in a synthetic medium containing galactosides as potential carbon sources. Methyl-β-D-thiogalactoside (TMG) is an inducer but not a substrate

Table 3. β-Galactosides as Inducers and Substrates of β-Galactosidase

	Induction at 10^{-2} M[b]	*Affinity for Enzyme*[c]	*Hydrolysis*
Lactose	2500	100	+
Methyl-β-D-galactoside	2800	10	+
Phenyl-β-D-galactoside	560	600	+
o-Nitrophenyl-β-D-galactoside (ONPG)	1060	1,000	+

Notes: [a] Data from Monod and Cohn (1952).
[b] Activity of β-galactosidase induced after 4 h growth in the presence of 10^{-3} M inducer: expressed as mμ M o-nitrophenyl-β-D-galactoside hydrolyzed/min/mg bacteria.
[c] Measured as competitive inhibition of hydrolysis of ONPG set at 1000.

Table 4. β-Galactosides as Growth Substrates[a]

Carbon Source	*Inducer*	*Substrate*	*Growth*
Methyl-β-D-thiogalactoside (TMG)	+	−	*0*
Phenyl-β-D-galactoside	*(+)*	+	*0*
Phenyl-β-D-galactoside + TMG	+	+	+

Note: [a] Data from Monod (1956).

and phenyl β-D-galactoside is a substrate but not an inducer. Growth occurs only when both galactosides are provided (Table 4). Later it was shown that induction of β-galactosidase involved *de novo* protein synthesis from amino acids rather than the induced enzyme being derived from a preexisting protein precursor (see below).

ENZYME ADAPTATION IN 1953

Substrates and Inducers

By replacing "enzyme adaptation" by "enzyme induction" and introducing the term "inducer" it became much easier to discuss the phenomenon. Much confusion had been caused by using "adaptation" indiscriminately to include both evolution by mutation and selection and a phenotypic response to specific environmental conditions. By this time most investigators were attempting to distinguish clearly between enzyme genotype and enzyme phenotype (Gale's biochemical potential and actual activities). The Hinshelwood view was still that adaptation of bacterial populations was mainly due to directed change in the chemical equilibria of cell components. A suggestion by Spiegelman that was considered for a while was that a gene produced a cytoplasmic self-replicating element, the "plasmagene", which in turn synthesized the adaptive enzyme (see Monod, 1947). Pollock, who was working with penicillinase, suggested that an "organizer" possibly formed by the combination between the inducer and a specific receptor, catalyzed the formation of the active enzyme from a nonspecific precursor.

Many of these ideas were aired and actively discussed at a Symposium on *Adaptation in Microorganisms* in 1953 (Gale and Davies). The clear message from this meeting was that experimental techniques were now available for detailed exploration of both genetic and phenotypic events in enzyme synthesis. Cohn and Monod pointed out that most enzymes were, in Karstrom's classification, "constitutive" but that there was no fundamental reason why they should be synthesized any differently from inducible enzymes. Indeed, some enzymes were known to be constitutive in some strains and inducible in others. In *E. coli* it appeared that a constitutive β-galactosidase mutation was due to a single gene change and no differences could be found between the physicochemical properties of the enzyme from the constitutive strain and that from the inducible strain. The implication of this unitary hypothesis was that investigation of the regulation of induced enzyme synthesis was, in effect, an investigation of protein synthesis in general.

Gratuitous Enzyme Synthesis

The existence of a nonsubstrate inducer meant that a bacterium growing on succinate plus TMG was synthesizing an enzyme, β-galactosidase, in response to a compound it could not metabolize. This may have been of no value to the bacterium but it was enormously valuable to the experimenter since it made it possible to study the kinetics of induced enzyme synthesis under conditions in which the enzyme was not required for growth. This gratuitous enzyme synthesis introduced a new way of measuring the kinetics of enzyme induction. By adding inducer to exponentially growing cultures it was possible to measure the differential rate of synthesis of the induced enzyme against the increase in cell mass. The ratio of increase of enzyme Z to the increase in bacterial mass M is expressed as:

$$P = \Delta Z \ / \ \Delta M$$

Figure 2 shows the differential rates of β-galactosidase synthesis induced by four nonsubstrate inducers added at saturating concentrations: isopropyl thiogalactoside, propyl thiogalactoside, methyl thiogalactoside, and butyl thiogalactoside. Washed suspensions had been superceded at last!

The hypothetical high-molecular weight precursors of induced enzymes were also banished as it became clear that induction involved the *de novo* synthesis of protein from amino acids. Several investigators had shown that amino acid analogues could prevent enzyme induction by reducing the available amino acid pool. Confirmation came from labeling experiments. It was shown that β-galactosidase made in response to the addition of inducer to exponentially growing cultures did not contain any of the radioactive label of previously formed cell proteins (Hogness et al.,1955).

1953 was a year full of promise for investigations on the regulation of gene expression. Many of the confusions that had surrounded the concept of enzyme adaptation could be set aside. The new and elegant studies on induced enzyme synthesis had already involved scientists from different disciplines. Biochemical and genetic studies were to become increasingly interdependent. The first gene transfer in bacteria had been the transformation of nonvirulent to virulent strains of pneumocci by free DNA (Avery et al., 1944). Hershey[2] and Chase

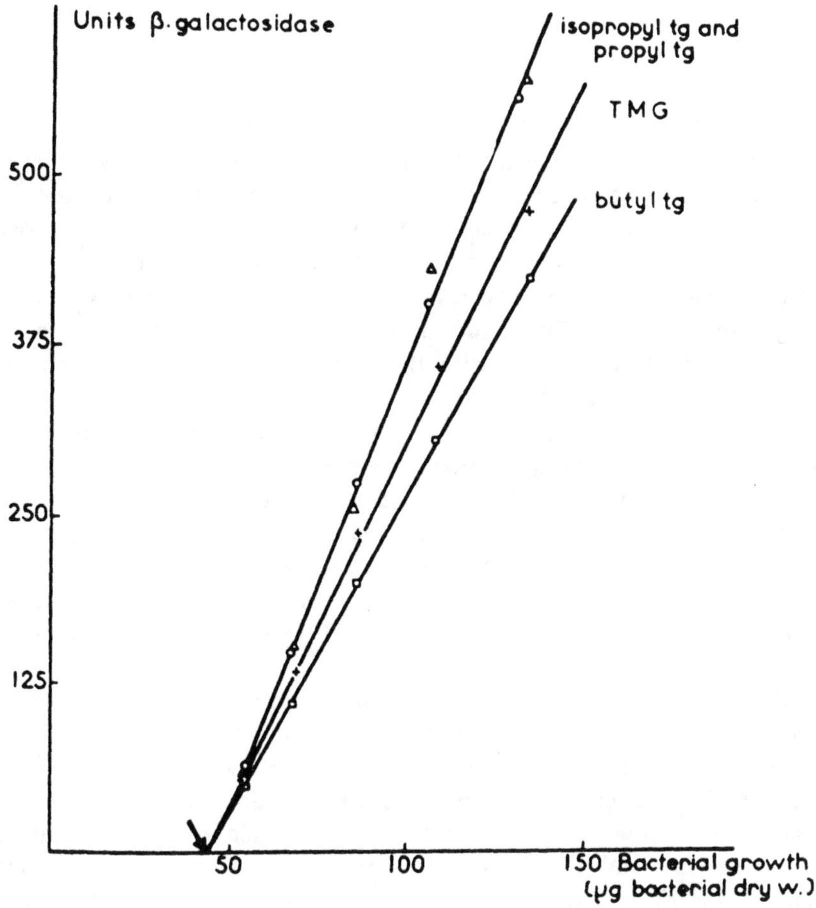

Figure 2. Synthesis of β-galactosidase in the presence of nonsubstrate inducers. *E. coli* ML30 was grown to the exponential phase on a mineral salts medium with succinate as sole carbon source. Methyl-, butyl- and isopropyl-thiogalactosides were added at the point indicated by the arrow at a concentration of 10^{-3} M (from Monod, 1956).

(1952) had shown that the basis of infection by lytic bacteriophage was the injection of DNA into the bacterial host. Transduction of short chromosomal segments of DNA by temperate bacteriophages had been demonstrated by Zinder and Lederberg (1952). Conjugation in bacteria had opened the way to mapping the bacterial chromosome (Hayes, 1953). Research on lysogenic bacteriophages (Jacob and Wollman, 1953) was about to provide some very important clues for elucidating the mechanism of enzyme induction. Furthermore, in 1953 Watson[2] and Crick[2] established the structure of DNA.

THE PERMEASE CONCEPT

Permeability

When inducers were added at saturating concentrations to exponentially growing cultures of *E. coli* the differential rate of β-galactosidase synthesis appeared to be constant and maximal from the start. Rather oddly, it seemed that with decreasing concentrations of the inducer, methyl-β-D-thiogalactoside (TMG), there were increasing lag periods before a steady rate of synthesis was achieved (Figure 3). When bacteria which had reached their maximum rate were washed free of inducer and allowed to grow again they were able to synthesize enzyme immediately after the inducer had been added. Bacteria that had been preinduced were able to accumulate ^{35}S-labeled TMG while noninduced bacteria were unable to do so. Later this effect was shown to be due to induction of a specific uptake system that became known as the lactose permease. This was the first clear-cut demonstration of such a system, but Cohen and Monod (1957), in reviewing the evidence for the lactose permease, discussed the evidence from earlier work for the existence of specific uptake systems for amino acids, sugars, and other compounds. Inducible permeases for tricarboxylic acid cycle intermediates had been detected in *Pseudomonas aeruginosa.*

The permease concept was readily accepted by pragmatic microbial biochemists. They realized that the presence, or absence, of specific inducible permeases could explain some of the anomalies in bacterial metabolism. For example, some bacterial strains appeared to possess all the enzymes for utilizing a particular organic compound, but were still unable to use that compound for growth. In many cases later work showed that such strains had defects in genes for specific uptake systems or in the regulator genes that controlled their synthesis. [A specific lactose-binding protein was identified later by Fox et al. (1967) and shown to be determined by the *lacy* gene.]

At the time there was considerable criticism of the permease concept among those who were interested in membrane transport. Peter Mitchell[1] objected to the term "permease" partly on the grounds that the ending "ase" belonged to enzymes and that complete uptake systems were not enzymes. It was suggested by some that his objection was merely semantic and that permease was a useful term even if the mechanism of uptake was not understood. However, it became clear that membrane transport was a complex process and that a specific binding protein was only one of the components. Mitchell himself was already involved in developing his chemi-osmotic theory of oxidative phosphorylation and membrane transport which led some years later to the award of a Nobel Prize.

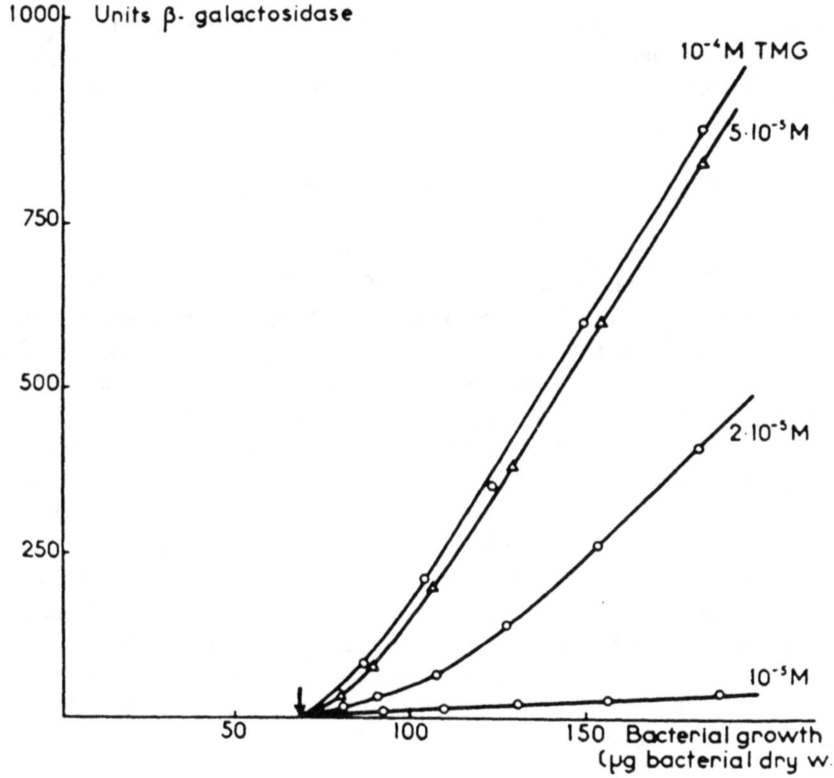

Figure 3. Induction of β-galactosidase at low concentrations of the nonsubstrate inducer: Methyl-thiogalactoside (TMG). Growth conditions were as for Figure 2. At low concentrations of inducer there is a long lag before the rate of synthesis of β-galactosidase reaches a steady state. The lag, ascribed to the y factor, was later shown to be related to the synthesis of the Lac permease (from Monod, 1956).

Lactose Mutants

The first lactose-negative mutants of *E. coli* had one of two phenotypes. Lac^- (absolute) could not be induced to produce β-galactosidase under any conditions but the permease could be induced. Lac^- (cryptic) produced normal amounts of β-galactosidase when induced with high concentrations of TMG but not with lactose (Monod, 1956). These were the *lacz* and *lacy* mutants that together with the *laci* (constitutive) mutants were to figure so largely in the next phase of research. In examining the accumulation of labeled TMG by induced and noninduced cultures, Georges Cohen isolated compound TMG-X which was later shown to be 6-acetyl-thiomethylgalactoside, the product of thiogalactoside transacetylase determined by another gene, *laca*. For a while this enzyme was wrongly thought to be a component of the lactose permease. Its role in metabolic activity is unclear but it turned out to be very useful to the experimenters since it was shown to be coordinately induced with β-galactosidase (i.e. simultaneous adaptation).

Mapping the *lac* Genes

The high frequency of genetic transfer from Hfr donors made it possible to construct a map of the chromosome of *E. coli* K12. With multiple markers it was found that the number of recombinants obtained from a conjugal cross could be arranged in decreasing order, suggesting polarity of transfer. This became clearer when the process of conjugation was interrupted at different times by separating the mating pairs by shaking the mixture in a Waring blender. This technique could be combined with killing off a sensitive donor by treatment with an antibiotic or a virulent bacteriophage. The interrupted mating experiments showed that genetic markers were transferred at a standard rate and could be ordered to give a time map of the bacterial chromosome (Wollman et al., 1956). Late markers were transferred with lower frequency due to breakage of the incoming DNA. By using a number of independent Hfr strains in which the sex factor had been inserted at different sites it became possible to build up a comprehensive chromosome map and to locate the position of the *lac* genes and their near neighbors as shown in Figure 4 (Jacob and Wollman, 1958).

TOWARDS THE OPERON THEORY

Regulation of Lactose and Bacteriophage λ-Genes

A major contribution to the development of the operon theory was the realization that induced enzyme synthesis was analogous to the release of bacteriophages from the lysogenic state. Prophages are maintained at specific sites on the bacterial chromosome and lytic particles are released spontaneously from a small proportion of the total population. Increased production of phage may be induced by UV irradiation or other treatments. The lysogenic character is perpetuated in the bacterial population and the prophage is normally maintained in its noninfectious state. *E. coli* K12 is lysogenic for λ-bacteriophage. Jacob and Wollman (1957) identified the chromosomal site of λ-prophage by an interrupted mating experiment involving crossing a nonlysogenic Hfr ly$^-$ strain with a F$^-$ ly$^+$ lysogenic strain. The nonlysogenic character was transferred in the same way as other genetic markers and the chromosomal site could be identified. In the reverse cross when the donor

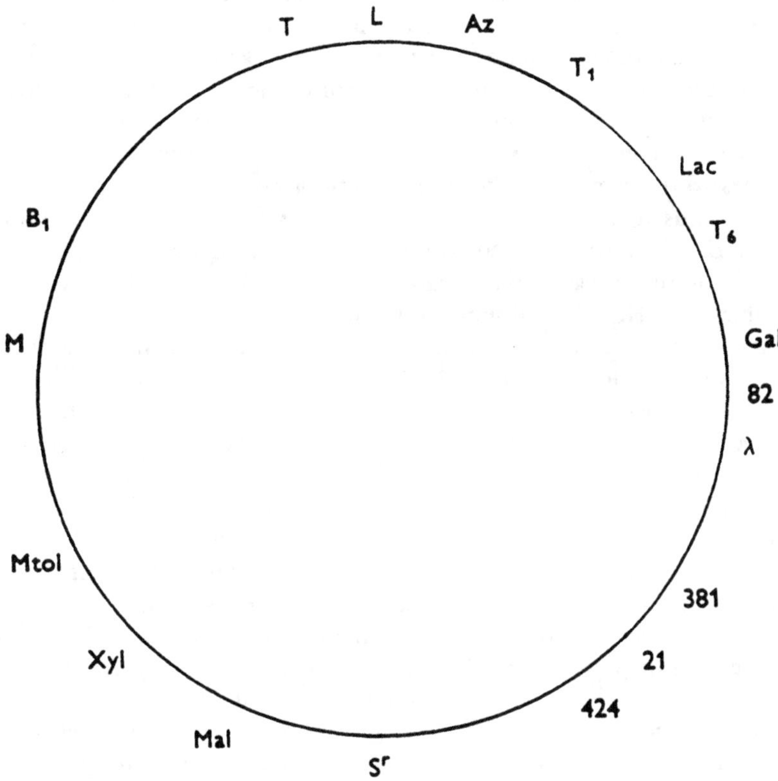

Figure 4. Map of *E. coli* chromosome in 1958. Diagrammatic representation of the K-12 chromosome, derived from a comparative study of the segments injected by different Hfr strains. The map represents only the sequence of characters, not the distances between them, measured in time intervals. The biosynthetic markers are T (threonine), L (leucine), M (methionine), and B_1 (thiamine). The catabolic markers are Lac (lactose), Mal (maltose, Gal (galactose), Xyl (xylose and Mtol (mannitol). Resistance characters are Az (sodium azide), Sr (streptomycin), and T_1 and T_6 (virulent bacteriophages). Inducible prophage sites are 82, λ, 381, 21, and 424 (from Jacob and Wollman, 1958).

carried the prophage they encountered the phenomenon of "zygotic induction". λ-bacteriophages were released soon after the time at which the λ-prophage had entered the nonlysogenic recipient bacterium. They concluded: (a) that the immunity of lysogenic bacteria is due to a cytoplasmic factor preventing phage development, and (b) that during conjugation the incoming gene could be expressed without recombination (see Jacob, 1979).

Analogous results were obtained in the so-called Pa-Ja-Ma experiments on β-galactosidase expression during conjugal transfer (Pardee et al., 1959). The *lac i, z, y,* and *a* genes controlling inducibility and the production of β-galactosidase, permease and transacetylase were shown by recombination frequencies and time mapping to be very closely linked but independent, since all possible phenotypes appeared among the recombinants. Mating experiments were carried out in the presence or absence of inducer between inducible and constitutive strains carrying positive or negative alleles of the *lac* genes to determine whether inducibility or constitutivity was dominant. With diploids it is relatively simple to demonstrate dominance. With the haploid *E. coli* it was decided to investigate the kinetics of β-galactosidase synthesis by the transient merozygotes produced in conjugal crosses. When inducible lac^+ (Hfr z^+ i^+) donors were mated with inducible lac^- (F^- z^- i^+) recipients in the presence of inducer they found that β-galactosidase synthesis began soon after the *lac* genes had been transferred (Figure 5a).

A second cross was made with strains Hfr z^+ i^+ and F^- z^- i^- in the absence of inducer. The donor was inducible so it could not synthesize β-galactosidase in the absence of inducer. Since the recipient had a defective *lac z* gene it was also unable to synthesize β-galactosidase. However, when the *lac z* gene from the donor entered the constitutive recipient cell, β-galactosidase synthesis began soon after transfer and continued during the first hour after mating. That is to say that the *lac z* gene was expressed constitutively soon after entry. Later, the rate of synthesis fell off unless the inducer (TMG) had been added (Figure 5b). These results were compatible with the absence of a cytoplasmic repressor in the i^- recipient. The conclusions from these experiments were that the *i* gene determining inducibility was distinct but linked to the *z* and *y* genes; that inducibility was dominant over constitutivity; and that constitutivity was due to the absence of a repressor of gene expression.

Another important discovery was that of F-prime plasmids or episomes. These originated from Hfr strains that had released the sex factor F attached to a small segment of the bacterial chromosome. F/*lac* plasmids can be transferred by conjugation and this made it possible to set up stable diploids and to confirm that the i^+ gene was trans-dominant. The experiments confirmed that the inducible i^+ allele was normally expressed in the cytoplasm and that the constitutive state resulted from the loss of capacity to synthesize an active repressor molecule (Pardee et al., 1959; Jacob et al., 1960).

Pardee et al. (1959) pointed out that there was already evidence for repression of enzymes of biosynthetic pathways by their end-products. Monod and Cohen-Bazire (1953) had shown that tryptophan synthetase was repressed by tryptophan and Cohn et al. (1953) had shown that the enzymes of the methionine biosynthetic pathway were repressed by methionine. Other examples of repression by amino acids of the enzymes of their biosynthetic pathways appeared later (see Vogel, 1957) so that it was reasonable to ask if biosynthetic and degradative enzymes were regulated in a similar manner. In 1959 the next two questions were:

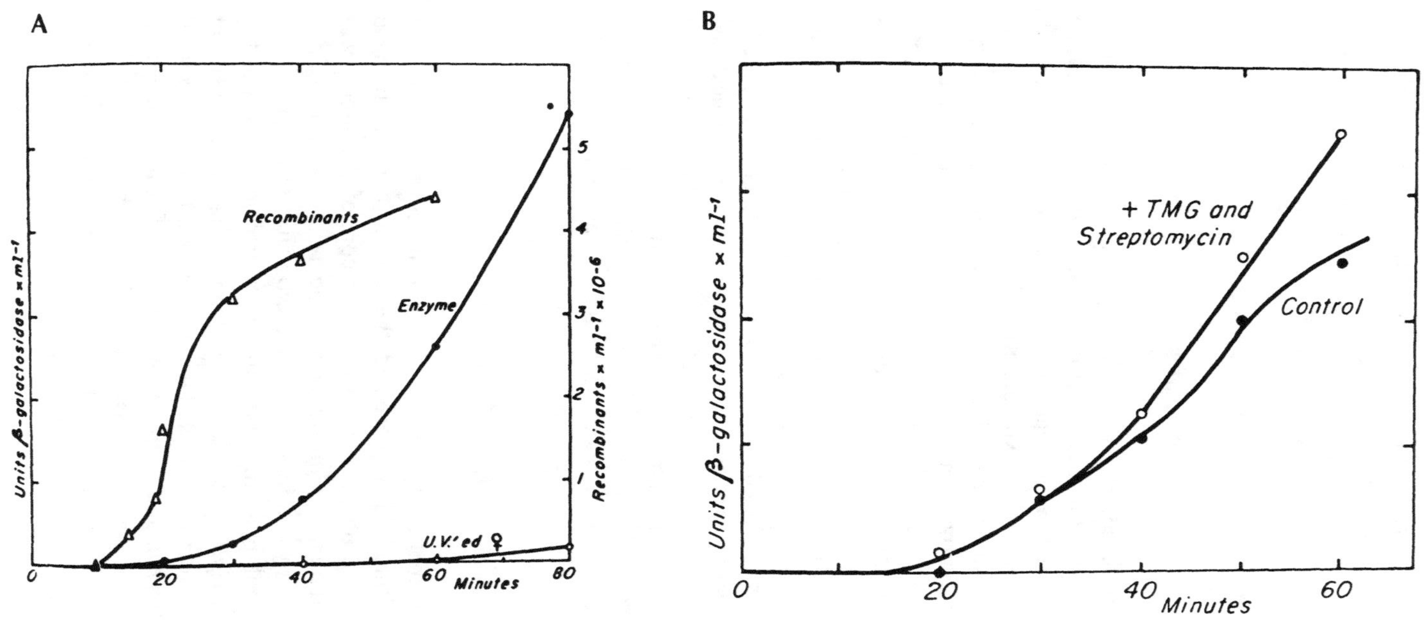

Figure 5. β-galactosidase synthesis during gene transfer. (**A**) Enzyme formation and appearance of recombinants during mating. An Hfr strain of *E. coli* K12 (z^+, i^+, Sm^s) was mated with an F^- recipient (z^-, i^+, Sm^r) strain in the presence of streptomycin (1 mg/mL) and isopropyl-thiogalactoside (10^{-3} M). The recombinants were selected by plating on streptomycin-lactose agar plates after separation in a blender. A control is shown as a separate experiment carried out with UV treated recipient cells (0.01% survival) (from Pardee et al., 1959). (**B**) Enzyme formation during the first hour of mating. An Hfr strain (z^+, i^+) was mated with a recipient (z^-, i^-) strain in the absence of inducer. To an aliquot was added streptomycin (0.8 mg/mL) at 20 min and methyl-thiogalactoside at 25 min to allow comparison of β-galactosidase synthesis with and without inducer. β-Galactosidase synthesis began almost immediately after gene transfer before recombination could have taken place (from Pardee et al., 1959).

1. What is the chemical nature of the repressor? Should it be considered a primary or secondary product of the gene?
2. Does the repressor act at the level of the gene itself, or at the level of the cytoplasmic gene product (enzyme-forming system)?

In a short time both these questions would be answered.

Protein Templates

Evidence had now accumulated, from research on enzyme induction and other studies that proteins were built on specific templates and not by polyenzymic reactions (see Siekevitz[2], 1996). It was accepted that DNA carried the information for protein sequence but there was no experimental evidence that it was directly involved. This left RNA. Studies with cell fractions had indicated that ribosomes were essential for protein synthesis but had shed no light on the synthesis of specific enzymes. The kinetic experiments introducing the *lacz* gene into a constitutive recipient had shown that the time before β-galactosidase was synthesized at the maximal rate was less than 2 min. If the template had a been a stable intermediate it was predicted that it would accumulate slowly in the cytoplasm and that enzyme synthesis would increase gradually.

When the donor strain was labeled with ^{32}P it was found that the radioactive zygotes suffered a progressive loss of enzyme-forming capacity due to the decay of the incoming genetic material. Riley et al. (1960) concluded that the template must be an unstable intermediate and that neither ribosomes nor the known soluble RNA fractions could fulfill this role. It is interesting to note that in 1956 Gale produced evidence that an unstable RNA was required for the synthesis of β-galactosidase by *Staphylococcus aureus.*

THE OPERON MODEL: 1961

In the historic review, "Genetic Regulatory Mechanisms in the Synthesis of Proteins", Jacob and Monod 1961 presented the genetic and biochemical evidence for negative control of gene expression. The core of this model was that: a *repressor* produced by a *regulator gene* prevented the expression of *one or more structural genes* by binding to an *operator* site; synthesis of the *structural gene product* took place in the cytoplasm; *messenger RNA* was the intermediate in the *transcription* of information from the gene to the site of protein synthesis; transcription was initiated from the *operator* site; and several linked genes formed an *operon* under common regulation. The main questions still to be decided were about the chemical nature of the repressor and whether it acted at the level of transcription or translation. They introduced the background to the operon model by saying that:

> The concept of the *structural gene* accounts for the multiplicity, specificity, and genetic stability of *protein structures*, and it implies that such structures are not controlled by environmental conditions or agents. It has been known for a long time, however, that the *synthesis* of individual proteins may be provoked or suppressed within a cell, under the influence of specific external agents.

Although the studies on β-galactosidase synthesis had been central to the development of these ideas, they suggested that there was enough evidence from other systems to provide a universal model for the control of protein synthesis. It was envisaged that in systems repressed by end-products, such as amino acid biosynthetic pathways, the regulator gene produced a molecule, the apo-repressor, which became an active repressor when combined with the end-product. For inducible systems, such as β-galactosidase, it was considered that the repressor itself prevented gene expression and was removed by combination with the inducer. They put forward the two alternative forms of the operon model shown in Figure 6.

In Model 1 the repressor acts at the gene level preventing transcription of the messenger RNA. In Model 2 the repressor acts at the translation level. They preferred Model 1 in which the repressor binds to an operator site adjacent to the structural genes. Synthesis of inducible enzymes, such as β-galactosidase, would occur when an inducer was bound to the repressor molecule removing it from the operator. In Model 2 the repressor binds to a polycistronic messenger at the level of translation. At that time it appeared that either model would be compatible with inducible systems, with λ-lysogeny and with end-product repression of biosynthetic enzymes. They concluded that negative control by repression was probably a general system for regulating protein synthesis.

In 1961 messenger RNAs and repressors were still hypothetical entities, but it was not long before they achieved physical reality. A rapidly renewed RNA fraction was identified in *E. coli* which fulfilled the requirement for the chemical messenger and became known as mRNA (Gros et al., 1961). Although there was some criticism of the experimental evidence from traditional biochemists, the significance of mRNA was not in doubt for long. This was indubitably the molecule that carried the information about protein sequence from the DNA of the gene via RNA to the site of protein synthesis in the cytoplasm.

Jacob said in 1979 that he thought that the model had "withstood the test of time and of deeper biochemical analysis with one exception." They had decided that the repressor was RNA and not protein, partly because some experiments had appeared to show that protein synthesis was not involved and partly because it seemed difficult to envisage the specific interaction required for DNA binding. The Lac repressor protein was isolated by Gilbert[2] and Müller-Hill in 1966. In *The Lactose Operon* they describe the ingenious methods they used. First, from extracts of inducible strains they obtained a

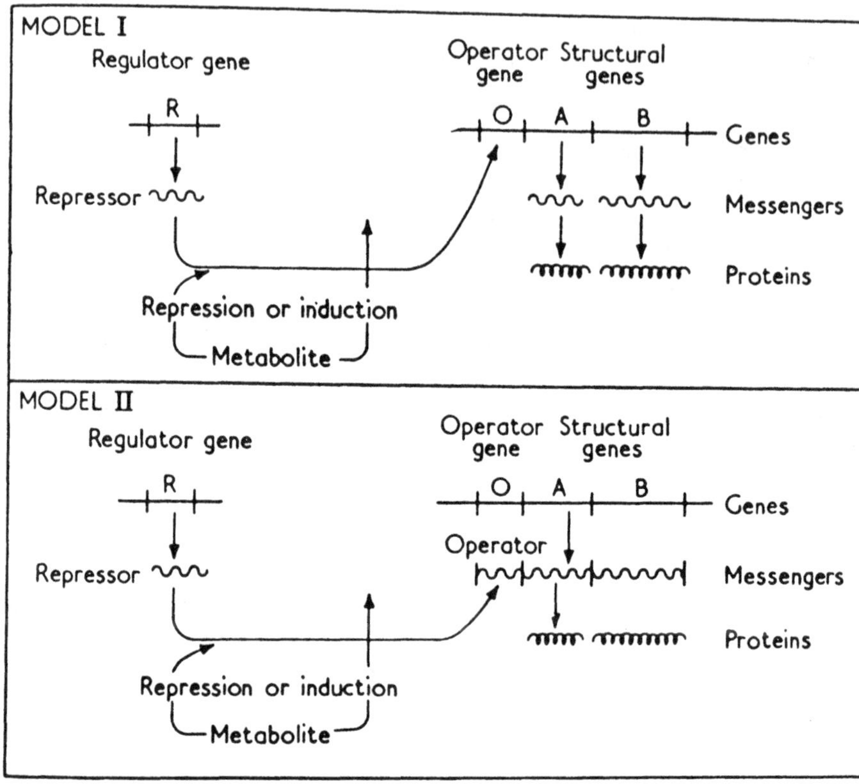

Figure 6. The operon models for the regulation of protein synthesis, 1961. Model I proposed that a regulator gene produced a repressor that acted directly at an operator site at the gene level and prevented the synthesis of cytoplasmic messengers and proteins. Model II proposed that the repressor acted at an operator site on a polycistronic messenger, coding for the proteins. Available evidence indicated a genetic site for the operator. It can be seen that both models indicate the active repressor is RNA and not protein (from Jacob and Monod, 1961).

protein that could bind to the inducer IPTG (isopropyl β-D-thiogalactoside). This protein was not found in constitutive strains. Wild-type cells produced very low amounts of repressor; about 0.002% of the cell protein (about 10 copies per genome). They then isolated i^{Q} mutants that produced 10-fold more repressor. To get even more repressor they attached the mutant *lac* genes to a defective phage that could be triggered by heat to multiply within the cell. By this means they raised the amount of repressor to 2.5% of cell protein.

These experiments provide an early example of a mutation within a weak promoter to give a stronger promoter and of increased yield of protein as a result of gene multiplication. It was very satisfactory to confirm that the Lac repressor molecule carried regions for specific inducer and DNA binding as had been predicted. The active form of the repressor is a tetramer and the inducer acts as an allosteric effector by changing the conformation in such a way that its affinity for the DNA binding site is decreased. Among the i^- mutants isolated at this time were some producing proteins that were unable to bind the inducer and they behaved as i^- dominant. Looking back, it seems just as well that these mutants had not turned up earlier when the experiments were carried out that indicated transdominance of the wild-type i^+. Shortly after the isolation of the *lac* repressor Ptashne (1967) isolated the λ-repressor.

An experiment in which part of the lactose operon was fused by deletion of intervening sequences illustrated the specificity of the DNA binding site for its regulator protein (Jacob et al., 1965). On one side the deletions extended into the *z* gene and on the other side into part of the operon for the biosynthesis of purines. These deletions resulted in new hybrid operons and with several of these the β-galactoside permease and the transacetylase were no longer induced by lactose but were repressed by purines. Gene fusions *in vitro* were to provide a valuable tool for testing gene expression. Figure 7 is a slightly modified version of the original Model 1 and it can be seen that while the operator remains the site of repressor binding the the site at which transcription is initiated is identified as the promoter.

The picture was rounded off by the isolation of purified *lac* operon DNA. (Shapiro et al., 1969). The *lac* genes were fused with bacteriophages λ and ϕ80 to increase the amount of DNA produced. The DNA of the two phages was separated into single strands, hybridized, and the nonhomologous single-stranded ends removed. The resulting double-stranded DNA was the purified *lac* operon that could be seen under the electron microscope (Figure 8).

OTHER OPERONS

The Lac operon model was so satisfying and elegant that it dominated thinking about the regulation of gene expression for a long time. Almost everyone interested in bacterial enzymes had accepted that negative control by repressors of structural genes could be a universal method for regulation of protein synthesis. It was tacitly assumed that most, if not all, inducible and repressible enzyme systems were similar to that for β-galactosidase. However, it is interesting to reflect that Monod had originally seen the inducer playing a positive role in enzyme induction. Before becoming committed to β-galactosidase he had been interested in amylomaltase (Monod and Torriani, 1950). In 1963 Maxime Schwartz returned to this problem. The structural genes

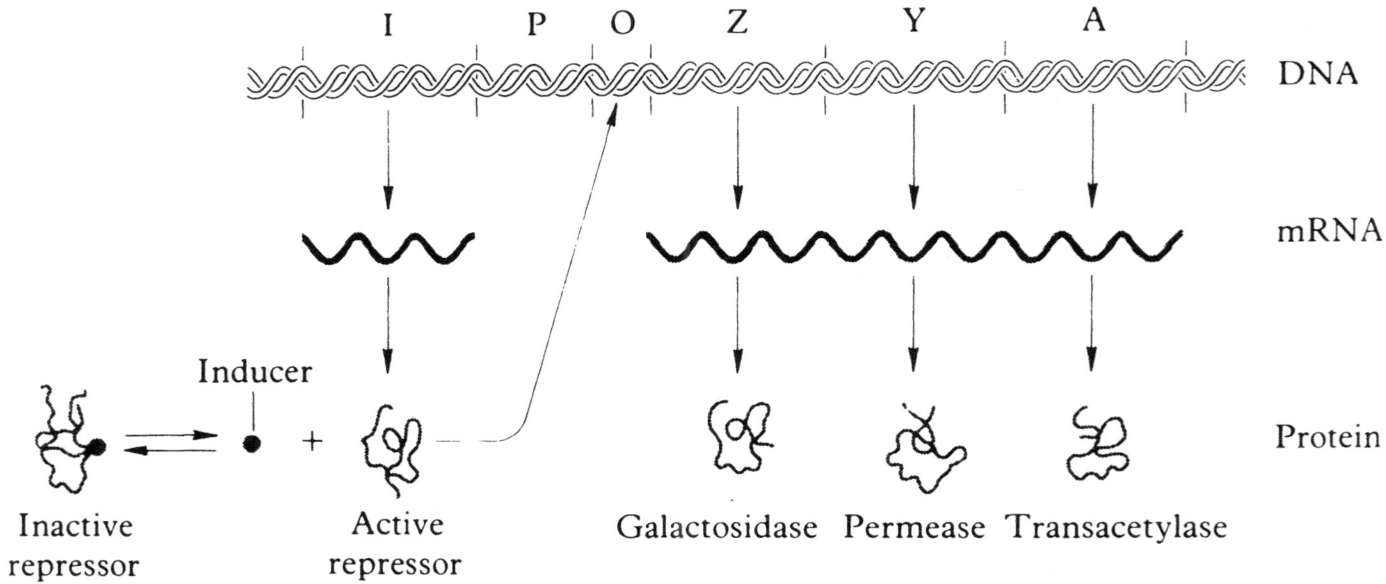

Figure 7. The classical *lac* operon. The model for the lactose operon, derived from the classical experiments of Jacob and Monod (1961). The *lac 1* gene produces a repressor protein that binds to the *lac* operator site and prevents transcription of the *lac* structural genes coding for the polycistronic messenger RNA for the synthesis of the β-galactosidase, permease, and transacetylase proteins. The inducer binds to the repressor and by removing it from the operator allows gene expression to occur (from Stent and Calendar, 1978).

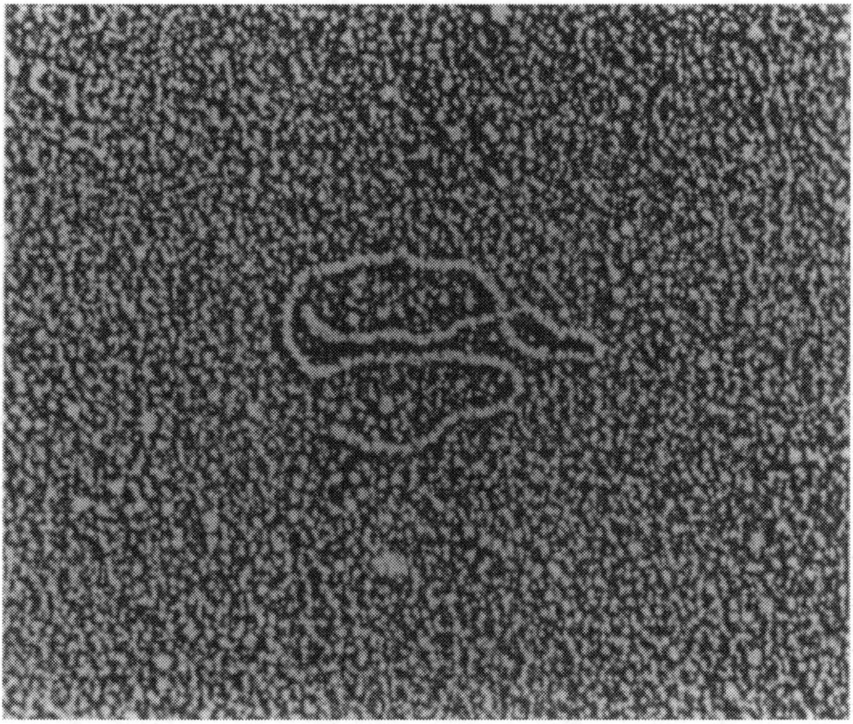

Figure 8. Electron micrograph of *lac* operon DNA. The DNA of *lac* genes, fused to bacteriophages λ and ϕ80 respectively was isolated and separated into single strands and allowed to rehybridize. The heteroduplex segments were treated with single-strand specific deoxyribonuclease from *Neurospora*. The single-stranded ends were completely degraded leaving pure *lac* duplex DNA (Shapiro et al., 1969) (Magnification not determined).

for amylomaltose, phosphorylase, and permease appeared to form a single operon but no constitutive i^- type mutants turned up. From the properties of the various *mal* mutants that did appear it was suspected that the *mal* operon might be under positive control. Since *laci*$^-$ mutants were pleiotropically constitutive it could be predicted that regulator mutants of an operon regulated by positive control would be pleiotropically negative. The problem was that there were many other classes of negative mutants possible and it took a long time before Monod was convinced of the existence of the positive regulator gene *malT* (Schwartz, 1979).

This was not the only operon out of step. At the same time Ellis Englesberg was working on the regulation of L-arabinose enzymes in *E. coli B*. He obtained

Table 5. Substrates and Inducers of Amidase of *Pseudomonas aeruginosa*[a,b]

Amide	*Inducer*	*Substrate*	*Repressor*	*Growth*
Formamide	+	+	+	0
Acetamide	+++	+++	·	+
Propionamide	+++	+++	·	+
Butyramide	-	(+)	+	0
Lactamide	+++	+	·	+
N-methylformamide	+++	-	·	0
N-acetylacetamide	+++	-	·	0
Cyanoacetamide	-	-	+	0
Acrylamide (I)[c]	+	++	·	0

Notes: [a] Data from Kelly and Clarke (1962) and Clarke and Drew (1988).
[b] The relative activities as inducers or substrates are shown as +, ++, +++. Some amides compete with inducers and repress enzyme synthesis.
[c] (I) Hydrolysis of acrylamide produces acrylic acid which inhibits growth.

evidence that the genes for epimerase, isomerase, and ribulokinase formed an operon under positive control by *araC*. Again it was some time before this gained general acceptance (Englesberg et al., 1969). It was a complex system since the *araC* protein has a dual role acting at one site as a repressor and being converted to an activator by combination with the inducer. *E. coli* had become the most widely used organism for studies on the regulation of enzyme synthesis and it was important to ask how other species of bacteria regulated enzyme synthesis.

The *Pseudomonas aeruginosa* enzyme that has been studied in most detail is the inducible aliphatic amidase that enables the organism to grow on acetamide (Kelly and Clarke, 1962). A large number of acetamide analogues were available for experimentation and it was soon established that substrate specificity was distinct from inducer specificity and that some amides competed with inducing amides to repress amidase synthesis (Table 5).

Since amides have the potential to provide both carbon and nitrogen sources for growth it was possible to devise a large variety of media for mutant selection. This produced various classes of constitutive mutants as well as mutants that were altered in substrate or inducer specificities. A trawl through several hundred regulator mutants did not identify any that were inducible at higher temperatures and constitutive at lower temperatures as had been found for β-galactosidase. Since it appeared that there were no mutants producing temperature-sensitive repressors it was decided to look for mutants with temperature-sensitive activator proteins. Among the collection of acetamide-negative mutants selected for inability to grow on acetamide at 37 °C were several that were acetamide-positive at 30 °C (Farin and Clarke, 1978). The amidase of these strains was no less thermostable than the wild-type enzyme. Kinetic studies on enzyme synthesis showed that only trace amounts of amidase

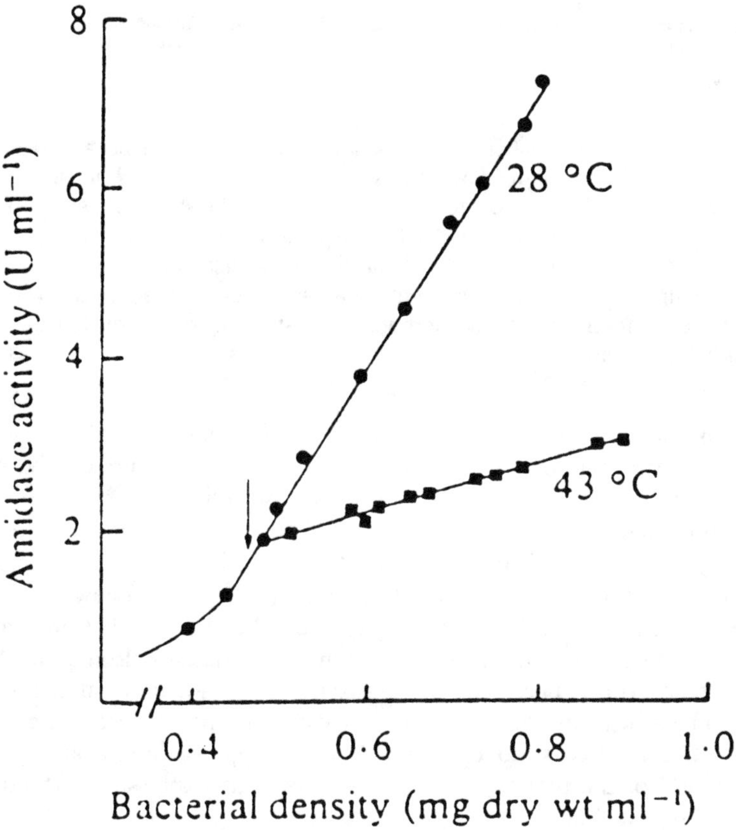

Figure 9. Positive control of *Pseudomonas aeruginosa* amidase. The *P. aeruginosa* temperature-sensitive mutant PAC438 was grown in a minimal medium at 28 °C. At the time indicated by the arrow the culture was divided and one part transferred to 43 °C. Growth continued but amidase synthesis ceased (data from Farin and Clarke, 1978).

were produced at high temperatures and that when the growth temperature was reduced there was short lag before the enzyme was synthesized at the normal rate (Figure 9). Genetic evidence, including the demonstration that constitutivity was transdominant over inducibility, confirmed that the structural gene *amiE* was controlled by an activator protein coded by the regulator gene *amiR* (Cousens et al., 1987). However, the amidase operon turned out to be more complex than this. Experiments with cloned gene segments, together with DNA sequence studies, identified an additional negative regulator gene, *amiC* (Figure 10). In the uninduced state the AmiC

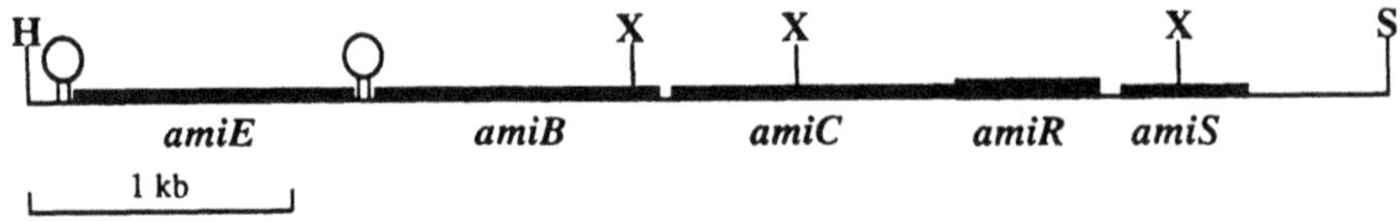

Figure 10. Genetic organization of the amidase operon of *P. aeruginosa.* The amidase structural gene, *amiE*, is controlled by the positive regulatory gene, *amiR*, and the negative regulator gene, *amiC*. The AmiC protein complexes with the AmiR protein to prevent activation of gene transcription. The AmiC protein carries the inducer-binding site and when an inducing amide is present the AmiR protein is released. The genes, *amiB* and *amiS*, are thought to code for components of an amide transporter system. The stem-loop structures represent transcription terminators. Restriction enzyme sites are as follows: *H* = *Hind*III; *X* = *Xho*I; *S* = *Sal*I (Wilson et al., 1995).

protein complexes with and inactivates the AmiR protein.The AmiC protein carries the inducer-binding site and when an inducing amide is present the AmiR protein is released. Positive control of transcription is mediated by a transcription–anti-terminating mechanism, whereby the AmiR protein allows the RNA polymerase to read through an upstream rho-independent transcription terminator and through into the structural genes (Wilson and Drew, 1995). The amidase operon includes two further genes, *amiB* and *amiS* that are thought to code for components of a membrane transport system (Wilson et al., 1995).

Positive control of bacterial gene transcription is not unusual and is now known to be at least as common as negative control, particularly for complex systems. The relative simplicity of the lactose operon contributed to the establishment of the first significant model for gene expression and all later systems were tested against it.

COMPLEX OPERON REGULATION

In the decades after the formulation of the operon theory there were many investigations of the regulation of protein synthesis in microorganisms and increasingly in higher organisms. The structure of complex operons and some of the molecular events in gene expression known in 1979 were described by contributors to *Biological Regulation and Development* edited by Goldberger (1979) and in *Regulation of Gene Expression–25 Years On* edited by Booth and Higgins (1986).

Bacterial Variants

In contrast to the relative simplicity of the regulatory control of the *lac* operon it soon became clear that in bacteria there existed many variants on the theme of induction and repression, transcription, and translation. Repression of tryptophan synthetase was one of the earliest examples of end-product repression of a biosynthetic enzyme. Further studies on the biochemistry and genetics of the tryptophan pathway in *E. coli* showed that all the genes for the enzymes converting chorismate to tryptophan formed a single linkage group. The structural genes are transcribed as a single messenger RNA under the control of the unlinked tryptophan regulator gene.

In addition to the primary promoter adjacent to *trpE* there is a secondary promoter that controls the last three enzymes. Different patterns of regulation of tryptophan biosynthesis occur in other bacterial genera. For example, in *Pseudomonas aeruginosa* the *trp* genes occur in three separate linkage groups widely scattered around the chromosome. Only the early enzymes are repressed by tryptophan. Tryptophan synthetase is induced by its substrate, indole glycerol phosphate, and the α-subunit of tryptophan synthetase acts as the repressor of both its own synthesis and the β-subunit.

Catabolic Pathways

Many complex regulatory systems have been identified for catabolic pathways, particularly in bacteria able to utilize a large number of organic compounds as growth substrates. For example, it was found that in several species the initial catabolic pathways for different aromatic compounds converge to produce a common intermediate such as catechol. These include salicylate, phenol, benzoate, anthranilate, mandelate, and benzene, all of which may occur in the natural environment. Stanier and colleagues established that in *Pseudomonas putida* catechol is converted to succinate and acetyl CoA via the β-ketoadipate pathway. This organism is also able to utilize intermediates of the pathways and the regulatory systems are correspondingly complex. D-mandelate is converted to benzoate by four enzymes induced by D- and L-mandelate and the intermediate phenylglyoxylate. Benzoate induces the enzyme benzoate oxidase which converts it to catechol. However, the inducer of catechol 1,2-oxygenase is its product, *cis,cis*-muconate, which also induces the next two enzymes of the pathway. The last two enzymes of the β-ketoadipate pathway are induced by the product, β-ketoadipate (Ornston, 1971). Although genes for the aromatic pathways may be located at different sites around the chromosome in the *Pseudomonas* species, in some cases they have common regulatory genes.

Many *Pseudomonas* species carry catabolic genes on plasmids, some of which are transmissible. These degradative plasmids were first identified when single mutations appeared to result in the loss of an entire pathway. Plasmid loss may be spontaneous and in nature it appears that at any one time plasmids may be present in only a small proportion of the bacterial population. This is an economical way of gene regulation since when the appropriate substrates appear in the environment the plasmid genes may be expressed and transferred to other bacteria. Among the plasmids that have been reported are CAM (camphor), OCT (octane), SAL (salicylate), NAH (naphthalene), and TOL (toluate). An alliance of plasmid and chromosomal genes can initiate a novel catabolic pathway.

Catabolite Repression

Another problem left over from earlier years was the *glucose effect*. Monod had observed diauxic growth of *E. coli* grown in the presence of both glucose and lactose. Gale had described ways in which glucose affected the synthesis of a number of different enzymes. There were many other reports of enzymes being repressed by the addition of various carbon compounds to growing cultures. This appeared to be a general phenomenon involving compounds other than glucose and Magasanik suggested it should be described as *catabolite repression*. He considered that in the regulation of enzyme synthesis there was a balance between induction and catabolite repression and that both acted at the level of transcription (Nakada and Magasanik, 1964). Catabolite repression is still not fully understood (see Booth and Higgins, 1986) but at least one specific mechanism has been established. The finding that cyclic AMP could overcome repression of β-galactosidase synthesis by glucose led to the identification of the cyclic AMP binding protein (CAP) as a positive agent for the transcription of many catabolic genes (Zubay et al., 1970).

MANIPULATING GENES

One of the unexpected results of the early research on enzyme adaptation was the establishment of biotechnology as an important industry. The key factor in most of those developments has been the ability to remove genes from one organism and to insert them into a suitable host in which they can be expressed at a high level. The early studies with *lac* genes fused to plasmids or prophages and the identification of derepressed and up-promoter mutants were obvious forerunners of these later developments. In some cases it is still useful to fuse the β-galactosidase gene to the structural and regulatory genes under study and to use a chromogenic galactoside to test for replication and expression.

Continued research on the regulation of enzyme synthesis, together with developments in cloning and DNA sequencing methods, has made it possible to examine gene structure and the initation and termination of gene expression at the molecular level. Discussions of gene expression in both prokaryotes and eukaryotes moved rapidly to the consideration of molecular interactions between cell components. It became possible to begin to fulfill the aim expressed in the 1961 review to provide "a valuable model for the interpretation of biochemical coordination within tissues and between organs in higher organisms." The insight gained from the study of microbial systems is now being applied to unravel the complexities of gene expression in the development and differentiation of plants and animals.

REFERENCES

Avery, O.T., MacCleod, C.M., & McCarty, M. (1944). Studies on the chemical nature of the substance inducing transformation of pneumococcal types. Induction of transformation by a desoxyribonucleic acid fraction isolated from Pneumococcus Type III. J. Exp. Med. 79, 137-158.

Beadle, G.W. (1974). Recollections. Annu. Rev. Biochem. 43, 1-13.

Beadle, G.W. & Ephrussi, B. (1936). Eye color in *Drosophila*. Genetics 21, 225-247.

Booth, I.R. & Higgins, C.F. (Eds.) (1986). Regulation of Gene Expression: 25 Years On. 39th Symposium of the Society for General Microbiology. Cambridge University Press.

Clarke, P.H. (1981). Adaptation. J. Gen. Microbiol. 126, 5-20.

Clarke, P.H. & Drew, R.E. (1988). An experiment in enzyme evolution. Studies with *Pseudomonas aeruginosa* amidase. Biosci. Rep. 8, 103-120.

Cohen, G.N. & Monod, J. (1957). Bacterial permeases. Bacteriol. Rev. 21, 169-194.

Cohn, M., Cohen, G., & Monod, J. (1953) The specific inhibitory effect of methionine in the formation of methionine synthetase of *Escherichia coli.* C.R.H. Acad. Sci. (Paris) 236, 746-748.

Cousens, D., Clarke, P.H., & Drew, R.E. (1987). The amidase regulatory gene (*amiR*) of *Pseudomonas aeruginosa.* J. Gen. Microbiol. 133, 2041-2052.

den Dooren de Jong. (1926). See Stephenson (1939). Bacterial Metabolism, pp. 190-198.

Dubos, R.J. (1945). The Bacterial Cell. Harvard University Press, Cambridge, MA.

Englesberg, E., Irr, J., Power, J., & Lee, N. (1969). Positive control of enzyme synthesis by gene C in the L-arabinose system. J. Bacteriol. 90, 946-957.

Farin, F. & Clarke, P.H. (1978). Positive regulation of amidase synthesis. J. Bacteriol. 135, 379-392.

Fildes, P., Gladstone, G.P., & Knight, B.C.J.G. (1933). The nitrogen and vitamin requirements of *B. typhosus.* Brit. J. Exp. Pathol. 14, 189-196.

Fox, C.F., Carter, J.R., & Kennedy, E.P. (1967). Genetic control of the membrane protein component of the lactose transport system of *Escherichia coli.* Proc. Nat. Acad. Sci. USA 57, 698-705.

Gale, E.F. (1943). Factors influencing the enzyme activities of bacteria. Bacteriol. Rev. 7, 139-173.

Gale, E.F. (1956). Nucleic acids and enzyme synthesis. In: Enzymes Units of Biological Structure and Function (Gaebler, O.H., Ed.). Academic Press, New York.

Gale, E.F. & Davies, R. (Eds.) (1953). Adaptation in Microorganisms. 3rd Symposium of the Society for General Microbiology. Cambridge University Press.

Garrod, A.E. (1923). Inborn Errors of Metabolism, 2nd ed. Henry Froude, London.

Gilbert, W. & Müller-Hill, B. (1966). Isolation of the Lac repressor. Proc. Nat. Acad. Sci. USA 56, 1891-1898.

Goldberger, R.F. (Ed.) (1979). Biological regulation and development. Gene Expression, Vol. 1. Plenum Press, New York.

Gros, F., Gilbert, W., Hiatt, H., Attardi, G., Spahr, P.F., & Watson, J.D. (1961). Molecular and biological characterization of messenger RNA. Cold Spring Harbor Symp. Quant. Biol. 26, 111-132.

Gutfreund, H. (Ed.) (1976). Enzymes: One hundred years. FEBS Lett. 62, Suppl. 4.
Hayes, W. (1953). The mechanism of genetic recombination in *Escherichia coli.* Cold Spring Harbor Symp. Quant. Biol. 18, 75-93.
Hayes, W. (1968). The Genetics of Bacteria and their Viruses, 2nd ed. Blackwell Scientific Publications, Oxford.
Hershey, A.D. & Chase, M. (1952). Independent functions of viral protein and nucleic acid in growth of bacteriophage. J. Gen. Physiol. 36, 39-56.
Hinshelwood, C. (1946). The Chemical Kinetics of the Bacterial Cell. Oxford University Press.
Hogness, D.S., Cohn, M., & Monod. J. (1955). Studies on the synthesis of β-galactosidase in *Escherichia coli:* The kinetics and mechanism of sulphur incorporation. Biochim. Biophys. Acta 16, 99-116.
Jacob, F. (1979). The switch. In: The Origins of Molecular Biology (Lwoff, A. & Ullman, A., Eds.). Academic Press, New York.
Jacob, F. & Monod, J. (1961). Genetic regulatory mechanisms in the synthesis of proteins. J. Mol. Biol. 3, 318-356.
Jacob, F. & Monod, J. (1970). In: The Lactose Operon (Beckwith, J.R. & Zipser, D., Eds). Cold Spring Harbor Press.
Jacob, F. & Wollman, E. (1953). Induction of phage development in lysogenic bacteria. Cold Spring Harbor Symp. Quant. Biol. 18, 101-121.
Jacob, F. & Wollman, E. (1957). Genetic aspects of lysogeny. In: The Chemical Basis of Heredity (McElroy, W.D & Glass, B., Eds.), pp. 468-499. Johns Hopkins Press, Baltimore.
Jacob, F. & Wollman, E. (1958). Genetic and physical determinations of chromosomal segments in *Escherichia coli.* Symp. Soc. Exp. Biol. 12, 75-92.
Jacob, F., Ullman, A., & Monod, J. (1965). Deletions fusing the lactose operon to a purine operon in *Escherichia coli.* J. Mol. Biol. 13, 704-719.
Jacob, F., Perrin, D., Sanchez, C., & Monod, J. (1960). The operon: A group of genes whose expression is coordinated by an operator. Compt. Rend. 250, 1727-1729.
Karlsson, J.L. & Barker, H.A. (1948). Evidence against the occurrence of a tricarboxylic acid cycle in *Azotobacter agilis.* J. Biol. Chem. 175, 913-921.
Karstrom, H. (1930). See Stephenson, M. (1939). Bacterial Metabolism, p. 304.
Kelly, M. & Clarke, P.H. (1962). An inducible amidase produced by a strain of *Pseudomonas aeruginosa.* J. Gen. Microbiol. 27, 305-316.
Kluyver, A.J. (1931). The Chemical Activities of Microorganisms. University of London, London, UK.
Lederberg, J. & Tatum, E.L. (1946). Gene recombination in *Escherichia coli.* Nature (London) 158, 558.
Lwoff, A. & Ullman, A. (Eds.) (1978). Selected papers in Molecular Biology by Jacques Monod. Academic Press, New York.
Lwoff, A. & Ullman, A. (Eds.) (1979). Origins of Molecular Biology: A Tribute to Jacques Monod. Academic Press, New York.
Massini, R. (1907). See Stephenson (1939). Bacterial Metabolism, p. 301.
Monod, J. (1941). Recherches sur la Croissance des Cultures Bactériennes. Thesis, Hermann Edition, Paris.
Monod, J. (1947). The phenomenon of enzymatic adaptation and its bearings on problems of genetic and cellular differentiation. Growth Symp. 11, 223-289.
Monod, J. (1949). The growth of bacterial cultures. Annu. Rev. Microbiol. 3, 371-394.
Monod, J. (1956). Remarks on the mechanism of enzyme induction. In: Enzymes Units of Biological Structure and Function, pp. 7-8. Academic Press, New York.
Monod, J. & Audureau, A. (1946). Mutation and enzymatic adaptation in *Escherichia coli mutabile.* Annales Institut Pasteur (Paris) 72, 868-878.

Monod, J. & Cohen-Bazire, G. (1953). The specific inhibition effect in the biosynthesis of tryptophan-desmolase in *Aerobacter aerogenes*. Compt. Rend. 236, 530-532.

Monod, J. & Cohn, M. (1952). The induced biosynthesis of enzymes (enzymatic adaptation). Adv. Enzymol. 13, 67-119.

Monod, J. & Torriani, A.M. (1950). On the amylomaltase of *Escherichia coli*. Annales Institut Pasteur (Paris) 78, 65-77.

Nakada, D. & Magasanik, B. (1964). The roles of inducer and catabolite repressor in the induction of β-galactosidase. Biochim. Biophys. Acta 61, 835-837.

Ornston, L.N. (1971). Regulation of catabolic pathways in *Pseudomonas*. Bacteriol. Rev. 35, 87-116.

Pardee, A., Jacob, F., & Monod, J. (1959). The genetic control and cytoplasmic expression of "inducibility" in the synthesis of β-galactosidase by *E. coli*. J. Mol. Biol. 1, 165-178.

Ptashne, M. (1967). Isolation of the λ phage repressor. Proc. Nat. Acad. Sci. USA 57, 306-313.

Riley, M., Pardee, A.B., Jacob, F., & Monod, J. (1960). On the expression of a structural gene. J. Mol. Biol. 2, 216-225.

Schwartz, M. (1979). In: The Origins of Molecular Biology (Lwoff, A. & Ullman, A., Eds.). Academic Press, New York.

Senez, J.C. (1968). General Microbiology. Deren et Cie, Paris.

Shapiro, J., MacHattie, L., Eron, L., Ihler, G., Ippen, K., & Beckwith, J. (1969). The isolation of pure *lac* operon DNA. Nature (London) 224, 768-774.

Siekevitz, P. (1996). In: Foundations of Modern Biochemistry (Ord, M.G. & Stocken, L., Eds.), Vol. 2, pp. 109-121. JAI Press, Greenwich, CT.

Stanier, R. (1947). Simultaneous adaptation: a new technique for the study of metabolic pathways. J. Bacteriol. 84, 339-348.

Stanier, R.Y., Palleroni, N., & Doudoroff, M. (1966). The aerobic pseudomonads; a taxonomic study. J. Gen. Microbiol. 43, 159-271.

Stent, G.S. & Calendar, R. (1978). Molecular Genetics, 2nd ed. Freeman, San Francisco.

Stephenson, M. (1939, 1947). Bacterial Metabolism. Longmans, Green, London.

Stephenson, M. & Stickland, L.H. (1932). Hydrogenlyases. Bacterial enzymes liberating molecular hydrogen. Biochem. J. 26, 712-724.

Stephenson, M. & Stickland, L.H. (1933). Hydrogenlyases. Further experiments on the formation of formic hydrogenlyase by *Bact. coli*. Biochem. J. 27, 1528-1532.

Vogel, H.J. (1957). Repression and induction as control mechanisms of enzyme biogenesis: the adaptive formation of acetylornithinase. In: The Chemical Basis of Heredity (McElroy, W.D. & Glass, B., Eds.). Johns Hopkins Press, Baltimore.

Watson, J.D. & Crick, F.H.C. (1953). Genetical implications of the structure of deoxyribonucleic acid. Nature (London) 171, 964-969.

Wilson, S.A. & Drew, R.E. (1995). Transcriptional analysis of the amidase operon from *Pseudomonas aeruginosa*. J. Bacteriol. 197, 3052-3057.

Wilson, S.A., Williams, R.J., Pearl, L.H., & Drew, R.E. (1995). Identification of two new genes in the *Pseudomonas aeruginosa* amidase operon, encoding an ATPase (AmiB) and a putative integral membrane protein (AmiS). J. Biol. Chem. 270, 18818-18824.

Wollman, E.L., Jacob, F., & Hayes, W. (1956). Conjugation and genetic recombination in *Escherichia coli* K-12. Cold Spring Harbor Symp. Quant. Biology 21, 141-162.

Yudkin, J. (1938). Enzyme variation in microorganisms. Biol. Rev. 13, 93-106.

Zinder, N.D. & Lederberg, J. (1952). Genetic exchange in *Salmonella*. J. Bacteriol. 64, 679-699.

Zubay, G., Schwartz, D., & Beckwith, J. (1970). Mechanism of activation of catabolic-sensitive genes: A positive control system. Proc. Nat. Acad. Sci. USA 66, 104-110.

Chapter 9

ANTIBODY SPECIFICITY AND DIVERSITY: THE PROTEINS (PART I)*

Lisa A. Steiner

* Part II, The Genes, will appear in Volume 4 of this series

INTRODUCTION

This chapter is concerned with the historical development of our understanding of the induced immune response, in particular with its central feature: the ability to make antibodies that react specifically with only one or a few among an indefinitely large number of naturally-occurring or artificial substances (antigens). Although the discriminating power of antibodies was well documented in the first half of this century, substantial progress in understanding the biological basis for their incredible specificity has only been made in the last half of the century. Beginning in the 1950s, the molecular nature of antibodies and structural features accounting for their diversity were described. These results formed the background for investigations of the genetic basis for antibody synthesis, which revealed a novel mechanism for generating the requisite diversity. More recently, the identity of the receptors for antigen on T lymphocytes was established, receptors that are remarkable both for their similarity to and their differences from antibodies or B-lymphocyte receptors.

The focus here is on studies of antibodies, their overall structure and the means by which recognition of diverse antigens is achieved. Mentioned only briefly or not included at all are other topics of much current interest such as antigen recognition by T cells[a], signalling from antigen to B or T cells, differentiation of B and T cells, and regulation of the immune response, including tolerance, to indicate just a few omissions. This review has been divided into two parts. Part I, in this Volume, is concerned with hypotheses about the origin of antibodies, with the structural basis for antibody specificity, and with ensuing speculations about possible means to account for the production of all the diverse antibodies. Part II, to be included in Volume 4 of this series, will focus on the extraordinary mechanism by which a limited number of genes can generate an apparently unlimited number of antibodies.

Further discussion of many of the subjects considered here can be found in the following sources. Debra Bibel (1988) presents invaluable excerpts, translated into English where appropriate, with commentary, of many of the classical papers by the pioneers in immunology. Arthur Silverstein's monograph (1989) consists of a number of thoughtful essays on the historical development of major concepts in immunology. Gallagher et al. (1995) have collected a series of essays written, with a historical slant, by some of the major contributors in the field; these lend a personal flavor and sense of excitement that can sometimes be hard to glean from the literature. The series, *Annual Review of Immunology* (1983-present), contains autobiographical essays that often present a body of work in a personal historical context.

The language of immunology can be an obstacle to a reader from another discipline. Many of the specialized terms are explained in the text; additional definitions and more details can be found in dictionaries and encyclopedias (e.g. Rosen et al., 1989; Roitt et al., 1992; Cruse and Lewis, 1995).

BEGINNINGS

Tracing the origin of a discipline, like that of a species, both problems in evolution, can be a perplexing task. Still, many immunologists would agree that the beginning of at least the modern chapter of their subject dates from work of a number of investigators in the latter years of the nineteenth century. Indeed, if one had to choose a single discovery that accelerated the transformation of the field from a collection of observations into a scientific discipline, a likely candidate would be the demonstration by von Behring[b] and Kitasato (1890) that an unimmunized animal could be made resistant to tetanus toxin by transfer of serum from an actively immunized donor. It would now be possible, at least in principle, to fractionate the serum and to identify and characterize the biochemical entity responsible for the resistance. In fact, it was almost 50 years before significant progress toward this goal was achieved. It was also at about this time that Elie Metchnikoff (1892) demonstrated that cells contribute to the state of immunity of an animal by virtue of their ability to phagocytize bacteria. These two discoveries launched the long-lasting debate whether cells or "humoral" factors are the more important weapons of the immune system. It is ironic that Robert Koch, whose institute in Berlin was the major center for advocates of the preeminence of humoral factors in immunity, discovered (in the guise of a cure) the skin test for tuberculosis (Koch, 1891). Although not appreciated by Koch, who attributed the reaction to heightened toxicity of tuberculin in patients with the disease, this test was ultimately to become the textbook example of the cell-mediated immune response, now known to be caused by T lymphocytes.

Other important discoveries in these years include cell lysis in the presence of immune serum and complement (Bordet, 1898), and anaphylaxis (Portier and Richet, 1902). But perhaps the signal event that marked the beginning of the modern era in immunology was the delivery, on March 22, 1900, of the Croonian Lecture to the Royal Society of London by Paul Ehrlich[21] (1900). In this lecture[c], Ehrlich outlined a theory for the origin of antibodies; he proposed that cells contain receptors or "side-chains" that have a normal function in the cell (taking in "food-stuffs") and that toxins (antigens) are also recognized by these receptors by virtue of their structural resemblance to the normal ligand (i.e., a cross-reaction). As a result of the interaction with antigen the cell is stimulated to produce, in fact to overproduce, more receptor and the excess is shed to become circulating "Antikörper" (antibody). Today, Ehrlich's proposal for the origin of specific antibodies would be called a selective theory in that it contains the essential, and by no means obvious, principle that receptors (or antibodies) exist in the animal before introduction of antigen, the antigen serving merely to select and thereby enhance production of those particular antibodies that bind the introduced antigen.

Ascribing the beginning of the modern era of immunology to advances that occurred at the turn of the century is not meant to ignore the landmark contributions of the earlier pioneers such as Jenner and Pasteur[1]. However, their observations and experiments, indispensable as they were in establishing practical methods for inducing immunity, were not accompanied by an appreciation of their biological basis.

STATEMENT OF PROBLEM: INSTRUCTION AND SELECTION

At the 1967 Cold Spring Harbor Symposium, Niels Jerne (1967) defined two classes of immunologists, *cis* and *trans*. The former are primarily concerned with initial events in the immune response, such as the interaction of antigens with cells, or the basis for tolerance. In contrast, the attention of the *trans*-immunologists is focused on the antibody molecule and the structural basis for specificity. The *cis*-immunolgists gradually worked forward and the *trans*-immunolgists worked backward, until the distinction between the two groups became largely blurred. However, the division in the field did exist for many years, exemplified in the debates between the French school of "cellularists", led by Metchnikoff at the Pasteur Institute and the German school of "humoralists", disciples of Koch in Berlin (see Silverstein, 1989). Moreover, until almost the time of the Cold Spring Harbor meeting, the schism in the field had also been reflected in a debate as to how to explain the capacity of each individual to respond to the seemingly limitless number of antigens by the production of specific antibodies.

Two theories, instructive and selective, had been advanced to account for the existence of the large and diverse universe of antibodies. These theories differ crucially in the role ascribed to antigen. According to selective theories, antigen does not directly participate in the synthesis of antibody, its role being "merely" to stimulate the production of those antibodies, and only those, that are complementary to the antigen. Although, as noted above, a form of selective theory had already been advanced by Ehrlich over half a century earlier, in perhaps the first systematic attempt to account for production of specific antibodies, the idea that an individual is naturally endowed with the capacity for making antibodies specific for any conceivable antigen, including ones never actually encountered, seemed far-fetched to most chemically-minded (*trans*) immunologists. Consequently, the alternative or instructive theory held sway during the first half of this century.

Specificity by Instruction

The essence of all forms of instruction theory is that the synthesis of antibody absolutely requires the presence of antigen, at least in its initial phase. Early forms of such theories have been reviewed by Silverstein (1989). In 1930, Breinl and Haurowitz proposed that deposition of antigen in antibody-forming cells somehow modulates globulin synthesis so that a modified protein, the antibody, is produced instead of the normal globulin. As antibodies and ordinary globins were found to be similar in their overall chemical composition and physical properties, it was supposed that they differed only in the arrangement and spatial position of the amino acids. Similar views about the effects of antigen on antibody production were put forward by Alexander (1931) and Mudd (1932). Subsequently, Rothen and Landsteiner (1939), referring to earlier work on protein folding by Mirsky and Pauling (1936), suggested that different antibodies could result from variable folding of the same polypeptide chain. This idea was pursued in depth by Linus Pauling[4] (1940) who proposed that all antibody molecules are identical in amino acid sequence, differing from one another only in conformation, the result of variable folding of the same precursor globulin molecule around the antigen as a physical template. This idea seemed plausible since nothing was then known about the relation between primary structure and conformation of proteins. An additional hypothesis, that variable pairing of disulfide bonds stabilizes these folded states, was advanced by Karush (1958, 1962).

Instruction theories seemed particularly attractive following the demonstration by Landsteiner that virtually any arbitrary chemical group could induce formation of specific antibodies. By qualitatively evaluating the precipitation of antibodies against diverse hapten–protein conjugates[d] Landsteiner clearly illustrated the range and discriminating power of the antibody response; even substances as similar as *o*-, *m*-, and *p*-aminobenzoate could be distinguished (Landsteiner and Lampl, 1918; Landsteiner, 1919, 1936). Landsteiner also showed that "simple substances lacking in antigenic power" (i.e., haptens) could specifically inhibit precipitation; further, he noted that inhibition "can result from weak affinities which would not be sufficient for causing specific precipitation" (Landsteiner, 1920, 1936; Landsteiner and van der Scheer, 1931). It seemed inconceivable that antibodies directed against compounds not normally found in nature could preexist the introduction of that antigen. The direct molding of the antibody on the antigen seemed more plausible. Indeed, in his 1940 paper Pauling, noting Landsteiner's observations, speculated that the number of configurations accessible to a polypeptide chain could provide "specificity for an apparently unlimited number of different antigens."

The specificity of antibodies was explored quantitatively by Linus Pauling, David Pressman, and their colleagues in a series of papers published in the early 1940s. The precipitation of antiserum and homologous antigen was inhibited by a variety of structurally related haptens; the results were expressed as the ratio of concentrations of homologous to heterologous ligand required to achieve a given level of inhibition. An example of the data obtained is summarized in Table 1, based on Pauling and Pressman (1945). Antigens for immunization of rabbits were prepared by conjugating, via diazo linkage, either *o*-, *m*-, or *p*-aminobenzenearsonate to sheep serum proteins. To assay reactivity for the hapten alone and not the carrier protein, each antiserum was precipitated with the hapten used for immunization diazotized to a different protein, ovalbumin. The inhibition of precipitation by *o*-, *m*-, and *p*-aminobenzenearsonate diazotized to phenol was determined. It is clear that each antiserum discriminates among haptens that differ only in the position of the arsonate group on the benzene ring relative to the diazo linkage. The results were interpreted in terms of the intermolecular forces operating between the hapten inhibitor and the antibody-combining site. The possible contributions of van der Waals attraction, hydrogen bond formation, and steric factors were evaluated and discussed as a basis for inferences about the structure of the site.

Table 1.[a]

Hapten Inhibitor[c]	Relative Inhibition[b]		
	Anti-o-aba-sh + o-aba-ova[d]	Anti-m-aba-sh + m-aba-ova[e]	Anti-p-aba-sh + p-aba-ova[f]
o-aba-phenol	1	0.16	0.021
m-aba-phenol	0.022	1	0.29
p-aba-phenol	0.0005	0.20	1

Notes: [a] Data abstracted from Tables I, III, and V in Pauling and Pressman (1945). The columns show the extent of inhibition, by each of the three haptens, of the antigen–antibody precipitates indicated.

[b] The ratio of concentrations of homologous to heterologous hapten inhibitor required to reduce the amount of precipitation by approximately one-half. The data have been normalized so that the relative inhibition by the homologous hapten has the value 1.

[c] Haptens are o-, m-, and p-azobenzenearsonate-phenol.

[d] Precipitate formed by antibodies to o-benzenearsonate-sheep serum proteins and o-benzenearsonate-ovalbumin.

[e] Precipitate formed by antibodies to m-benzenearsonate-sheep serum proteins and m-benzenearsonate-ovalbumin.

[f] Precipitate formed by antibodies to p-benzenearsonate-sheep serum proteins and p-benzenearsonate-ovalbumin.

Nevertheless, simple and appealing as it was, the Pauling template hypothesis posed problems for both *trans*- and *cis*-immunologists. The former puzzled over the inability of the theory to explain why both (or all) the antigen-combining sites of an antibody molecule have the same specificity. It was generally (although not universally) assumed that the formation of antigen–antibody precipitates indicated that both antigen and antibody must be at least bivalent. Indeed, it had been established by equilibrium dialysis that purified rabbit antibodies directed against a hapten have two combining sites for that hapten (Eisen and Karush, 1949; Lerman, 1949). However, these experiments did not address Pauling's prediction that immunization with an antigen bearing at least two distinctive determinants should lead to the production of some antibody molecules having dual specificity; there was no reason that *both* combining sites on one molecule should be shaped by only one determinant group of an antigen that has many such groups. Early attempts to use the precipitin reaction to find such antibodies were not generally successful (Landsteiner and van der Scheer, 1938; Haurowitz and Schwerin, 1943; Eisen et al., 1954); their absence was convincingly documented in a later, more quantitative, study by Nisonoff et al. (1959). Thus, it was concluded that both sites of an antibody molecule have the same specificity, inconsistent with the Pauling model.

Toward Clonal Selection

In 1941, in a monograph entitled "The Production of Antibodies", Macfarlane Burnet (the quintessential *cis*-immunologist), expressed concern about the lack of biological precedent for the instruction/template theory. In addition, he cited four characteristics of antibody production that could not easily be accounted for by this theory: (1) the enhanced and more rapid booster response to a second antigenic exposure; (2) the change in "character" of the antibody following repeated antigenic stimulation; (3) the continuation of antibody production long after antigen appears to have disappeared from the body; and (4) the exponential increase in antibody titer during the initial period after immunization, a feature that led Burnet to conclude that antibody production is a function not only of the cell originally stimulated, but also of its descendants.

In this monograph, Burnet proposed a new theory, which was revised and developed in 1949 with Frank Fenner. Inspired by the discovery of adaptive (i.e. inducible) enzymes in bacteria, it was suggested that antigen induces formation of a "self-replicating system," possibly similar to an enzyme, which is caused to multiply by an appropriate antigenic stimulus. The function of the first contact with antigen is to produce an "adaptive modification" of the "enzyme", while subsequent contacts stimulate its replication; circulating antibodies are partial replicas of the enzyme, carrying its "specifically modified adsorptive pattern", but not its enzymatic activity. Once the critical changes are effected in the cell, antigen is no longer required for further antibody production, and daughter cells inherit the modification. Although these proposals assigned an essential role to antigen in initiating antibody formation and so were a form of instruction theory, the emphasis on cellular mechanisms, as well as the recognition that the capacity for continuing antibody synthesis did not require the presence of antigen and could be passed on to daughter cells, was a step along the road to the eventual formulation of the clonal selection theory.

In their 1949 monograph and in an earlier paper (1948), Burnet and Fenner put forward the notion that body components are identified by "self-markers" and that antibodies to any component carrying such a marker cannot be produced. The ABO blood group substances are examples of self-markers. Moreover, the process for recognizing these markers occurs early in ontogeny and such recognition, once established, brings about a life-long lack of response to the self-antigen. As support for this idea, these authors cited the observation of Owen (1945) that nonidentical cattle twins, who share circulatory systems *in utero*, are blood cell chimeras as adults, unable to reject each other's red blood cells. The imaginative idea that foreign antigens introduced at a critical early period in embryonic life can be adopted as "self", thereby inducing a permanent state of tolerance, was experimentally verified by Peter Medawar and colleagues who showed, first, that the chimeric cows are unable to reject each other's skin (Anderson et al., 1951; Billingham et al., 1952) and, second, that a similar state of long-lasting tolerance could also be induced in neonatal mice (Billingham et al., 1953).

These experiments demonstrated that the distinction between self and nonself can be learned. This concept was eventually incorporated by Burnet into the clonal selection theory (see below), taking the form that self-reactive clones are deleted during a critical period early in ontogeny (negative selection). Moreover, Burnet suggested that autoimmunity might result from the anomalous presence of self-reactive clones. These might arise in several ways: the release of antigens that are normally sequestered from the immune system with consequent production of antibodies; the pathological modification of a self-antigen giving rise to antibodies that cross-react with the normal antigen; or a change, perhaps the result of somatic mutation, in an antibody-forming clone so that it now reacts with a self-antigen.

In 1955, Niels Jerne, responding to the perceived deficiencies in the antigen–template idea, proposed "the natural-selection theory of antibody formation". Jerne suggested that "natural antibodies" are always present in the circulation in small amounts and that some of these are able to bind to any introduced antigen. The antigen then carries the complementary antibodies to cells in which the antibodies are reproduced, the cells acting merely as a sort of nonspecific copying machine. The process is initiated by the spontaneous production of small numbers of antibody molecules of random specificity, either early in life or continuously. Jerne also considered the possible analogy between induction of antibody formation and of adaptive (inducible) enzyme synthesis, but his arguments differed significantly from those of Burnet in that the adaptation was induced by introduction of antibody, not antigen, into the cell.

The natural selection theory was able to deal with most of the objections that had been raised to the antigen-template idea. Thus, the secondary response was readily explained by Jerne's theory: antigen encounters an increased amount of antibody resulting from the primary stimulus and therefore more antibody is brought back to the globulin-reproducing cells. Similar reasoning, as well as multiplication of cells, could explain the exponential increase in amount of antibody early in the response. An increase in antibody "character" or "avidity" in response to restimulation with antigen could be explained by selection and subsequent replication of those antibody molecules best able to bind to the antigen. In addition, antigen does not have to be present in order for antibodies to be produced once it has carried the antibody molecules to the cell that will reproduce them. The theory also provided a simple explanation for the absence of antibodies against antigens either naturally present or artificially introduced during early ontogeny. If, at this time, antibodies are produced in limited quantities, they will be removed when they encounter the corresponding antigen, their disappearance leading to the permanent loss of that specificity.

Clonal (Cell) Selection Proposed

The similarity between the ideas of Jerne and those proposed by Ehrlich over 50 years previously was noted by David Talmage (1957a). He pointed out that Ehrlich's notion, which postulated that a replica is made of an intrinsic cellular receptor rather than of an extracellular protein, was more nearly in accord with contemporary views of protein synthesis. Indeed, the antigen–template idea, Burnet's adaptive enzyme hypothesis, and Jerne's natural selection theory all require that external substances brought into the cell influence biosynthetic processes, which seemed implausible. Moreover, Jerne's hypothesis seemed at odds with well established observations that cells, not molecules, can transfer the ability to respond to antigenic stimulation from one individual to another. Accordingly, Talmage concluded that the unit selected is the cell itself—only those cells having receptors with affinity for the antigen are selected and multiply.

Also in 1957, Burnet published a paper entitled "A Modification of Jerne's Theory of Antibody Production Using the Concept of Clonal Selection". While Burnet acknowledged that most immunological phenomena were entirely consistent with the natural selection theory, he agreed with Talmage that it seemed implausible that an antibody molecule could stimulate a cell to produce replicas. Accordingly, Burnet proposed a theory that retained the advantages of Jerne's but overcame this difficulty by ascribing the recognition of antigen not to circulating natural antibody but to clones of lymphatic cells. Thus, the clonal selection theory was born. Further details were elaborated in the Abraham Flexner lectures given at Vanderbilt University the following year and subsequently published (Burnet, 1959).

The origin of diverse antibody-forming cells was assumed by Burnet to be the result of selection after undirected random somatic mutation. He further supposed that this diversification is accomplished early in embryonic life. Joshua Lederberg (1959) subsequently introduced the notion that a variety of antibody-producing cells, arising from somatic mutation accompanying cell proliferation, continue to be generated throughout the life of the animal. Burnet (1964), in a "Darwinian modification" of the clonal selection theory, agreed that it was not necessary to assume that all antibody specificities are present at the time of birth, but that diversification by somatic mutation and antigen-driven selection could be an ongoing process. It is now clear that unusually rapid somatic mutation of antibody genes in cells undergoing antigen-driven proliferation substantially increases the extent of diversity established by other mechanisms, will be discussed in Part II.

According to clonal selection, prior to antigenic exposure there already exist in the body "multiple clones of globulin-producing cells, each responsible for one genetically determined type of antibody globulin." (In context it is clear that by genetically determined, Burnet did not mean encoded in the germline, but encoded in the DNA of somatic cells.) The role of antigen is to stimulate the expansion of any cell carrying a complementary receptor, which is identical in specificity to the antibody that the cell and its progeny will produce. Like natural or "antibody-based" selection, the clonal theory could explain the basic features of the immune response (e.g., the increase in quantity and avidity of antibody following repeated antigenic exposure). As noted previously, it could also be adapted to explain the phenomenon of self-tolerance. However, crucial predictions of the theory remained to be addressed:

1. How many different specificities can be made by a single antibody-forming cell or clone? "One (or possibly a small number)", according to clonal selection (Burnet, 1959). Unspecified, but presumably a large number, according to the template theory.
2. Does an antibody-forming cell have receptors identical in specificity to the antibodies ultimately produced by that cell and its descendants? Yes, "the essence of the hypothesis", according to clonal selection (Burnet, 1957). Irrelevant for the template theory.
3. Is antigen necessary or even present in antibody-forming cells? Not necessarily, according to clonal selection; any antigen present in the initially stimulated cell would likely be diluted out during clonal expansion. Absolutely, according to the template theory.

Burnet's idea that each cell usually produces antibodies corresponding to only one potential antigenic determinant would seem to be an efficient way for a selective immune system to operate, as multispecificity would introduce complications. If cells produce antibodies/receptors having two or more unrelated combining sites, stimulation by one antigen might result in the production of some antibodies not at all complementary to the immunogen. And worse, a cell that simultaneously produced antibodies to foreign and to self-antigens might either be selected against early in life with loss of ability to react with the foreign antigen or, if such a cell survived, stimulation by the foreign antigen might lead to self-destruction.

Clonal Selection Prevails

Once the current of thought had swept immunologists toward clonal selection, a flurry of experiments followed to test its main precepts. Support for the concept that each cell or clone is specific for only one antigen was provided by a large number of experimental approaches. Thus, individual lymph node cells from rats immunized with two different bacteria, and manipulated into microdrops, produced antibody reacting against one or the other, but never both (Nossal and Lederberg, 1958; Nossal, 1960). Consistent with this result, Coons (1958) and White (1958), using the fluorescent antibody technique, also reported that single cells did not generally contain antibodies of two different specificities. Green et al. (1967), also using fluorescent antibodies, showed that after immunization with a hapten-carrier conjugate single cells produced antibody against the hapten or the carrier, but not both. The hemolytic plaque technique, developed by Jerne and Nordin (1963), was an efficient frequently used method for evaluating the presence of "double producers". Evidence for the homogeneity in affinity of antibodies produced by a single clone of cells was obtained by Klinman (1969). Askonas et al. (1970) were able to select a single antibody-forming clone by sequential transfer of limited numbers of spleen cells from primed mice into irradiated recipients who also received the same antigen; evidence for the monoclonality of the antibodies in the sera of the recipients was obtained by isoelectric focusing.

In contradistinction to the foregoing results, Attardi et al. (1959, 1964a,b) found that a significant proportion of single lymph node cells from hyperimmunized rabbits inactivated two immunologically unrelated bacteriophages; in one experiment, of 280 total cells, 44 inactivated one or the other phage and 10 inactivated two. This finding stimulated considerable discussion since the results differed substantially from those obtained by most investigators. One attempt to repeat the experiment exactly did not reproduce the results, i.e., no double producers were found (Mäkelä, 1967).

It has been argued that the intensity of the immunization schedule (multiple inoculations over 13 months), which had been designed to elicit the maximum possible antibody response, may have allowed efficient selection for double producers, even if they do not occur under ordinary conditions of immunization. Selection for double producers might be the result of "leakiness" in the system that ordinarily prevents the expression in one cell of antibodies with two different combining sites (allelic exclusion, to be discussed in Part II). Lively and personal accounts, by two of the chief protagonists in the experiments and debates about this issue can be found in biographical chapters in recent volumes of the *Annual Review of Immunology* (Cohn, 1994; Nossal, 1995).

A number of other studies also reported that antibodies having more than one specificity were sometimes produced by a single cell or clone. These results, however, were the exception in what was coming to be the consensus view: one cell, one antibody [see Sigal and Klinman (1978) for review and discussion of some of the exceptions]. This view found additional support from analysis of, first, myeloma tumors, and then hybridomas (see below); each of these monoclonal cell lines produces homogeneous antibody-like proteins derived from the progeny of a single B lymphocyte.

The possible presence of antibody-like molecules on the surfaces of lymphoid cells from unimmunized mice was suggested by the adherence of bacteria (Mäkelä and Nossal, 1961) and by the binding of a radiolabeled antigen (Naor and Sulitzeanu, 1967) to a small proportion of such cells (less than ~1%). It did not seem likely that these results were due to the passive adsorption of external antibody (e.g., from the medium), but the possibility that they are a consequence of antibody in the process of being secreted from the cells could not be ruled out. A clear distinction between secreted and "cell-bound" or receptor antibody would not be possible until tools for analysis at the molecular level became available (as will be discussed in Part II). Nevertheless, a variety of experiments pointed toward a similarity in antigen binding properties between the putative receptor and the secreted antibody, suggesting that their combining sites must be closely related, if not identical.

Talmage (1957a,b) had observed that antibodies produced 50 days after immunization form complexes that are more stable (dissociate more slowly) than complexes formed by antibodies produced 12 days after immunization. He related this finding to the low dose of antigen required to induce a secondary, relative to a primary, response and suggested that antibody responses are stimulated by an equilibrium reaction between antigen and preexisting cellular receptors with affinity for a ligand similar to that of the antibody ultimately formed.

Subsequently, Herman Eisen and colleagues showed that during the period after immunization of a rabbit with a hapten–protein conjugate, antibodies increase progressively in average association constant (affinity) for the hapten (Eisen and Siskind, 1964). Although other possibilities were considered, this was demonstrated to be a direct consequence of an increase in the affinity of antibodies secreted by lymph node cells (Steiner and Eisen, 1966, 1967a). A second injection of the same conjugate resulted in the rapid synthesis of large amounts of high-affinity antibody (Steiner and Eisen, 1967b). These findings were consistent with predictions of clonal selection. As antigen decreases with time after immunization, cells with receptors of relatively high affinity are selectively stimulated and the population of such cells expands. Repeated immunization results in the enhanced production of antibodies of correspondingly high affinity. Indeed, a change in the type of antibody produced during the course of an immune response and in the secondary response was one of the observations that led to the formulation of selective theories of antibody formation. (The basis for changes in antibody affinity after immunization will be discussed in more detail in Part II.)

That the receptor has antibody-like properties was also suggested by experiments of N. A. Mitchison and co-workers. After immunization of mice with a hapten–protein conjugate, spleen cells were restimulated *in vitro* with the same conjugate and then transferred into irradiated mice; antibody production in the recipient was determined. The response was completely inhibited when restimulation with the conjugate occurred in the presence of excess hapten, consistent with the idea that the receptors on lymphocytes behave similarly to the antibodies produced by these cells (Brownstone et al., 1966). If, instead, the restimulation took place in the presence of a free hapten that differed slightly from that in the immunizing conjugate, antibody production was not abrogated, but shifted in specificity away from the hapten inhibitor. Thus, cells from mice originally immunized with hapten A conjugated to a protein and then restimulated with the same conjugate in the presence of a related hapten A′ produced, after transfer, antibodies with diminished binding of A′ relative to A (Mitchison, 1967). This suggested that the range of binding properties of the receptors and the antibodies produced is similar.

A different approach to the same question was taken by Ada and Byrt (1969). A population of lymphocytes was exposed *in vitro* to a highly radioactive antigen; the capacity to respond to that antigen was found to be diminished without impairing the response to an unrelated antigen. Presumably, the labeled antigen was bound only to cells with receptors of complementary specificity and these cells were thereby inactivated. Although this experiment indicated that antibody-producing cells have antibody-like receptors of the same specificity as the antibody to be produced, it did not demonstrate that the cells have receptors of *only* that specificity. However, Raff et al. (1973) subsequently showed that naive as well as primed B cells appear to have receptors with specificity for only one antigen. Virtually all of the antibody-like receptors on the surface of single cells were aggregated into a "cap" by cross-linking with a multivalent antigen. Strictly speaking, this experiment showed only that all the receptors bound the same antigen, but not necessarily the identical determinant on that antigen. However, it seems quite unlikely that if several different receptors were present on a cell, all would react with different determinants on the very same antigen.

A number of experiments demonstrated that populations of lymphocytes could be depleted or enriched for a particular specificity by selectively binding cells, via their surface receptors, to an antigen-containing matrix and, in some cases, recovering the adherent cells from the matrix (e.g. Wigzell and Andersson, 1969; Wofsy et al., 1971; Rutishauser et al., 1973). The depleted or enriched populations were prepared from unimmunized mice as well as from immunized mice. Evidently, the specificity of the secreted antibody must match that of the receptor. The culmination of this type of experiment was the demonstration that even a single lymphocyte from an enriched population of naive lymphocytes could be stimulated to proliferate into a clone that made the expected antibody (Nossal and Pike, 1976). A significant advance in isolating viable cells of a particular specificity was the development of the fluorescence-activated cell sorter, which allowed the efficient separation of cells having a surface antigen labeled with a fluorescent marker (Bonner et al., 1972; Julius et al., 1974).

As to the necessity for antigen at the site of antibody production, the experimental approach adopted was to introduce into an animal an antigen that was sufficiently highly radiolabeled that a very small quantity of antigen within an antibody-forming cell would be detectable. The definitive result would be that the number of antigen molecules found is less than some upper limit judged to be too small to be compatible with the rate of antibody synthesis. Indeed, Nossal et al. (1965) and McDevitt et al. (1966) reported that the upper limit on the number of antigen molecules in each antibody-producing cell was about 5–15, respectively. Even allowing for as many as 30 "antigenic sites" per molecule, the maximum number of such sites is far smaller than the number of polyribosomes per cell and, assuming that no loss of label from antigenic sites had occurred, it was concluded that there was probably no requirement for antigen at the site of antibody formation.

Many of the experiments cited in the preceding paragraphs were actually carried out for the sake of "completeness," long after any form of instruction-template theory was considered viable by most immunologists. Indeed, by the 1950s it had become clear that proteins are encoded by specific genes and the details of protein biosynthesis were becoming known. Accordingly, the idea that antibodies are synthesized in a fundamentally different way from other proteins became increasingly unattractive. The experimental demonstration that the specificity and presumably the three-dimensional structure of proteins (Epstein et al., 1963), including antibodies (Haber, 1964; Whitney and Tanford, 1965), is determined by their amino acid sequence was the final blow to the antigen–template idea. Thus, by the early to mid-1960s it was generally accepted that no hypothesis depending on instruction by antigen could be correct, and that some form of selective theory must be invoked to account for antibody production (of course, *cis*-immunologists were quicker off the mark to accept clonal selection than were *trans*-immunologists). The numerous experiments supporting the clonality of antibody production served to confirm what was by this time generally believed. Clonal selection is consistent with the modern biological world view and its principal features are central to our current understanding of the operation of the immune system.

Role of the Lymphocyte

Clonal selection gives center stage to the lymphocyte, the most common of the white blood cells, as the key cell in immune responses. Although there had been hints that lymphocytes participate in antibody responses (Ehrich and Harris, 1942), their role in immunity was firmly established in the early 1960s largely through the work of Gowans and collaborators (Gowans et al., 1962; see also recent review by Gowans, 1996). Also at this time it became clear that the debate about the relative importance of antibodies and cells in immune responses, which had engaged immunologists from the time of von Behring (a *trans*-immunologist) and Metchnikoff (a *cis*-immunologist), was coming to resolution. There are two major classes of lymphocytes, the B-lineage cells that mature in the *b*ursa of Fabricius in birds or the *b*one marrow of mammals and produce antibodies, and the T cells that mature in the *t*hymus and are responsible for cell-mediated immunity (e.g. delayed hypersensitivity) and whose cooperation with B lymphocytes is required for antibody production (Claman et al., 1966; Mitchell and Miller, 1968; Jacobson et al., 1970; Mitchison et al., 1970; Raff, 1970). Both B and T cells recognize and interact specifically with antigen.

Antibodies are modified forms of the receptors for antigen on B lymphocytes, reminiscent of Ehrlich's early prediction. The nature of the antibody-like receptor is considered in more detail in Part II. The identity of the T-cell antigen receptor was the subject of considerable debate until the matter was resolved in the early 1980s and the receptor was found to be a molecule different from, but with considerable resemblance to, the antibody molecule. Critically, the T-cell receptor differs from antibody in its recognition pattern: it recognizes peptides derived from the antigen that are bound to molecules of the major histocompatibility complex (MHC). The clonal selection theory applies equally to T cells as it does to B cells.

THE ANTIBODY MOLECULE: STRUCTURE AND SPECIFICITY

As noted in the preceding section, the demonstration that immunity can be passively transferred by serum paved the way for eventually establishing the molecular nature of antibodies. What was needed were methods for fractionating serum and purifying its various constituents. It is indeed one of the striking aspects of the gradually evolving understanding of antibody structure that it closely followed advances in methods for purifying and characterizing proteins. In the 1930s, a center for developing such methodology was the Svedberg[1] laboratory in Uppsala and it was there that the first steps were taken toward determining which serum component carries antibody activity. Analysis in the ultracentrifuge indicated that antibodies sedimented either at 17 to 19 S or at 6 to 7 S, corresponding to molecular weights of about one million and 160,000, respectively (Heidelberger et al., 1936; Heidelberger and Pedersen, 1937; Kabat[24], 1939).

With the technique of free boundary electrophoresis, Tiselius and Kabat (1939) demonstrated that the antibody activity in a rabbit antiserum to ovalbumin was confined to the γ-globulin region (fraction migrating slowest toward the anode). However, antibodies were sometimes found to migrate faster on electrophoresis. In addition, proteins having no antibody activity (e.g., properdin, a protein of the alternative complement pathway) might also migrate in the γ-globulin fraction. Subsequently, it was realized that all antibodies, even those belonging to different classes (e.g., IgG, IgM, IgA, see below), share many basic structural features despite differences in size or in electrophoretic properties. Eventually, the term "immunoglobulin" was introduced to include the set of all proteins that share antigenic determinants, and hence essential structural features with antibodies (Ceppellini et al., 1964). (An antigenic determinant, or "epitope" in current terminology, is the portion of an antigen that makes contact with the combining site of a particular antibody; an antibody, like any protein, can also function as an antigen and so has a number of epitopes.) The term immunoglobulin is preferentially used when emphasis is on the antibody as a protein, regardless of its antigen-binding activity.

The Papain Fragments

In the 1950s the pioneering work of Fred Sanger[2] in working out methods to establish the amino acid sequence of insulin (Sanger and Tuppy, 1951; Sanger and Thompson, 1953) ushered in an era of rapid advances in the determination of the sequences of proteins. Rodney Porter[26], a Ph.D. student of Sanger, became interested in the question of the chemical basis for antibody activity. At that time, the only proteins whose sequence had been determined (insulin, ribonuclease, lysozyme) were at least an order of magnitude smaller than antibody molecules. An even more formidable problem was that different preparations of antibody, known to bind to the same antigenic determinant, varied measurably in molecular properties (e.g., as demonstrated by electrophoresis) and were impossible to fractionate into homogeneous constituents. Nevertheless, the heterogeneity did not obscure the substantial similarity among all immunoglobulin molecules, similarity that had contributed to Pauling's formulation of the variable folding model of antibody specificity. The major distinction among antibodies was in their recognition of different antigens, not in their overall molecular structure. As Porter (1973a) noted in his Nobel lecture, "This combination of an apparently infinite range of antibody combining specificity associated with what appeared to be a nearly homogeneous group of proteins astonished me and indeed still does." In this respect, antibodies stand in contrast to enzymes, which typically differ substantially from one another in structure as well as in specificity.

Porter's plan was to reduce the magnitude of the sequencing problem by breaking the antibody molecule into fragments, hoping that one or more of the smaller pieces would retain specificity for antigen. He was influenced by the work of Landsteiner, who had shown that in many cases, only a small part of an antigen is needed to bind to antibody, suggesting that the combining site of the antibody may also be smaller than the whole antibody molecule (Porter, 1950a). The plan also depended on the supposition that the heterogeneity of the antibody preparations would not prevent the isolation of constituent pieces.

The specific approach adopted by Porter was digestion of antibody with papain. Initial efforts along these lines, by himself and others, had shown that active products of lower molecular weight could be produced by treatment with proteolytic enzymes, but these products had not been isolated or characterized in detail (Parfentjev, 1936; Petermann and Pappenheimer, 1941; Northrop,

1941–1942; Petermann, 1946; Porter, 1950a). These early experiments were hampered by technical limitations in availability of pure enzymes and in effective methods for fractionating mixtures of proteins. However, by the late 1950s, the requisite materials and techniques had improved significantly. Thus, the availability of a method for preparing highly purified papain (Kimmel and Smith, 1954) meant that the specific digestion products would not be substantially contaminated by enzyme. In addition, the newly introduced carboxymethylcellulose ion exchange resins (Peterson and Sober, 1956) provided efficient means for separating the products of digestion.

In a renewed effort, Porter treated several rabbit antibodies, each specific for a different antigen, with crystalline papain; the resulting digests were fractionated on columns of carboxymethylcellulose. In each case, three fractions of approximately equal size were obtained and named, in order of their elution—fractions I, II, and III; these fractions together accounted for almost all of the starting material and were resistant to further digestion with papain (Porter, 1958, 1959). None of the fractions precipitated with the corresponding antigen, but fractions I and II, which were very similar in size and amino acid composition, specifically inhibited the precipitation of the antigen by the homologous antiserum. Fraction III, which had no inhibitory activity, was shown to contain structural features responsible for transmission of the antibody across the placenta[e] (Brambell et al., 1960), for binding to guinea pig skin, permitting anaphylactic reactions upon antigenic challenge (Ovary and Karush, 1961; Ishizaka et al., 1962), and for activating complement (Taranta and Franklin, 1961; Ishizaka et al., 1962).

Fraction III crystallized readily when dialyzed against buffers of neutral pH, an unexpected result since the starting antibody preparation did not crystallize. This finding suggested that a fragment of the protein might be more homogeneous than the starting material. Another inference from these experiments was that the IgG molecule consists of three tightly folded globular segments that are resistant to further digestion by papain, whereas the polypeptide(s) connecting these segments are more exposed to proteolytic digestion.

In later years, Porter enjoyed recalling that initially he believed that the crystals, which appeared upon dialysis of the papain digest in the cold, consisted of cystine, the oxidation product of the cysteine that had been used to activate papain. Accordingly for several months he poured the crystals down the sink. As he noted in his Nobel lecture (1973a), it was fortunate that his neighbor at the National Institute for Medical Research was Olga Kennard, a crystallographer. When he finally asked her opinion about the crystals, she remarked that they looked like crystals of protein, not cystine. They were then identified as the material in the third peak obtained by fractionation of the products of digestion on carboxymethylcellulose.

In the original experiments, fraction I was obtained in the pass-through of the ion-exchange column; fractions II and III were eluted with the gradient. The yields of the three fractions were very similar, and it was initially thought that the IgG molecule consisted of one each of fragments I, II, and III. It was subsequently shown, however, that the similar yield of fractions I and II was fortuitous, the result of their chemical heterogeneity, as reflected in overall charge, and choice of column elution conditions (Palmer et al., 1962). Thus, more negatively charged IgG molecules were found to contain two fragments of type I, whereas more positively charged molecules contained two fragments of type II. Fragments I/II were later renamed Fab (fragment antigen-binding) and the crystallizable piece III was renamed Fc (fragment crystallizable); see Ceppellini et al. (1964) for a summary of immunoglobulin nomenclature. The Fc piece is shared by all IgG molecules, whereas the two Fab fragments differ from one molecule to another and contain the combining sites for antigen.

The Four-Chain Model

Another approach to the determination of the structure of antibodies was taken by Gerald Edelman[25], while he was still a graduate student at the Rockefeller Institute. Edelman found that when IgG was reduced with a mercaptan, in the presence of dissociating solvents such as 6 M urea, its molecular weight dropped significantly, demonstrating that it consists of a number of polypeptide chains cross-linked by disulfide bridges (Edelman, 1959; Edelman and Poulik, 1961). These findings disagreed with earlier end-group analyses, which had indicated that rabbit antibodies have approximately one free amino-terminal residue, consistent with a single-chain protein (Porter, 1950b). To resolve this discrepancy and to obtain products that might retain some biological activity, Porter and colleagues carried out the reduction and chain separation in aqueous solution. A key step was the use of mild conditions of reduction, which had been shown by Cecil and Wake (1962) to cleave interchain disulfide bonds, in general, more readily than intrachain bonds. The chains prepared in this way remained soluble and antigenically active after separation by gel filtration in 1 N acetic acid (Fleischman et al., 1962).

By good luck, Julian Fleischman, a postdoctoral fellow, had joined the Porter laboratory bringing antisera to the Fab and Fc fragments, which he had helped prepare in Melvin Cohn's laboratory at Stanford. A simple double immunodiffusion experiment established the relation between the chains and fragments. There was initially some uncertainty about the molecular weights of the separated chains since the heavy chain in particular tends to aggregate. The four-chain model for IgG was first proposed by Porter at a meeting in New York, but the possibility that the molecule consists of only two or even three chains was also considered (Porter, 1962; Fleischman et al., 1962). However, redetermination of the molecular weights (Pain, 1963) was consistent with the presence of two heavy and two light chains per molecule. Additional data suppporting the now-familiar four-chain model were presented by Fleischman et al. (1963).

As pointed out by Fleischman (1981), a critical finding, which contributed to the final model, was the observation of Alfred Nisonoff and colleagues (1960) that pepsin digestion of the IgG molecule at pH ~4.5 degrades the Fc fragment but leaves two Fab-like fragments intact and linked to one another by a disulfide bond; the resulting bivalent fragment is called $F(ab')_2$. Therefore, previous models, in which the combining sites were placed at the distal ends of a cigar-shaped molecule, were no longer tenable.

The IgG model, with minor modification, mainly in the number and location of interchain disulfide bridges, has stood the test of time. The molecule consists of two identical heavy chains and two identical light chains of mol wt 50,000 and 23,000, respectively. Each Fab fragment consists of one entire light chain plus the amino-terminal half of one heavy chain; the Fc fragment consists of the remainder of both heavy chains (Figure 1). The presence of two Fab fragments, each of which is active in binding antigen, was consistent with the results mentioned in the preceding section that IgG antibodies are bivalent. The model was reconciled with Porter's end-group analyses when it was realized that the heavy chains have a blocked amino-terminus and the light chains are heterogeneous at the amino-terminus; under these circumstances, the yield of a single end-group per molecule did not reflect the actual number of chains.

There are two major types of light chains, κ and λ, which were originally differentiated by their antigenicity (Korngold and Lipari, 1956). In addition to IgG, there are several other immunoglobulin classes, the number and type varying in different species. The classes are defined according to their heavy chains. The light chains in a molecule of any class can be either κ or λ. The classes in humans and mice are IgG, IgM, IgA, IgD, and IgE; the heavy chains in these classes are γ, μ, α, δ, and ϵ. Thus, an IgG molecule consists of two heavy (γ) chains and either two κ chains or two λ chains. The IgM class had previously been identified as the high molecular weight antibodies, sedimenting at ~19 S. Some of these classes exist as several variants, called subclasses. [See monograph by Nisonoff et al. (1975) for a comprehensive review of studies on immunoglobulin structure from the late 1950s until 1975.] The genes encoding different classes of heavy chains are linked (Herzenberg, 1964; Lieberman and Potter, 1966).

Studies of the inheritance of the genes encoding κ and λ chains in the rabbit indicated that these genes are not linked either to each other or to genes encoding heavy chains (reviewed by Kelus and Gell, 1967; Mage et al., 1973). It was later shown that in humans (Erikson et al., 1981; Kirsch et al., 1982; Malcolm et al., 1982; McBride et al., 1982) and mice (Swan et al., 1979; D'Eustachio et al., 1980; D'Eustachio et al., 1981) the genes for κ, λ, and heavy chains map to different chromosomes.

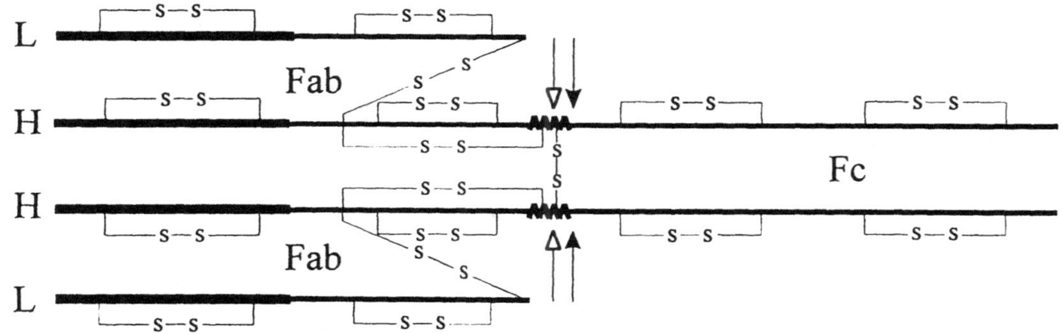

Figure 1. Diagrammatic representation of polypeptide chain structure of rabbit IgG (adapted from Porter, 1973a,b). Heavy (H) and light (L) chains, and Fab and Fc fragments are indicated. The jagged segments within the heavy chains represent the hinge region. The sites of papain and pepsin digestion are indicated by the open and closed arrows, respectively. Digestion with papain yields two Fab fragments and one Fc fragment. Digestion with pepsin yields a bivalent $F(ab')_2$ fragment, held together by the disulfide bridge between the heavy chains. Pepsin degrades the section of the molecule corresponding to Fc into a number of fragments, and does not yield a piece similar in size to Fc. There are two domains in each light chain, one variable (V_L, thicker line) and one constant (C_L, thinner line). In addition to the hinge region, there are four domains in each heavy chain: one V_H (thicker line) and three C_H domains (thinner line). The interchain and intrachain disulfide bridges are shown. In addition to the conserved intrachain disulfide bonds within each domain, the heavy chain of rabbit IgG has a disulfide bond linking the amino- and carboxyl-terminal ends of the first C domain (C_H1, the constant part of the heavy chain in Fab), as indicated here (O'Donnell et al., 1970); this bond is not found in human or mouse IgG heavy chains, but several other immunoglobulin or immunoglobulin-like domains have an intrachain bond in similar positions, as discussed by Steiner (1985).

V and C Regions

The unraveling of the molecular structure of the antibody molecule was a most significant achievement and laid the foundation for further work to determine the structural basis for specificity. These next studies required sequence analysis of the antibody heavy and light chains. Rather than use antibody preparations with their inherent heterogeneity, many investigators turned to myeloma proteins (the products of plasma cell tumors), which appeared to be homogeneous counterparts of induced antibodies. Patients with multiple myeloma often produce large amounts of so-called Bence-Jones proteins, which are secreted in the urine and had been shown to be identical to the light chains of the myeloma proteins circulating in the blood of the same patient (Edelman and Gally, 1962). The smaller size and the ready availability of large quantities of Bence-Jones proteins facilitated sequence analysis.

Examination of Bence-Jones proteins by tryptic peptide mapping and by determining the amino acid composition of isolated peptides, indicated that the κ and λ types are very different in primary structure (Putnam, 1962; Putnam and Easley, 1965). Comparison of a number of Bence-Jones proteins of either type showed that some peptides were common to all the proteins of that type, whereas others were found only in an individual protein. In particular, within each type the carboxyl-terminal peptide was uniform, whereas the amino-terminal peptide varied (Putnam and Easley, 1965; Titani and Putnam, 1965).

In 1965, Norbert Hilschmann and Lyman Craig determined the partial sequence of two κ Bence-Jones proteins. On the basis of overlaps obtained by analyzing peptides from chymotryptic and peptic digests, they ordered the tryptic peptides of one of these proteins, and partially ordered and aligned most of the tryptic peptides from the other one, as well as some of the peptides from a third κ protein that had been studied by Putnam and co-workers (Putnam et al., 1963; Titani and Putnam, 1965). From this analysis, they drew the important conclusion that the variability in κ chains is not distributed throughout the sequence, but is confined to approximately the amino-terminal half (~110 amino acid residues), the carboxyl-terminal half being uniform or nearly uniform. This was the first demonstration that an immunoglobulin polypeptide chain consists of one segment that is variable (V) in sequence and another that is constant (C).

The existence of V and C segments in κ chains was reminiscent of the presence of variable and constant parts of the intact IgG molecule, as indicated by Porter's fractionation of rabbit IgG into heterogeneous Fab and homogeneous Fc. Data from sequence analysis of the λ type of light chains, and also of heavy chains, indicated that they too consist of V and C regions. Further, it was determined that the set of V regions found on κ chains differs from the set on λ chains and both differ from the set on heavy chains. These three sets or "groups" of V regions (i.e., V_κ, V_λ, and V_H) are distinguished by certain features of sequence that can be discerned despite the variability. (For nomenclature of V-region "groups" and "subgroups", see Asofsky et al., 1969; see also section "Two Genes, One Polypeptide".)

The presence of V and C regions in immunoglobulins was established largely by analysis of myeloma proteins. However, studies of normal heterogeneous rabbit IgG, carried out largely by the Porter laboratory, led to similar conclusions. This was reassuring as the myeloma proteins, although thought probably to be valid models for antibodies, are the products of abnormal cells. Additional evidence in support of the supposition that myeloma proteins are essentially equivalent to homogeneous antibodies was the observation that some of them could be shown to have antibody-like activity (Eisen et al., 1970; Kunkel, 1970). For example, an IgG myeloma protein was found to bind ε-dinitrophenyl-L-lysine with an association constant of $\sim 2 \times 10^4\ M^{-1}$; like IgG antibodies, there were two ligand binding sites per molecule, but unlike induced antibodies, which consist of a population of molecules that differ in binding constant, the binding sites in the myeloma were found to be homogeneous with respect to affinity (Eisen et al., 1967). [See review by Fahey (1962) for a summary of early views about the proposition that myeloma proteins are or are not "normal" immunoglobulins.]

The Immunoglobulin Domain

A prominent feature of antibodies that emerged from the first studies of the amino acid sequence of heavy and light chains was that the basic structural unit is a domain of approximately 110 amino acid residues, which contains a disulfide bridge linking half-cystines 60 or 70 residues apart in the linear sequence. Light chains consist of two such domains (V_κ and C_κ, or V_λ and C_λ), whereas heavy chains consist of one V_H domain and, usually, three or four C_H domains. For example, the heavy (μ) chain of IgM contains four C_μ domains; the heavy (γ) chain of IgG contains three C_γ domains, as well as an extra short segment called the "hinge region", which is thought to impart flexibility to the molecule and is the region that is susceptible to proteolysis by enzymes such as papain and pepsin. The domains were recognized because of the conserved disulfide bridge at nearly the same position within each domain and because of other similarities in amino acid sequence.

These features of the immunoglobulin domain led early workers in the field to propose that the immunoglobulin molecule evolved from an ancestral domain by successive rounds of gene duplication (Hill et al., 1966; Singer and Doolittle, 1966). Further evidence in support of this hypothesis was provided by Edelman and colleagues (1969), who determined the complete amino acid sequence of both heavy and light chains of an IgG myeloma protein; the results clearly revealed the repeating domain structure of both chains. Immunoglobulins composed of these domains and having basically very similar structures have been identified in representatives of all vertebrate classes, with the exception of the primitive agnatha, or jawless vertebrates (reviewed by Du Pasquier, 1993).

In recent years it has become evident that many non-immunoglobulin proteins contain domains having significant similarity in sequence to immunoglobulin domains. A number of these proteins have important roles in the immune response: the T-cell receptors; the T-cell membrane accessory molecules, CD3, CD4 and CD8; class I and class II molecules of the MHC; and most Fc receptors. Other proteins containing immunoglobulin-like domains are found on the surfaces of a variety of cells, often in the nervous system, and have no known role in immunity, but appear to mediate cell–cell interactions [e.g., the neural cell adhesion molecule, N-CAM (Cunningham et al., 1987)]. These immunoglobulin-like domains are considered to be members of a large family that has been designated the immunoglobulin superfamily (Williams and Barclay, 1988). Studies of three-dimensional structure have largely confirmed the predictions based on the sequences; domains in the immunoglobulin superfamily typically have immunoglobulin-like folds (see below). There may be some variation in number and length of strands, but the core of the fold is conserved. Although not an absolutely conserved feature, most of these domains have the characteristic intradomain disulfide bridge linking half-cystine residues about 60 residues apart in the linear sequence.

Although most of the domains that were originally identified as belonging to the immunoglobulin superfamily were found in proteins expressed on cell surfaces, more recently certain intracellular muscle proteins, e.g. titin (Labeit et al., 1990), twitchin (Benian et al., 1989), and telokin (Holden et al., 1992) have also been found to contain domains belonging to this family. Many proteins are composed of a mixture of domain types. Thus, the membrane-proximal domains of the MHC class I and class II molecules are immunoglobulin-like, but the peptide binding domains are not (Bjorkman et al., 1987; Brown et al., 1993). For this reason, it may be preferable to confine the term, superfamily, to domains rather than proteins. Immunoglobulin-like domains have also been identified in a number of proteins from invertebrates (e.g. Harrelson and Goodman, 1988; Seeger et al., 1988; Benian et al., 1989). It has been proposed that the immunoglobulin-like domain evolved in early metazoans from a primitive cell adhesion molecule (Williams, 1982, 1987; Edelman, 1987). When the adaptive immune system evolved in vertebrates, the domain was utilized for the specific B- and T-cell receptors as well as in other molecules, some of which interact with these receptors.

Complementarity-Determining Regions

The major objective of the studies on the polypeptide arrangement and sequence of antibodies had been to elucidate the structural basis for specificity.

By the end of the 1960s it had become clear that the combining region for antigen must lie in the V region of the heavy and/or light chain. Numerous attempts were made to localize the site to one chain or the other by a variety of experimental approaches (reviewed by Nisonoff et al., 1975). These included determining the activity of isolated heavy and light chains and of recombined molecules having chains derived from different antibodies or from antibodies and nonspecific immunoglobulins. The general conclusion was that both chains are required for optimal antigen-binding activity. In addition, the technique of affinity labeling was developed (Wofsy et al., 1962) to localize the site to a part of the heavy and/or light chain. Although impeded by the heterogeneity of the combining-site region in the pooled antibody preparations, it was found that in most cases residues in both heavy and light chains contribute to the site.

Beginning in the mid-1960s, the amino acid sequence of many V regions of heavy and light chains was determined. The immunoglobulins used for these studies were mostly myeloma proteins derived from plasma cell tumors. The human tumors were from patients with the disease, multiple myeloma; the mouse plasmacytomas were induced in BALB/c mice by methods developed by Michael Potter (1982). The results provided clues about the extent of variability and the probable location of the combining site.

Analysis of the amino acid sequences of a large group of light-chain V regions led T. T. Wu and Elvin Kabat (1970) to conclude that variability is not distributed uniformly throughout the V region. They distinguished between framework regions with relatively little variability from one molecule to another and hypervariable regions. For light chains, the hypervariable regions were considered to be positions 24–34, 50–56, and 89–97. Heavy chains also have three hypervariable regions in similar positions. It was predicted that in the folded configuration of the antibody molecule, residues from the six hypervariable regions would be near each other and would form the major part of the antigen-combining site. This prediction turned out to be correct, and the hypervariable regions are now generally referred to as complementarity-determining regions (CDRs).

Three-Dimensional Structure

Details about the combining site, as well as the overall three-dimensional structure of antibody molecules, were provided by X-ray crystallographic studies. It proved to be difficult to crystallize intact immunoglobulins and the first structures determined at high resolution were of isolated Fab fragments of myeloma proteins (Poljak et al., 1973; Segal et al., 1974), of the dimer of a Bence-Jones protein (Schiffer et al. 1973) and of Fc derived from pooled human IgG (Deisenhofer et al., 1981). These studies delineated the basic structure of the immunoglobulin domain or immunoglobulin fold, as it is often called: ~110 amino acids arranged in two approximately parallel sheets, each formed by segments of anti-parallel β-strands that are connected by loops of polypeptide chain. The two sheets are held together by the conserved intradomain disulfide bridge; side chains of amino acid residues fill the space between the sheets and stabilize the bilayer structure. These structural features are also characteristic of other domains in the immunoglobulin superfamily. In the case of V domains, the framework regions are in the β-strands and the CDRs are in the connecting loops. As predicted, the six CDRs are clustered and form the combining site for antigen. When different Fabs are compared, the framework segments are nearly superimposable, and the CDRs differ.

In a few cases it has been possible to crystallize an entire immunoglobulin molecule. The first two such IgGs examined, myeloma proteins Dob and Mcg, turned out to have deletions of the hinge region, an anomaly that is probably related to their ease of crystallization (Fett et al., 1973; Steiner and Lopes, 1979). Both proteins were found to be T-shaped, the Fc forming the stem and the two Fabs the arms of the T (Sarma et al., 1971; Silverton et al. 1977; Rajan et al., 1983; Guddat et al., 1993). Two other IgGs, Kol and Zie, with intact hinges, were also crystallized, but in both cases no electron density corresponding to Fc was seen, presumably because flexibility in the hinge allows the Fc to assume more than one conformation with respect to the Fab arms in the crystal lattice (Ely et al., 1978; Marquart et al., 1980).

More recently, a structure, at 3.5 Å, of an intact mouse IgG antibody with no structural defects was at last obtained (Harris et al., 1992). The monoclonal antibody (see below) used in this study was directed against a canine lymphoma. The molecule is asymmetric; the "hinge angles" between Fc and the two Fabs are substantially different and the "elbow angles" between the V_L/V_H and C_L/C_H axes in the two Fabs are also different. The authors indicated that the asymmetric conformation observed in the crystals should not be considered as a static structure in solution, but represents only one of many possible transient conformations. The hinge appears to have an extended and open configuration, in contrast to the compact globular Fab and Fc fragments. This observation is consistent with early predictions about the structure of the IgG molecule, as discussed above, that were based on the susceptibility of the hinge to proteolytic digestion. The structure of the hinge would appear to allow the Fabs considerable freedom of movement.

To date, a considerable number of Fab structures has been determined and in some cases complexes of Fab or Fv with antigen as well [see reviews by Wilson and Stanfield (1993), Padlan (1994); Braden and Poljak (1995), and Davies and Cohen (1996)]. [Fv is the portion of Fab containing the V regions of one heavy and one light chain, and therefore the combining site (Inbar et al., 1972)]. In antibodies that bind small antigens or haptens, the binding site appears as a crevice, whereas in antibodies that bind globular proteins the site is much flatter with an undulating surface complementary to the surface of the antigen. In each complex, at least four of the six CDRs make contact with antigen, not necessarily the same ones in each case, but always the third CDR in both chains. Both of the third CDRs are located near the center of the antigen-binding surface. The total number of residues in the antibody forming contacts with antigen varies from 8 to 10 for smaller ligands to about 20 for larger ones. Many of the aromatic residues in the CDRs are involved in these contacts; occasional framework residues can also make contact with antigen. Conformational changes in both antibody and antigen may accompany complex formation.

Anti-Antibodies (Idiotypy)

When antibodies are used as antigens, the induced anti-antibodies recognize a variety of epitopes on the immunizing antibody. Some of these epitopes, called "isotypic[f] determinants" are uniform in all individuals in one species. Isotypic determinants differentiate among immunoglobulin classes, one set expressed on IgG, another on IgM, etc.; they may also distinguish different light-chain types (κ vs. λ). Other epitopes, "allotypic determinants", are different in different groups of individuals within the same species. Allotypic determinants differentiate among products of allelic genes at one locus.

Other epitopes, called "idiotypic determinants", are unique to antibodies against one antigen in one individual or, perhaps in a group of individuals. The idiotypic determinants are associated with the CDRs and overlap, to a greater or lesser extent, with the antibody combining site. The presence of unique epitopes on immunoglobulins was first demonstrated by Henry Kunkel and colleagues who showed that each of a battery of human myeloma proteins was "individually specific" (Slater et al., 1955). That this individuality is also characteristic of induced antibodies was subsequently shown in the human (Kunkel et al., 1963) and rabbit (Oudin and Michel, 1963; Gell and Kelus, 1964). In current usage, idiotypic determinant refers to these individually specific epitopes whether they ocur on myeloma proteins or on antibodies. The term "idiotype" refers to the set of all idiotypic determinants on an immunoglobulin molecule.

Since anti-idiotypic antibodies and the epitope of the immunogen may bind to the same region of the antibody, they sometimes compete in binding. Thus, Brient and Nisonoff (1970) demonstrated that a hapten can partially inhibit the interaction of anti-hapten antibodies with anti-idiotypic antibodies raised against them. However, not all idiotype:anti-idiotype reactions can be inhibited by the relevant antigen or hapten. Other experiments demonstrated that sometimes the expression of an antibody bearing a particular idiotypic determinant can be suppressed by exposure of antibody-producing cells *in vivo* or *in vitro* to antibodies directed against these determinants (reviewed by Nisonoff, 1991; see also Part II).

The possibility of complex interactions, both stimulatory and inhibitory, between antibodies and antigens, on the one hand, and antibodies and anti-idiotypic antibodies on the other, were among the considerations that led Jerne in 1974 to put forward "a network theory of the immune system". Jerne suggested that an antibody molecule not only recognizes an epitope of an introduced antigen, but the idiotypic determinants of that antibody are recognized in turn by other antibody molecules, the anti-idiotypic antibodies. Jerne also proposed that this interacting network of idiotypic determinants and anti-idiotypes has a dominant role in regulating antibody production in the immune response. He suggested that sometimes the structure of the anti-idiotypic combining site resembles the structure of the epitope, and referred to such a site as the "internal image" of the foreign epitope. Robert Schwartz (1984) and Arthur Silverstein (1986) have pointed out that Ehrlich, among others, had already, at the turn of the century, speculated about the existence and possible function of anti-antibodies. However, the subject then remained largely dormant until work in the 1950s rekindled interest in this subject (see additional citations in Silverstein's article).

Support for some degree of structural similarity between an antigenic epitope and the combining site of certain anti-idiotypic antibodies was provided by Sege and Peterson (1978) who showed that an IgG fraction of an antiserum raised against purified insulin antibodies can inhibit the binding of insulin to its receptor and also has an insulin-like effect on cells. Similar approaches have been applied in studies of other ligand–receptor systems and have in fact been used to isolate a number of receptors. The ability of some anti-idiotypic antibodies to serve as antigenic mimics has also led to efforts to use them as vaccines in cases where it may be undesirable to use the antigen itself for immunization. Although there has been some success with this approach in experimental systems, a practical anti-idiotypic vaccine for human use has not yet been developed. (For a discussion of concepts and practical uses of anti-idiotypes, see reviews by Nisonoff, 1991 and Greenspan and Bona, 1993.) It has proven to be difficult to substantiate the idea that "network" interactions have an important regulatory role in the immune response.

The conjectural relationship between anti-idiotypes and antigens has been explored by examining the crystal structure of complexes of a monoclonal anti-lysozyme Fab or Fv fragment with lysozyme (Amit et al., 1986; Bhat et al., 1990) and of the same antibody fragment with Fab or Fv fragments of two different anti-idiotypes. The complexes were evaluated in terms of such criteria as sharing of contact residues and interactions with solvent. The authors concluded that in one case (Bentley et al., 1990) there was little similarity in the interactions of antibody with antigen and with anti-idiotype, but in the other case (Fields et al., 1995) there was considerable similarity. In the latter, most of the residues in anti-lysozyme that form contacts with lysozyme are also in contact with the anti-idiotype. Significantly, the atoms in lysozyme and the anti-idiotype that form these contacts are in similar positions, although the actual residues differ. In addition, about half of the hydrogen bonds in the two complexes are superimposable (Fields et al., 1995).

Monoclonal Antibodies

As mentioned previously, many of the early studies of antibody sequence and three-dimensional structure were carried out with mouse or human myeloma proteins. In 1975, Georges Köhler and Cesar Milstein[27] described a technique for "the manufacture of predefined specific antibodies by means of permanent tissue culture lines." These monoclonal antibodies, like myeloma proteins, are products of the clonal expansion of a single B cell, in this case a B cell that has been fused with a myeloma cell. However, monoclonal antibodies have a distinct advantage in that they can be directed against designated antigens, and they soon replaced myeloma proteins in structural studies. The hybrid cell line or hybridoma retains the unlimited growth characteristics of the myeloma parent, and continues to secrete the antibody product of the B cell. A critical step in the development of the hybridoma technique was to establish conditions for selecting rare hybrid cells from the overwhelming majority of unfused myeloma cells. This was accomplished by a modification of the selection technique that had been introduced by Littlefield (1964). In the presence of appropriate selective medium, only the hybrids and not the mutant parental cells survive. The possibility of producing unlimited quantities of homogeneous antibody directed against any antigen or epitope of choice has had a major impact on immunology and many other areas of biomedical research and biotechnology.

TWO GENES, ONE POLYPEPTIDE

The notion that immunoglobulins consist of Fc fragments that are invariant and Fab fragments that vary from one molecule to another prompted early speculation that Fc and Fab represent distinct biosynthetic units. However, experimental evidence did not support this possibility (Porter, 1959; Fleischman, 1963). These experiments were carried out before the polypeptide structure of the molecule was understood. The elucidation of the actual molecular structure, as well as the discovery of V and C regions in the light and heavy chains (see preceding section), prompted additional debate about the genetic control of the antibody polypeptide chains. If the chains are encoded by a single germline gene, one would have to assume that GOD [the "generator of diversity" (Lennox and Cohn, 1967)] modifies only the portion encoding the V region and not the portion encoding the C region. On the other hand, if the chains are encoded by multiple germline genes, there would have to be similar constraints on diversification of the constant region, now occurring over evolutionary time, rather than somatically. It seemed that the best way out of this dilemma was to imagine that the two parts of the chain are actually encoded by separate genes and that these genes, or their RNA or protein products, are then joined. Even so, it would be necessary to ensure that only the *V* and not the *C* gene(s) are diversified.

No evidence could be obtained that V and C regions are joined by disulfide, ester, or other non-peptide linkages. In addition, pulse-labeling experiments, carried out with both light and heavy chains, did not support the independent synthesis of V and C regions (Fleischman, 1967; Knopf et al., 1967). Indeed, it was later shown by sequence analysis of immunoglobulin light-chain mRNA that the V and C regions are encoded by a single mRNA molecule (Brownlee et al., 1973). Although fusion did not appear to take place posttranslationally, it might occur at either the RNA or DNA level. In principle, fusion of DNA would seem more economical since the somatic change would then be inherited by all daughter cells in the clone and fusion would not be required after each cell division (Lennox et al., 1967).

In 1965, following the demonstration that light chains are composed of V and C regions, Dreyer and Bennett formally proposed that each light chain is actually encoded by two distinct genes, one for the C region and another for the V region, i.e. that joining takes place at the DNA level. The mechanism they proposed was based on the insertion of temperate viruses, such as λ, into specific loci in the bacterial genome. It was imagined that genes for the V regions are rings of DNA stacked along the chromosome and that one or another of them inserts into a specific nucleotide sequence in the common region. It was assumed that the multiple V-region rings evolved from a common ancestor by gene duplication, which would account for the observed homology among V regions. The main feature of this proposal—the combination of distinct segments of DNA to form a gene encoding an antibody chain—turned out to be correct, although this was not to be demonstrated for more than 10 years (see Part II).

In the next several years, evidence accumulated that the V and C region of each heavy and light chain are encoded by separate genes. Some of this evidence emerged from the sequence analysis of the V regions of a large number of human and mouse light and heavy chains. Thus, when sequences of V_κ regions were compared, it became evident that they fell into related sets, designated subgroups (Gray et al., 1967, Hood et al., 1967, Milstein, 1967, Niall and Edman, 1967; also reviews by Edelman and Gall, 1969, Hood and Talmage, 1970; Milstein and Pink, 1970; see also Asofsky et al., 1969). The sequences within each subgroup are more similar to one another than to sequences in the other subgroups; certain residues are conserved at particular framework positions and there are conserved sequence gaps. These data were generally interpreted as indicating that each of the subgroups is encoded by at least one germline gene; otherwise, parallel mutations would have to occur repeatedly to generate identical replacements within each subgroup. In humans, an allelic (allotypic) phenotype in immunoglobulins was shown to correspond to sequence differences in the C region of κ chains, indicating that the C_κ region is encoded by a single germline gene (Terry et al., 1969). This result could be rationalized with the subgroup data if the V and C regions of each κ chain are encoded by separate genes.

The classification of many V_κ regions into subgroups was facilitated by the introduction of the automated protein sequenator (Edman and Begg, 1967). Unlike many λ and heavy chains, κ chains have free amino-terminal end groups and are amenable to Edman degradation. Consequently the sequence of the amino-terminal ~20–25 amino acid residues of a large number of κ chains was determined and used for subgroup classification. Although residues characteristic of the different subgroups occur at scattered positions in the V region[g], the sequence of the amino-terminal segment was usually sufficient for subgroup assignment.

The λ and heavy chain V-region sequences were also classified into a number of subgroups and the same reasoning applied. However, in the case of the heavy chains, an additional argument for the existence of separate *V* and *C* genes was put forward. When sequences of μ and γ heavy chains were first determined, it was observed that the V-region subgroup classification did not correlate with the class of heavy chain. Thus, the V regions of two γ chains were found to be very similar to the V region of a μ chain, but different from the V region of another γ chain (Press and Hogg, 1969; Wikler et al., 1969). As additional sequence data accumulated, it became clear that this observation could be generalized: the different heavy chain classes share the same V_H subgroups (reviewed by Putnam, 1977).

Even more convincing was the discovery of two myeloma proteins in the same patient: one IgG and the other IgM, in which the μ and γ chains had *identical* V regions (Wang et al., 1970; Wang et al., 1977). The two light chains were also identical (Wang et al., 1969). Thus, only the C regions of the two proteins differed. The C regions of the μ and γ heavy chains are products of distinct genes, and the shared V region is presumably the product of a single gene. Again, the simplest explanation for these observations was to assume that the V and C regions are encoded by separate genes. Individual plasma cells in the bone marrow synthesized the IgG protein or the IgM, but not both (Wang et al., 1969). It was proposed that a clone originally producing one of the myeloma proteins (e.g., IgM) generated a subclone producing the other, the same V_H gene being expressed in association with either C_H gene. Both clones remained viable and continued to produce their immunoglobulin products. Evidence that a single lymphoid cell can synthesize both IgM and IgG antibody, presumably the result of a similar class switch, was presented by Nossal et al. (1964). The basis for the switch in immunoglobulin class will be considered in Part II. These two myeloma heavy chains (different C regions, same V region) represent the converse of the situation found for the sequence of κ light chains (different V-region subgroups, same C region). Both observations are consistent with separate genetic control of V and C regions.

Additional support came from studies of the expression of rabbit allotypic determinants. An allotypic determinant is an epitope reflecting allelic variation of a particular immunoglobulin gene (Oudin, 1956a,b; Dray and Young, 1958; Dubiski et al., 1959; see also reviews by Kelus and Gell, 1967; Mage et al., 1973). In rabbits, there are three alleles at the *a* locus: *a*1, *a*2, and *a*3 (Oudin, 1960b; Dray et al., 1962; Dubiski et al., 1962). In 1963, Charles Todd reported that in an individual rabbit, both IgG and IgM, which have different heavy-chain C regions, express the same *a*-locus allotypic determinants. Shortly thereafter, Arnold Feinstein (1963) observed that immunoglobulins of the IgA class also express the same allotypic determinants as found on IgG.

The *a*-locus allotypic determinants behave genetically like products of classical Mendelian alleles at a single genetic locus. Moreover, as discussed below, they were shown to be associated with heavy chains. However, the C regions of the IgM, IgG, and IgA heavy chains are products of distinct genes. The observation that three (or more) immunoglobulin classes can express the same set of allotypic determinants, which became known, after its discoverer, as the "Todd phenomenon", posed a severe dilemma for understanding the genetic control of heavy chains. How can these polypeptides behave, on the one hand like the products of several genes, and on the other like the products of a single gene? The dilemma would easily be resolved if it were indeed the case that the rabbit heavy chain is encoded by two distinct genes, one for the V region and another for the C region *and* if the allotypic determinants are associated with the V region. In this case, the putative V_H gene, which has three alleles (allotypes), could be expressed in association with *any* of the C_H genes encoding the C regions of the heavy chains of the different immunoglobulin classes.

Also in 1963, the basic four-chain structure of rabbit IgG became known and methods for separating heavy and light chains were available. Accordingly it was now possible to determine the location of the *a*-locus allotypic determinants on the immunoglobulin molecule. They were soon shown to be present on heavy, not light chains, and on the Fab, not the Fc fragment (Kelus et al., 1961; Feinstein et al., 1963; Stemke, 1964; Dray and Nisonoff, 1965; Wilhelm and Lamm, 1966). Therefore, they had to be associated with the heavy-chain segment in Fab, which is called Fd and consists of C_H1 (the first constant domain of the heavy chain) and V_H. Additional evidence consistent with this assignment was obtained by analyzing the amino acid composition and sequence of Fd derived from IgG of restricted allotypy (from partially inbred rabbits). Although heterogeneity of the V regions presented difficulties for these studies, it appeared nonetheless that the overall amino acid composition of

Fd reflected allotypic differences (Koshland, 1967). Moreover, sequence analysis suggested that certain V_H, but no C_H1, positions correlated with the *a*-locus allotype (Wilkinson, 1969; Fleischman, 1971, 1973; Mole et al., 1971), thereby further localizing the allotypic differences to the V_H portion of Fd. Presumably, conserved features of framework regions in V_H account for the allotype-related features.

By the early 1970s, most immunologists were firmly convinced that the "two genes–one polypeptide" hypothesis must be correct, but this idea had not penetrated into broader biological circles and, where it had, skepticism prevailed. The debate in the immunology community now centered on the number of genes in the germline required to encode all the V regions.

MANY GERMLINE GENES OR FEW?

It had long been realized that the number of different antibody specificities must be enormous, although it has never been possible to estimate this number with any degree of confidence. As discussed previously, a persuasive argument that the repertoire is large was provided by Landsteiner who demonstrated that antibodies specific for many arbitrary chemical groups could be produced. Thus, the number of specificities would appear to exceed even the number of antigens found in nature. That most antigens have many immunogenic epitopes further compounds the number of possible antibody combining sites. Indeed, even the response to a single haptenic group attached to a unique residue in a protein can be heterogeneous. This was shown by preparing a derivative of ribonuclease with the 2,4-dinitrophenyl (DNP) group attached only to the ϵ -amino sidechain of the lysyl residue at position 41. The anti-DNP response to this immunogen appeared to be as heterogeneous as the response to proteins having many DNP groups attached to different residues (Eisen et al., 1964).

As soon as it became accepted that antibodies, like other proteins, are encoded by genes, immunologists began to grapple with the question of the number of genes that would be required to account for all the specificities. The possibility of forming antibody sites by diverse combinations of heavy and light-chain V regions might effect some reduction in the total number of genes required, but it was nonetheless agreed that this number must be "large". The fundamental question that was then asked was: are all of the genes necessary for antibody formation present in the germline of each individual or are one or a few germline genes diversified extensively by somatic processes during the lifetime of each individual? This question had indeed been framed by Lederberg (1959) soon after Burnet's proposal of the clonal selection theory.

The strength of the "germline position", as summarized for example by Hood and Talmage (1970), was that no special biological mechanisms are required to generate all the different antibodies. Multiple *V* genes arise through evolutionary time by the accepted means of gene duplication, mutation, and selection, just like the genes encoding any other protein. On the other hand, the "somatic position" posited that these same genetic events occur, but in the lifetime of each individual. Selection acts on cells producing different antibodies rather than on individuals (or species). This view had been proposed by Burnet (1959) and Lederberg (1959) and was subsequently championed by Milstein (Brenner and Milstein, 1966), Jerne (1971), and Cohn and Weigert (Cohn et al., 1974), among others. Although not usually discussed in these terms, it is interesting to consider that the instructionists were actually the ultimate proponents of a somatic position since they proposed that antigen acts either on the cellular antibody-forming "machinery" or directly on the antibody itself. If the Pauling model is viewed in a genetic light, then a single globulin-producing gene would suffice for all the antibodies.

Impassioned debates about the two extreme positions and assorted intermediate views raged through the early 1970s (see monograph by Kindt and Capra, 1984). Opponents of the germline position pointed out that a substantial fraction of the genome would have to be devoted to all the *V* genes; to this argument proponents responded essentially "so what?". Further, was it reasonable to suppose that forces of mutation and selection, acting through evolutionary time, would yield V regions specific for artificial determinants of the sort Landsteiner had shown could elicit specific responses? Possibly not, but neither did it seem obvious that somatic processes would select for such antibodies. The existence of nonfunctional genes posed something of a problem for the germline theory since it was not obvious how such genes would be selected against to prevent their accumulation over eons of evolutionary time; the presence of nonfunctional genes arising as a result of somatic processes in the lifetime of each individual seemed less of a problem. If a somatic theory was to be accepted, did mutations occur uniformly throughout the V region, selection resulting in the concentration of observed variation in CDRs, or was the mutation process itself directed by some means to the CDRs?

It was generally agreed that the existence of V-region subgroups, as discussed in the preceding section, was strong evidence in favor of the existence of at least one germline gene for the V regions in each subgroup. However, there was no agreement on the exact definition of a subgroup. A relatively stringent definition would place fewer sequences in each subgroup and hence require more subgroups and more germline *V* genes. Moreover, as the number of sequences analyzed increased, so did the number of subgroups and, consequently, the minimum number of germline genes. Nonetheless, an extreme somatic model, e.g.., only one germline gene for all V_κ was not tenable in the face of even the most lax definition for a subgroup.

An observation that seemed consistent with somatic mutation as a source of antibody diversification was the pattern of sequence variation in mouse λ light chains. Most of the V regions of these light chains were found to be identical in amino acid sequence; the minority of V_λ that differed in sequence had only one to three amino acid replacements and these were usually in the CDRs. It was argued that these data suggest that the V_λ are encoded by a single germline gene that is varied to a limited extent by somatic mutation (Weigert et al., 1970). These data will be discussed in more detail in Part II.

A major difficulty for the position that V regions are encoded by multiple germline genes was the existence of the *a*-locus allotypic determinants associated with the variable regions of rabbit heavy chains, as described in the preceding section. The genes encoding the V regions expressing these epitopes behave as if they are alleles at a single genetic locus. A variety of alternatives were considered to explain the inheritance pattern. Perhaps the allotypes are alleles of a regulatory, not a structural gene. However, no evidence for such a possibility was forthcoming and the inheritance of the *a*-locus allotypic determinants remained a strong argument for the existence of only a single V_H gene in the germline, at least of rabbits. (An explanation for the allelic behavior of rabbit V_H genes will be discussed in Part II.)

Another argument that was advanced to favor relatively few germline genes was the presence, at specific positions in V regions, of "species specific" or "phylogenetically associated" residues. A residue at such a position is found, for example, in all (or most) V_κ regions of a particular species, and differs from the residue at that same position in the V_κ of other species. If all V_κ regions were encoded by separate germline genes, then during the evolution of a new species, each of these genes would have to undergo the same mutation at the position in question. Again, the occurrence of many parallel mutations seemed unlikely. But in this case also, the correlation of species with specific residues weakened as more sequences were examined.

So the arguments continued through the early 1970s, the difficulties with each theory being countered by *ad hoc* solutions. Although both the somatic and germline camps had their die-hard adherents, most immunologists adopted a "wait and see" attitude. They did not have long to wait.

ACKNOWLEDGMENTS

I would like to acknowledge those who have indirectly or directly contributed to the writing of this article. I am particularly indebted to Fred Richards, Herman Eisen and the late Rodney Porter, in whose laboratories I was introduced to research in protein chemistry and immunology. The Helen Hay Whitney Foundation and the American Heart Association supported my transition from medicine into basic science. I thank the Institute of Allergy and Infectious Diseases of the National Institutes of Health for supporting my research for the last 30 years, via grant AI-08054.

In the preparation of this chapter, I have benefited from discussions with a number of colleagues, who suggested particular items for inclusion, corrected errors, and provided support during a process that neither I nor they (nor my editors) thought would ever end. These colleagues include Neil Barclay, Herman Eisen, Julian Fleischman, John Kimball, Norman Klinman, Leonard Lerman, Alfred Nisonoff, and Carol Warner. I received invaluable assistance from the staff of the MIT Libraries, in particular from Louisa Worthington, Anne Battis, and Paul Vermouth, in obtaining books and articles not in our collection. I thank Catherine Willett, whose artistic sense and computer skills are responsible for Figure 1, and Edwin Kim and Hilda Harris-Ransom for collecting and typing innumerable references.

NOTES

[a] The terms B lymphocyte and B cell are used interchangeably, as are T lymphocyte and T cell.

[b] The first Nobel Prize in physiology or medicine was awarded to Emil von Behring for his work on "serum therapy."

[c] In Ehrlich's words, "food-stuffs" are "those substances which are able to enter into the composition of the protoplasm, and so are readily assimilated." Toxins (which can induce antibody formation) have "haptophore groups" allowing the toxin to be recognized by the cell's "side-chains". "The *antitoxines represent nothing more than side-chains reproduced in excess during regeneration, and therefore pushed off from the protoplasm, and so coming to exist in a free state*" (italics Ehrlich's). Another "group ... designated toxophore, is the cause of the toxic action." Other substances, "alkaloids, aromatic amines, antipyretics, aniline dyes" do not contain "haptophore groups" and cannot be "assimilated;" hence, they do not cause antibody formation, a belief that would, before two decades were out, be countered by the studies of Landsteiner who showed that such groups, if coupled to suitable carrier proteins, could induce specific antibodies. "The relationship of the corresponding groups," i.e., the "side-chains" (antitoxins) and "food-stuffs" (toxins) "must be specific... adapted to one another, as, e.g., male and female screw (Pasteur), or as lock and key (E. Fischer)". Further, "the cells become, so to say, educated or trained to reproduce the necessary side-chains in ever-increasing quantity."

[d] Many of the ligands used by Landsteiner and others to explore the range of antibody specificity were haptens coupled to proteins. The term hapten was introduced by Landsteiner (1921) to designate substances that can react specifically with antibodies, but that cannot alone induce antibody formation. To induce antibody formation, haptens must be conjugated to immunogenic carriers (usually proteins). Molecular size is not part of the definition; many, but not all, haptens are relatively low in molecular weight.

[e] Silverstein (1996) has recently pointed out that the passive transfer of maternal antibody to fetus (placental transfer) and newborn (via maternal milk) had already been described by Ehrlich in a series of papers in 1892–1893.

[f] The terms isotypic, allotypic, and idiotypic, referring to epitopes (or "specificities") of immunoglobulins, were introduced by Oudin (1956b, 1960a,b, 1966). See also Dray et al., 1962.

[g] The carboxyl-terminal 12 or 13 residues of the V region do not conform to the subgroup classification that is evident in the remainder of the region; the reason for this became clear when the genes encoding the V region were identified, as will be discussed in Part II of this series.

REFERENCES

Ada, G.L. & Byrt, P. (1969). Specific inactivation of antigen-reactive cells with ^{125}I-labelled antigen. Nature 222, 1291-1292.

Alexander, J. (1931). Some intracellular aspects of life and disease. Protoplasma 14, 296-306.

Amit, A.G., Mariuzza, R.A., Phillips, S.E.V., & Poljak, R.J. (1986). Three-dimensional structure of an antigen-antibody complex at 2.8 Å resolution. Science 233, 747-753.

Anderson, D., Billingham, R.E., Lampkin, G.H., & Medawar, P.B. (1951). The use of skin grafting to distinguish between monozygotic and dizygotic twins in cattle. Heredity 5, 379-397.

Askonas, B.A., Williamson, A.R., & Wright, B.E.G. (1970). Selection of a single antibody-forming cell clone and its propagation in syngeneic mice. Proc. Natl. Acad. Sci. USA 67, 1398-1403.

Asofsky, R., Binaghi, R.A., Edelman, G.M., Goodman, H.C., Heremans, J.F., Hood, L., Kabat, E.A., Rejnek, J., Rowe, D.S., Small, P.A., & Trnka, Z. (1969). An extension of the nomenclature for human immunoglobulins. Bull. Wld. Hlth. Org. 41, 975-978.

Attardi, G., Cohn, M., Horibata, K., & Lennox, E.S. (1959). Symposium on the biology of cells modified by viruses or antigens. II. On the analysis of antibody synthesis at the cellular level. Bacteriol. Revs. 23, 213-223.

Attardi, G., Cohn, M., Horibata, K., & Lennox, E.S. (1964a). Antibody formation by rabbit lymph node cells. I. Single cell responses to several antigens. J. Immunol. 92, 335-345.

Attardi, G., Cohn, M., Horibata, K., & Lennox, E.S. (1964b). Antibody formation by rabbit lymph node cells. III. The controls for microdrop and micropipet experiments. J. Immunol. 92, 356-371.

Benian, G.M., Kiff, J.E., Neckelmann, N., Moerman, D.G., & Waterston, R.H. (1989). Sequence of an unusually large protein implicated in regulation of myosin activity in *C. elegans*. Nature 342, 45-50.

Bentley, G.A., Boulot, G., Riottot, M.M., & Poljak, R.J. (1990). Three-dimensional structure of an idiotope-anti-idiotope complex. Nature 348, 254-257.

Bhat, T.N., Bentley, G.A., Fischmann, T.O., Boulot, G., & Poljak, R.J. (1990). Small rearrangements in structures of Fv and Fab fragments of antibody D1.3 on antigen binding. Nature, 347 483-485.

Bibel, D.J. (1988). Milestones in Immunology. A Historical Exploration. Science Tech Publishers, Madison.

Billingham, R.E., Brent, L., & Medawar, P.B. (1953). 'Actively acquired tolerance' of foreign cells. Nature 172, 603-606.

Billingham, R.E., Lampkin, G.H., Medawar, P.B., & Williams, H.L.L. (1952). Tolerance to homografts, twin. Diagnosis and the freemartin condition in cattle. Heredity 6, 201-212.

Bjorkman, P.J., Saper, M.A., Samraoui, B., Bennett, W.S., Strominger, J.L., & Wiley, D.C. (1987). Structure of the human class I histocompatibility antigen, HLA-A2. Nature 329, 506-512.

Bonner, W.A., Hulett, H.R., Sweet, R.G., & Herzenberg, L.A. (1972). Fluorescence activated cell sorting. Rev. Sci. Instrum. 43, 404-409.

Bordet, J. (1898). Sur l'agglutination et la dissolution des globules rouges par le sérum d'animaux injectés de sang défibriné. Annales de l'Institute Pasteur 12, 688-695.

Braden, B.C. & Poljak, R.J. (1995). Structural features of the reactions between antibodies and protein antigens. FASEB J. 9, 9-15.

Brambell, F.W.R., Hemmings, W.A., Oakley, C.L., & Porter, R.R. (1960). The relative transmission of the fractions of papain hydrolyzed homologous γ-globulin from the uterine cavity to the foetal circulation in the rabbit. Proc. R. Soc. Lond. B 151, 478-482.

Breinl, F. & Haurowitz, F. (1930). Chemische Untersuchung des Präzipitates aus Hämoglobin und Anti-Hämoglobin-Serum und Bemerkungen über die Natur der Antikörper. Z. Physiol. Chem. 192, 45-57.

Brenner, S. & Milstein, C. (1966). Origin of antibody variation. Nature 211, 242-243.

Brient, B.W. & Nisonoff, A. (1970). Quantitative investigations of idiotypic antibodies. IV. Inhibition by specific haptens of the reaction of anti-hapten antibody with its anti-idiotypic antibody. J. Exp. Med. 132, 951-962.

Brown, J.H., Jardetzky, T.S., Gorga, J.C., Stern, L.J., Urban, R.G., Strominger, J.L., & Wiley, D.C. (1993). Three-dimensional structure of the human class II histocompatibility antigen HLA-DR1. Nature 364, 33-39.

Brownlee, G.G., Cartwright, E.M., Cowan, N.J., Jarvis, J.M., & Milstein, C. (1973). Purification and sequence of messenger RNA for immunoglobulin light chains. Nature New Biology 244, 236-240.

Brownstone, A., Mitchison, N.A., & Pitt-Rivers, R. (1966). Biological studies with an iodine-containing synthetic immunological determinant 4-hydroxy-3-iodo-5-nitrophenylacetic acid (NIP) and related compounds. Immunology 10, 481-492.

Burnet, F.M. (1941). The Production of Antibodies. MacMillan, Melbourne.

Burnet, F.M. (1957). A modification of Jerne's theory of antibody production using the concept of clonal selection. Aust. J. Sci. 20, 67-69.

Burnet, F.M. (1959). The clonal selection theory of acquired immunity. Vanderbilt University, Nashville.

Burnet, M. (1964). The clonal selection theory of immunity—A Darwinian modification. Aust. J. Sci. 27, 6-7.

Burnet, F.M. & Fenner, F. (1948). Genetics and immunology. Heredity 2, 289-324.

Burnet, F.M. & Fenner, F. (1949). The Production of Antibodies. MacMillan, Melbourne.

Cecil, R. & Wake, R.G. (1962). The reactions of inter- and intra-chain disulphide bonds in proteins with sulphite. Biochem. J. 82, 401-406.

Ceppellini, R., Dray, S., Edelman, G., Fahey, J., Franěk, F., Franklin, E., Goodman, H.C., Grabar, P., Gurvich, A.E., Heremans, J.F., Isliker, H., Karush, F., Press, E., & Trnka, Z. (1964). Nomenclature for human immunoglobulins. Bull. Wld. Hlth. Org. 30, 447-450.

Claman, H.N., Chaperon, E.A., & Triplett, R.F. (1966). Thymus-marrow cell combinations. Synergism in antibody production. Proc. Soc. Exp. Biol. Med. 122, 1167-1171.

Cohn, M. (1994). The wisdom of hindsight. Annu. Rev. Immunol. 12, 1-62.

Cohn, M., Blomberg, B., Geckeler, W., Raschke, W., Riblet, R., & Weigert, M. (1974). First order considerations in analyzing the generator of diversity. In: The Immune System: Genes, Receptors, Signals (Sercarz, E.E., Williamson, A.R., & Fox, C.F., Eds.), pp. 89-117. Academic Press, New York.

Coons, A.H. (1958). The cytology of antibody formation. J. Cell. Comp. Physiol. 52 (Suppl. 1), 55-67.

Cruse, J.M. & Lewis, R.E. (1995). Illustrated Dictionary of Immunology. CRC Press, Boca Raton.

Cunningham, B.A., Hemperly, J.J., Murray, B.A., Prediger, E.A., Brackenbury, R., & Edelman, G.M. (1987). Neural cell adhesion molecule: structure, immunoglobulin-like domains, cell surface modulation, and alternative RNA splicing. Science 236, 799-806.

D'Eustachio, P., Bothwell, A.L.M., Takaro, T.K., Baltimore, D., & Ruddle, F.H. (1981). Chromosomal location of structural genes encoding murine immunoglobulin λ light chains. J. Exp. Med. 153, 793-800.

D'Eustachio, P., Pravtcheva, D., Marcu, K., & Ruddle, F.H. (1980). Chromosomal location of the structural gene cluster encoding murine immunoglobulin heavy chains. J. Exp. Med. 151, 1545-1550.

Davies, D.R. & Cohen, G.H. (1996). Interactions of protein antigens with antibodies. Proc. Natl. Acad. Sci. USA 93, 7-12.

Deisenhofer, J. (1981). Crystallographic refinement and atomic methods of a human Fc fragment and its complex with fragment B of protein A from *Staphylococcus aureus* at 2.9- and 2.8-Å resolution. Biochemistry 20, 2361-2370.

Dray, S., Dubiski, S., Kelus, A., Lennox, E.S., & Oudin, J. (1962). A notation for allotypy. Nature 195, 785-786.

Dray, S. & Nisonoff, A. (1965). Relationship of genetic control of allotypic specificities to the structure and biosynthesis of rabbit γ-globulin. In: Molecular and Cellular Basis of Antibody Formation (Sterzl, J., Ed.), pp. 175-191. Academic Press, New York.

Dray, S. & Young, G.O. (1958). Differences in the antigenic components of sera of individual rabbits as shown by induced isoprecipitins. J. Immunol. 81, 142-149.

Dreyer, W.J. & Bennett, J.C. (1965). The molecular basis of antibody formation: a paradox. Proc. Natl. Acad. Sci. USA 54, 864-869.

Dubiski, S., Dudziak, Z., Skalba, D., & Dubiska, A. (1959). Serum groups in rabbits. Immunology 2, 84-92.

Dubiski, S., Rapacz, J., & Dubiska, A. (1962). Heredity of rabbit gamma-globulin iso-antigens. Acta Genet. 12, 136-155.

Du Pasquier, L. (1993). Evolution of the immune system. In: Fundamental Immunology, 3rd ed. (Paul, W.E., Ed.), pp. 199-233. Raven Press, New York.

Edelman, G.M. (1959). Dissociation of γ-globulin. J. Am. Chem. Soc. 81, 3155-3156.

Edelman, G.M. (1987). CAMS and Igs: cell adhesion and the evolutionary origins of immunity. Immunol. Rev. 100, 11-45.

Edelman, G.M., Cunningham, B.A., Gall, W.E., Gottlieb, P.D., Rutishauser, U., & Waxdal, M.J. (1969). The covalent structure of an entire g immunoglobulin molecule. Proc. Natl. Acad. Sci. USA 63, 78-85.

Edelman, G.M. & Gall, W.E. (1969). The antibody problem. Annu. Rev. Biochem. 38, 415-466.

Edelman, G.M. & Gally, J.A. (1962). The nature of Bence-Jones proteins. Chemical similarities to polypeptide chains of myeloma globulins and normal γ-globulins. J. Exp. Med. 116, 207-227.

Edelman, G.M. & Poulik, M.D. (1961). Studies on structural units of the γ-globulins. J. Exp. Med. 113, 861-884.

Edman, P. and Begg, G. (1967). A protein sequenator. European J. Biochem. 1, 80-91.

Ehrich, W.E. & Harris, T.N. (1942). The formation of antibodies in the popliteal lymph node in rabbits. J. Exp. Med. 76, 335-348.

Ehrlich, P. (1900). On immunity with special reference to cell life. Proc. R. Soc. Lond. 66, 424-448.

Eisen, H.N., Carsten, M.E., & Belman, S. (1954). Studies of hypersensitivity to low molecular weight substances. III. The 2,4-dinitrophenyl group as a determinant in the precipitin reaction. J. Immunol. 73, 296-308.

Eisen, H.N. & Karush, F. (1949). The interaction of purified antibody with homologous hapten. Antibody valence and binding constant. J. Amer. Chem. Soc. 11, 363-364.

Eisen, H.N., Little, J.R., Osterland, C.K., & Simms, E.S. (1967). A myeloma protein with antibody activity. Cold Spring Harbor Symp. Quant. Biol. 32, 75-81.

Eisen, H.N., Michaelidis, M.C., Underdown, B.J., Schulenburg, E.P., & Simms, E.S. (1970). Myeloma proteins with antihapten activity. Fed. Proc. 29, 78-84.

Eisen, H.N., Simms, E.S., Little, J.R., Jr., & Steiner, L.A. (1964). Fed. Proc. 23, 559.

Eisen, H.N. & Siskind, G.W. (1964). Variations in affinities of antibodies during the immune response. Biochemistry 3, 996-1008.

Ely, K.R., Colman, P.M., Abola, E.E., Hess, A.C., Peabody, D.S., Parr, D.M., Connell, G.E., Laschinger, C.A., & Edmundson, A.B. (1978). Mobile Fc region in the Zie IgG2 cryoglobulin: comparison of crystals of the $F(ab')_2$ fragment and the intact immunoglobulin. Biochemistry 17, 820-823.

Epstein, C.J., Goldberger, R.F., & Anfinsen, C.B. (1963). The genetic control of tertiary protein structure: studies with model systems. Cold Spring Harbor Symp. Quant. Biol. 28, 439-449.

Erikson, J., Martinis, J., & Croce, C.M. (1981). Assignment of the genes for human λ immunoglobulin chains to chromosome 22. Nature 294, 173-175.

Fahey, J.L. (1962). Heterogeneity of γ-globulins. Adv. Immunol. 2, 41-109.

Feinstein, A. (1963). Character and allotypy of an immune globulin in rabbit colostrum. Nature 199, 1197-1199.

Feinstein, A., Gell, P.G.H., & Kelus, A.S. (1963). Immunochemical analysis of rabbit gamma-globulin allotypes. Nature 200, 653-654.

Fett, J.W., Deutsch, H.F., & Smithies, O. (1973). Hinge-region deletion localized in the IgG1-globulin Mcg. Immunochemistry 10, 115-118.

Fields, B.A., Goldbaum, F.A., Ysern, X., Poljak, R.J., & Mariuzza, R.A. (1995). Molecular basis of antigen mimicry by an anti-idiotope. Nature 374, 739-742.

Fleischman. J.B. (1963). The synthesis of γ-globulin by rabbit lymph node cells. J. Immunol. 91, 163-169.

Fleischman, J.B. (1967). Synthesis of the γG heavy chain in rabbit lymph node cells. Biochemistry 6, 1311-1320.

Fleischman, J.B. (1971). A partial amino acid sequence in the heavy chain of a rabbit antibody to group C streptococcal carbohydrate. Biochemistry 10, 2753-2761.

Fleischman, J.B. (1973). Amino acid sequences in the Fd of a rabbit antibody heavy chain. Immunochemistry 10, 401-407.

Fleischman, J.B. (1981). Citation classic: commentary on citations of "the arrangement of the peptide chains in γ-globulin". Current Contents 24, no. 11, 17.

Fleischman, J.B., Pain, R.H., & Porter, R.R. (1962) Reduction of γ-globulins. Arch. Biochem., Biophys., Suppl. 1, 174-180.

Fleischman, J.B., Porter, R.R., & Press, E.M. (1963). The arrangement of the peptide chains in γ-globulin. Biochem. J. 88, 220-228.

Gallagher, R.B., Gilder, J., Nossal, G.J.V., & Salvatore, G. (1995). Immunology: The Making of Modern Science. Academic Press, London.

Gell, P.G.H. & Kelus, A. (1964). Anti-antibody or clone-product? Nature 201, 687-689.

Gowans, J.L. (1996). The lymphocyte—a disgraceful gap in medical knowledge. Immunol. Today 17, 288-291.

Gowans, J.L., McGregor, D.D., Cowen, D.M., & Ford, C.E. (1962). Initiation of immune responses by small lymphocytes. Nature 196, 651-655.

Gray, W.R., Dreyer, W.J., & Hood, L. (1967). Mechanism of antibody synthesis: size differences between mouse kappa chains. Science 155, 465-467.

Green, I., Vassalli, P., Nussenzweig, V., & Benacerraf, B. (1967). Specificity of the antibodies produced by single cells following immunization with antigens bearing two types of antigenic determinants. J. Exp. Med. 125, 511-526.

Greenspan, N.S. & Bona, C.A. (1993). Idiotypes: structure and immunogenicity. FASEB J. 7, 437-444.

Guddat, L.W., Herron, J.N., & Edmundson, A.B. (1993). Three-dimensional structure of a human immunoglobulin with a hinge deletion. Proc. Natl. Acad. Sci. USA 90, 4271-4275.

Haber, E. (1964). Recovery of antigenic specificity after denaturation and complete reduction of disulfides in a papain fragment of antibody. Proc. Natl. Acad. Sci. USA 52, 1099-1106.

Harrelson, A.L. & Goodman, C.S. (1988). Growth cone guidance in insects: fasciclin II is a member of the immunoglobulin superfamily. Science 242, 700-708.

Harris, L.J., Larson, S.B., Hasel, K.W., Day, J., Greenwood, A., & McPherson, A. (1992). The three-dimensional structure of an intact monoclonal antibody for canine lymphoma. Nature 360, 369-372.

Haurowitz, F. & Schwerin, P. (1943). The specificity of antibodies to antigens containing two different determinant groups. J. Immunol. 47, 111-119.

Heidelberger, M. & Pedersen, K.O. (1937). The molecular weight of antibodies. J. Exp. Med. 65, 393-414.

Heidelberger, M. & Pedersen, K.O. & Tiselius, A. (1936). Ultracentrifugal and electrophoretic studies on antibodies. Nature 138, 165.

Herzenberg, L.A. (1964). A chromosome region for $gamma_{2a}$ and $beta_{2A}$ globulin H chain isoantigens in the mouse. Cold Spring Harbor Symp. Quant. Biol. 29, 455-462.

Hill, R.L., Delaney, R., Fellows, R.E., Jr., & Lebovitz, H.E. (1966). The evolutionary origins of the immunoglobulins. Proc. Natl. Acad. Sci. USA 56, 1762-1769.

Hilschmann, N. & Craig, L.C. (1965). Amino acid sequence studies with Bence-Jones proteins. Proc. Natl. Acad. Sci. USA 53, 1403-1409.

Holden, H.M., Ito, M., Hartshorne, D.J., & Rayment, I. (1992). X-Ray structure determination of telokin, the C-terminal domain of myosin light chain kinase, at 2.8 Å resolution. J. Mol. Biol. 227, 840-851.

Hood, L., Gray, W.R., Sanders, B.G., & Dreyer, W.J. (1967). Light chain evolution. Cold Spring Harbor Symp. Quant. Biol. 32, 133-146.

Hood, L. & Talmage, D.W. (1970). Mechanism of antibody diversity: germ line basis for variability. Science 168, 325-334.

Inbar, D., Hochman, J., & Givol, D. (1972). Localization of antibody-combining sites within the variable portions of heavy and light chains. Proc. Natl. Acad. Sci. USA 69, 2659-2662.

Ishizaka, K., Ishizaka, T., & Sugahara, T. (1962). Biological activity of soluble antigen-antibody complexes. VII. Role of an antibody fragment in the induction of biological activities. J. Immunol. 88, 690-701.

Jacobson, E.B., L'Age-Stehr, J., & Herzenberg, L.A. (1970). Immunological memory in mice. II. Cell interactions in the secondary immune response studied by means of immunoglobulin allotype markers. J. Exp. Med. 131, 1109-1120.

Jerne, N.K. (1955). The natural-selection theory of antibody formation. Proc. Natl. Acad. Sci. USA 41, 849-857.

Jerne, N.K. (1967). Summary: Waiting for the end. Cold Spring Harbor Symp. Quant. Biol. 32, 591-603.

Jerne, N.K. (1971). The somatic generation of immune recognition. Eur. J. Immunol. 1, 1-9.

Jerne, N.K. (1974). Towards a network theory of the immune system. Ann. Immunol. (Inst. Pasteur) 125C, 373-389.

Jerne, N.K. & Nordin, A.A. (1963). Plaque formation in agar by single antibody-producing cells. Science 140, 405.

Julius, M.H. & Herzenberg, L.A. (1974). Isolation of antigen-binding cells from unprimed mice. J. Exp. Med. 140, 904-920.

Kabat, E.A. (1939). The molecular weight of antibodies. J. Exp. Med. 69, 103-118.

Karush, F. (1958). Specificity of antibodies. Trans. N.Y. Acad. Sci. 20, 581-592.

Karush, F. (1962). The role of disulfide bonds in the acquisition of immunologic specificity. J. Pediatrics 60, 103-113.

Kelus, A.S. & Gell, P.G.H. (1967). Immunoglobulin allotypes of experimental animals. Prog. Allergy 11, 141-184.

Kelus, A., Marrack, J.R., & Richards, C.B. (1961). The antigenic specificity of the pieces of the γ-globulin molecules of individual rabbits. In: Protides of the Biological Fluids, Proc. of the 8th Colloquium, Bruges (Peeters, H., Ed.), pp. 176-179. Elsevier, Amsterdam.

Kimmel, J.R. & Smith, E.L. (1954). Crystalline papain. I. Preparation, specificity, and activation. J. Biol. Chem. 207, 515-531.

Kindt, T.J. & Capra, J.D. (1984). The Antibody Enigma. Plenum, New York.

Kirsch, I.R., Morton, C.C., Nakahara, K. & Leder, P. (1982). Human immunoglobulin heavy chain genes map to a region of translocations in malignant B lymphocytes. Science 216, 301-303.

Klinman, N.R. (1969). Antibody with homogeneous antigen binding produced by splenic foci in organ culture. Immunochemistry 6, 757-759.

Knopf, P.M., Parkhouse, R.M.E., & Lennox, E.S. (1967). Biosynthetic units of an immunoglobulin heavy chain. Proc. Natl. Acad. Sci. USA 58, 2288-2295.

Koch, R. (1891). Fortsetzung der Mittheilungen über ein Heilmittel gegen Tuberculose. Deutsche Medizinische Wochenschrift 17, 101-102.

Köhler, G. & Milstein, C. (1975). Continuous cultures of fused cells secreting antibody of predefined specificity. Nature 256, 495-497.

Korngold, L. & Lipari, R. (1956). Multiple-myeloma protein. III. The antigenic relationship of Bence Jones proteins to normal gamma-globulin and multiple-myeloma serum proteins. Cancer 9, 262-272.

Koshland, M.E. (1967). Location of specificity and allotypic amino acid residues in antibody Fd fragments. Cold Spring Harbor Symp. Quant. Biol. 32, 119-127.

Kunkel. H.G. (1970). Individual antigenic specificity, cross specificity and diversity of human antibodies. Fed. Proc. 29, 55-58.

Kunkel, H.G., Mannik, M., & Williams, R.C. (1963). Individual antigenic specificity of isolated antibodies. Science 140, 1218-1219.

Labeit, S., Barlow, D.P., Gautel, M., Gibson, T., Holt, J., Hsieh, C.-L., Francke, U., Leonard, K., Wardale, J., Whiting, A., & Trinick, J. (1990). A regular pattern of two types of 100 residue motif in the sequence of titin. Nature 345, 273-276.

Landsteiner, K. (1919). Über die Bedeutung der Proteinkomponente bei den Präcipitinreaktionen der Azoproteine. XIII. Mitteilung über Antigene. Biochem. Z. 93, 106-118.

Landsteiner, K. (1920). Spezifische Serumreaktionen mit einfach zusammengesesetzten Substanzen von bekannter Konstitution (organischen säuren). XIV. Mitteilung über Antigene und serologische Spezifität. Biochem. Z. 104, 280-299.

Landsteiner, K. (1921). Über heterogenetische Antigen und Hapten XV. Mitteilung über Antigene. Biochem. Z. 119, 294-306.

Landsteiner, K. (1936). The specificity of serological reactions. Charles C. Thomas, Springfield, IL. This was the first English edition of the original German edition, Die Spezifizität der serologischen Reaktionen. Julius Springer, Berlin, 1933. The second English edition was published by Harvard University Press, Cambridge, MA in 1945.

Landsteiner, K. & Lampl, H. (1918). Über die Abhängigkeit der Serologischen spezifizität von der chemischen Struktur. (Darstellung von Antigenen mit bekannter chemischer Konstitution der spezifischen Gruppen.) XII. Mitteilung über Antigene. Biochem. Z. 86, 343-394.

Landsteiner, K. & van der Scheer, J. (1931). On the specificity of serological reactions with simple chemical compounds (inhibition reactions). J. Exp. Med. 54, 295-305.

Landsteiner, K. & van der Scheer, J. (1938). On cross reactions of immune sera to azoproteins. II. Antigens with azocomponents containing two determinant groups. J. Exp. Med. 67, 709-723.

Lederberg, J. (1959). Genes and antibodies. Do antigens bear instructions for antibody specificity or do they select cells lines that arise by mutation? Science 129, 1649-1653.

Lennox, E.S. & Cohn, M. (1967). Immunoglobulins. Annu. Rev. Biochem. 36, 365-406.

Lennox, E.S., Knopf, P.M., Munro, A.J., & Parkhouse, R.M.E. (1967). A search for biosynthetic subunits of light and heavy chains of immunoglobulins. Cold Spring Harbor Symp. Quant. Biol. 32, 249-254.

Lerman, L.S. (1949). Studies on the Reaction of Antibody with Simple Substances. Ph.D. Thesis, California Institute of Technology.

Lieberman, R. & Potter, M. (1966). Close linkage in genes controlling γA and γG heavy chain structure in BALB/c mice. J. Mol. Biol. 18, 516-528.

Littlefield, J.W. (1964). Selection of hybrids from matings of fibroblasts *in vitro* and their presumed recombinants. Science 145, 709-710.

Mage, R. , Lieberman, R., Potter, M., & Terry, W.D. (1973). Immunoglobulin allotypes. In: The Antigens (Sela, M., Ed.), Vol. 4, pp. 299-376. Academic Press, New York.

Mäkelä, O. (1967). The specificity of antibodies produced by single cells. Cold Spring Harbor Symp. Quant. Biol. 32, 423-430.

Mäkelä, O. & Nossal, G.J.V. (1961). Bacterial adherence: a method for detecting antibody production by single cells. J. Immunol. 87, 447-456.

Malcolm, S., Barton, P., Murphy, C., Ferguson-Smith, M.A., Bentley, D.L., & Rabbits, T.H. (1982). Localization of human immunoglobulin κ light chain variable region genes to the short arm of chromosome 2 by *in situ* hybridization. Proc. Natl. Acad. Sci. USA 79, 4957-4961.

Marquart, M., Deisenhofer, J., Huber, R., & Palm, W. (1980). Crystallographic refinement and atomic models of the intact immunoglobulin molecule Kol and its antigen-binding fragment at 3.0 Å and 1.9 Å resolution. J. Mol. Biol. 141, 369-391.

McBride, O.W., Hieter, P.A., Hollis, G.F., Swan, D., Otey, M.C., & Leder, P. (1982). Chromosomal location of human kappa and lambda immunoglobulin light chain constant region genes. J. Exp. Med. 155, 1480-1490.

McDevitt, H.O., Askonas, B.A., Humphrey, J.H., Schechter, I., & Sela, M. (1966). The localization of antigen in relation to specific antibody-producing cells. I. Use of a synthetic polypeptide [(T,G)-A—L] labelled with iodine-125. Immunol. 11, 337-351.

Metchnikoff, E. (1892). Lecons sur la pathologie comparée de l'inflamation faites a l'Institute Pasteur in Avril et Mai 1891. Masson, Paris.

Milstein, C. (1967). Linked groups of residues in immunoglobulin κ chains. Nature 216, 330-332.

Milstein, C. & Pink, J.R.L. (1970). Structure and evolution of immunoglobulins. Prog. Biophys. Mol. Biol. 21, 209-263.

Mirsky, A.E. & Pauling, L. (1936). On the structure of native, denatured, and coagulated proteins. Proc. Natl. Acad. Sci. USA 22, 439-447.

Mitchell, G.F. & Miller, J.F.A.P. (1968). Cell to cell interaction in the immune response. II. The source of hemolysin-forming cells in irradiated mice given bone marrow and thymus or thoracic duct lymphocytes. J. Exp. Med. 128, 821-837.

Mitchison, N.A. (1967). Antigen recognition responsible for the induction *in vitro* of the secondary response. Cold Spring Harbor Symp. Quant. Biol. 32, 431-439.

Mitchison, N.A., Rajewsky, K., & Taylor, R.B. (1970). Cooperation of antigenic determinants and of cells in the induction of antibodies. In: Developmental Aspects of Antibody Formation and Structure (Sterzl, J. & Riha, I. Eds.), Vol. II, pp. 547-561. Academia/Academic Press, New York.

Mole, L.E., Jackson, S.A., Porter, R.R., & Wilkinson, J.M. (1971). Allotypically related sequences in the Fd fragment of rabbit immunoglobulin heavy chains. Biochem. J. 124, 301-318.

Mudd, S. (1932). A hypothetical mechanism of antibody formation. J. Immunol. 23, 423-427.

Naor, D. & Sulitzeanu, D. (1967). Binding of radioiodinated bovine serum albumin to mouse spleen cells. Nature 214, 687-688.

Niall, H.D. & Edman, P. (1967). Two structurally distinct classes of kappa-chains in human immunoglobulins. Nature 216, 262-263.

Nisonoff, A. (1991). Idiotypes: concepts and applications. J. Immunol. 147, 2429-2438.

Nisonoff, A., Winkler, M.H., & Pressman, D. (1959). The similar specificity of the combining sites of an individual antibody molecule. J. Immunol. 82, 201-208.

Nisonoff, A., Wissler, F.C., Lipman, L.N., & Woernley, D.L. (1960). Separation of univalent fragments from the bivalent rabbit antibody molecule by reduction of disulfide bonds. Arch. Biochem. Biophys. 89, 230-244.

Nisonoff, A., Hopper, J.E., & Spring, S.B. (1975). The Antibody Molecule. Academic Press, New York.

Northrop, J.H. (1941-1942). Purification and crystallization of diphtheria antitoxin. J. Gen. Phys. 25, 465-485.

Nossal, G.J.V. (1960). Antibody production by single cells. IV. Further studies on multiply immunized animals. Br. J. Exp. Pathol. 41, 89-96.

Nossal, G.J.V. (1995). Choices following antigen entry: antibody formation or immunologic tolerance? Annu. Rev. Immunol. 13, 1-27.

Nossal, G.J.V. & Lederberg, J. (1958). Antibody production by single cells. Nature 181, 1419-1420.

Nossal, G.J.V, Ada, G.L., & Austin, C.M. (1965). Antigens in immunity. IX. The antigen content of single antibody-forming cells. J. Exp. Med. 121, 945-954.

Nossal, G.J.V. & Pike, B.L. (1976). Single cell studies on the antibody-forming potential of fractionated, hapten-specific B lymphocytes. Immunology 30, 189-202.

Nossal, G.J.V., Szenberg, A., Ada, G.L., & Austin, C.M. (1964). Single cell studies on 19S antibody production. J. Exp. Med. 119, 485-502.

O'Donnell, I.J., Frangione, B., & Porter, R.R. (1970). The disulfide bonds of the heavy chain of rabbit immunoglobulin G. Biochem. J. 116, 261-268.

Oudin, J. (1956a). Sérologie. Réaction de précipitation spécifique entre des sérums d'animaux de même espèce. C. R. Acad. Sci. (Paris) 242, 2489-2490.

Oudin, J. (1956b). L' "allotypie" de certains antigènes protéidiques du sérum. C. R. Acad. Sci. (Paris) 242, 2606-2608.

Oudin, J. (1960a). Allotypy of rabbit serum proteins. I. Immunochemical analysis leading to the individualization of seven main allotypes. J. Exp. Med. 112, 107-124.

Oudin, J. (1960b). Allotypy of rabbit serum proteins. II. Relationships between various allotypes: their common antigenic specificity, their distribution in a sample population; genetic implications. J. Exp. Med. 112, 125-142.

Oudin, J. (1966). The genetic control of immunoglobulin synthesis. Proc. R. Soc. Lond. B 166, 207-219.

Oudin, J. & Michel, M. (1963). Une nouvelle forme d'allotypie des globulines γ du sérum de lapin, apparemment liée à la fonction et à la spécificité anticorps. C. R. Acad. Sci. (Paris) 257, 805-808.

Ovary, Z. & Karush, F. (1961). Studies on the immunologic mechanism of anaphylaxis. II. Sensitizing and combining capacity *in vivo* of fractions separated from papain digests of antihapten antibody. J. Immunol. 86, 146-150.

Owen, R.D. (1945). Immunogenetic consequences of vascular anastomoses between bovine twins. Science 102, 400-401.

Padlan, E.A. (1994). Anatomy of the antibody molecule. Mol. Immunol. 31, 169-217.

Pain, R.H. (1963) The molecular weights of the peptide chains of γ-globuln. Biochem. J. 88, 234-239.

Palmer, J.L., Mandy, W.J., & Nisonoff, A. (1962). Heterogeneity of rabbit antibody and its subunits. Proc. Natl. Acad. Sci. USA 48, 49-53.

Parfentjev, I.A. (1936). Method for purification of antitoxins and the like. U.S. Patent No. 2,065,196.

Pauling, L. (1940). A theory of the structure and process of formation of antibodies. J. Am. Chem. Soc. 62, 2643-2657.

Pauling, L. & Pressman, D. (1945). The serological properties of simple substances. IX. Hapten inhibition of precipitation of antisera homologous to the *o*-, *m*-, and *p*-azophenylarsonic acid groups. J. Amer. Chem. Soc. 67, 1003-1012.

Petermann, M.L. (1946). The splitting of human gamma globulin antibodies by papain and bromelin. J. Amer. Chem. Soc. 68, 106-113.

Petermann, M.L. & Pappenheimer, A.M., Jr. (1941). The ultracentrifugal analysis of diphtheria proteins. J. Phys. Chem. 45, 1-9.

Peterson, E.A. & Sober, H.A. (1956). Chromatography of proteins. I. Cellulose ion-exchange adsorbents. J. Amer. Chem. Soc. 78, 751-755.

Poljak, R.J., Amzel, L.M., Avey, H.P., Chen, B.L., Phizackerley, R.P., & Saul, F. (1973). Three-dimensional structure of the Fab' fragment of a human immunoglobulin at 2.8-Å resolution. Proc. Natl. Acad. Sci. USA 70, 3305-3310.

Porter, R.R. (1950a). The formation of a specific inhibitor by hydrolysis of rabbit antiovalbumin. Biochem. J. 46, 479-484.

Porter, R.R. (1950b). A chemical study of rabbit antiovalbumin. Biochem. J. 46, 473-478.

Porter, R.R. (1958). Separation and isolation of fractions of rabbit gamma-globulin containing the antibody and antigenic combining sites. Nature 182, 670-671.

Porter, R.R. (1959). The hydrolysis of rabbit γ-globulin and antibodies with crystalline papain. Biochem. J. 73, 119-126.

Porter, R.R. (1962). The structure of gamma-globulin and antibodies. In: Basic Problems in Neoplastic Disease (Gellhorn, A. & Hirschberg, E., Eds.), pp. 177-194. Columbia University Press, New York.

Porter, R.R. (1973a). Structural studies of immunoglobulins. Les Prix Nobel en 1972, pp 173-183. Stockholm: The Nobel Foundation (also Science 180, 713-716).

Porter, R.R. (1973b). Immunoglobulin structure. In: Defense and Recognition. (MTP International Review of Science), Biochemistry (Porter, R.R., Ed.), Series 1, Vol. 10, pp. 159-197. Butterworth, London.

Portier, P. & Richet, C. (1902). De l'action anaphylactique de certains venins. Comptes Rendus la Société de Biologie (Paris) 54, 170-172.

Potter, M. (1982). Pathogenesis of plasmacytomas in mice. In: Cancer—A Comprehensive Treatise, 2nd ed. (Becker, F.I., Ed.), pp. 135-159. Plenum, New York.

Press, E.M. & Hogg, N.M. (1969). Comparative study of two immunoglobulin G Fd-fragments. Nature 223, 807-810.

Putnam, F.W. (1962). Structural relationships among normal human γ-globulin, myeloma globulins, and Bence-Jones proteins. Biochim. Biophys. Acta 63, 539-541.

Putnam, F.W. (1977). Immunoglobulins I. Structure. In: The Plasma Proteins, 2nd ed. (Putnam, F.W. Ed.), Vol. III. Academic Press, New York.

Putnam, F.W. & Easley, C.W. (1965). Structural studies of the immunoglobulins. I. The tryptic peptides of Bence-Jones proteins. J. Biol. Chem. 240, 1626-1638.

Putnam, F.W., Easley, C.W. & Helling, J.W. (1963). Structural study of human γ-globulin through the analysis of the tryptic peptides of Bence-Jones proteins. Biochim. Biophys. Acta 78, 231-233.

Raff, M.C. (1970). Role of thymus-derived lymphocytes in the secondary humoral immune response in mice. Nature 226, 1257-1258.

Raff, M.C., Feldmann, M., & de Petris, S. (1973). Monospecificity of bone marrow-derived lymphocytes. J. Exp. Med. 137, 1024-1030.

Rajan, S.S., Ely, K.R., Abola, E.E., Wood, M.K., Colman, P.M., Athay, R.J., & Edmundson, A.B. (1983). Three-dimensional structure of the Mcg IgG1 immunoglobulin. Mol. Immunol. 20, 787-799.

Roitt, I.M. & Delves, P.J. (1992). Encyclopedia of Immunology (3 volumes). Academic Press, London.

Rosen, F.S., Steiner, L.A., & Unanue, E.A. (1989). Macmillan Dictionary of Immunology. Macmillan Press, London.

Rothen, A. & Landsteiner, K. (1939). Adsorption of antibodies by egg albumin films. Science 90, 65-66.

Rutishauser, U., D'Eustachio, P., & Edelman, G.M. (1973). Immunological function of lymphocytes fractionated with antigen-derivatized fibers. Proc. Natl. Acad. Sci. USA 70, 3894-3898.

Sanger, F. & Thompson, E.O.P. (1953). The amino acid sequence in the glycyl chain of insulin. Biochem. J. 53, 353-374.

Sanger, F. & Tuppy, H. (1951). The amino acid sequence in the phenylalanyl chain of insulin. Biochem. J. 49, 463-490.

Sarma, V.R., Silverton, E.W., Davies, D.R., & Terry, W.D. (1971). The three-dimensional structure at 6 Å resolution of a human γG1 immunoglobulin molecule. J. Biol. Chem. 246, 3753-3759.

Schiffer, M., Girling, R.L., Ely, K.R., & Edmundson, A.B. (1973). Structure of a λ-type Bence-Jones protein at 3.5 Å resolution. Biochemistry 12, 4620-4631.

Schwartz, R.S. (1984). Immunological specificity, anergy and immunoregulation. Prog. Allergy 35, 1-14.

Seeger, M.A., Haffley, L., & Kaufman, T.C. (1988). Characterization of *amalgam*: a member of the immunoglobulin superfamily from Drosophila. Cell 55, 589-600.

Segal, D.M., Padlan, E.A., Cohen, G.H., Rudikoff, S., Potter, M., & Davies, D.R. (1974). The three-dimensional structure of a phosphorylcholine-binding mouse immunoglobulin Fab and the nature of the antigen binding site. Proc. Natl. Acad. Sci. USA 71, 4298-4302.

Sege, K. & Peterson, P.A. (1978). Use of anti-idiotypic antibodies as cell-surface receptor probes. Proc. Natl. Acad. Sci. USA 75, 2443-2447.

Sigal, N.H. & Klinman, N.R. (1978). The B-cell clonotype repertoire. Adv. Immunol. 26, 255-337.

Silverstein, A.M. (1986). History of immunology. Anti-antibodies and anti-idiotypic immunoregulation, 1899-1904: the inexorable logic of Paul Ehrlich. Cell. Immunol. 99, 507-522.

Silverstein, A.M. (1989). A History of Immunology. Academic Press, New York.

Silverstein, A.M. (1996). Paul Ehrlich: the founding of pediatric immunology. Cell. Immunol. 174, 1-6.

Silverton, E.W., Navia, M.A., & Davies, D.R. (1977). Three-dimensional structure of an intact human immunoglobulin. Proc. Natl. Acad. Sci. USA 74, 5140-5144.

Singer, S.J. & Doolittle, R.F. (1966). Antibody active sites and immunoglobulin molecules. Science 153, 13-25.

Slater, R.J., Ward, S.M., & Kunkel, H.G. (1955). Immunological relationships among the myeloma proteins. J. Exp. Med. 101, 85-108.

Steiner, L.A. (1985). Immunoglobulin disulfide bridges: Theme and variations. Biosci. Rep. 5, 973-989.

Steiner, L.A. & Eisen, H.N. (1966). Symposium on *in vitro* studies of the immune response. I. Variations in the immune response to a simple determinant. Bact. Revs. 30, 383-396.

Steiner, L.A. & Eisen, H.N. (1967a). Sequential changes in the relative affinity of antibodies synthesized during the immune response. J. Exp. Med. 126, 1161-1183.

Steiner, L.A. & Eisen, H.N. (1967b). The relative affinity of antibodies synthesized in the secondary response. J. Exp. Med. 126, 1185-1205.

Steiner, L.A. & Lopes, A.D. (1979). The crystallizable human myeloma protein Dob has a hinge-region deletion. Biochem. 18, 4054-4067.

Stemke, G.W. (1964). Allotypic specificities of A- and B-chains of rabbit gamma globulin. Science 145, 403-405.

Swan, D., D'Eustachio, P., Leinwand, L., Seidman, J., Keithley, D., & Ruddle, F.H. (1979). Chromosomal assignment of the mouse κ light chain genes. Proc. Natl. Acad. Sci. USA 76, 2735-2739.

Talmage, D.W. (1957a). Allergy and immunology. Annu. Rev. Med. 8, 239-256.

Talmage, D.W. (1957b). The primary equilibrium between antigen and antibody. Ann. N.Y. Acad. Sci. 70, 82-93.

Taranta, A. & Franklin, E.C. (1961). Complement fixation by antibody fragments. Science 134, 1981-1982.

Terry, W.D., Hood, L.E., & Steinberg, A.G. (1969). Genetics of immunoglobulin κ-chains: chemical analysis of normal human light chains of differing Inv types. Proc. Natl. Acad. Sci. USA 63, 71-77.

Tiselius, A. & Kabat, E.A. (1939). An electrophoretic study of immune sera and purified antibody preparations. J. Exp. Med. 69, 119-131.

Titani, K. & Putnam, F.W. (1965). Immunoglobulin structure: amino- and carboxyl-terminal peptides of type I Bence Jones proteins. Science 147, 1304-1305.

Todd, C.W. (1963). Allotypy in rabbit 19S protein. Biochem. Biophys. Res. Commun. 11, 170-175.

von Behring, E. & Kitasato, S. (1890). Über das Zustandekommen der Diphtherie-Immunität und der Tetanus-Immunität bei Thieren. Deutsche Medizinische Wochenschrift 16, 1113-1114.

Wang, A.-C., Wang, I.Y., & Fudenberg, H.H. (1977). Immunoglobulin structure and genetics. Identity between variable regions of a μ and a $\gamma 2$ chain. J. Biol. Chem. 252, 7192-7199.

Wang, A.-C., Wang, I.Y.F., McCormick, J.N., & Fudenberg, H.H. (1969). The identity of light chains of monoclonal IgG and monoclonal IgM in one patient. Immunochemistry 6, 451-459.

Wang, A.C., Wilson, S.K., Hopper, J.E., Fudenberg, H.H., & Nisonoff, A. (1970). Evidence for control of synthesis of the variable regions of the heavy chains of immunoglobulins G and M by the same gene. Proc. Natl. Acad. Sci. USA 66, 337-343.

Weigert, M.G., Cesari, I.M., Yonkovich, S.J., & Cohn, M. (1970). Variability in the lambda light chain sequences of mouse antibody. Nature 228, 1045-1047.

White, R.G. (1958). Antibody production by single cells. Nature 182, 1383-1384.

Whitney, P.L. & Tanford, C. (1965). Recovery of specific activity after complete unfolding and reduction of an antibody fragment. Proc. Natl. Acad. Sci. USA 53, 524-532.

Wigzell, H. & Andersson, B. (1969). Cell separation on antigen-coated columns. Elimination of high rate antibody-forming cells and immunological memory cells. J. Exp. Med. 129, 23-36.

Wikler, M., Köhler, H., Shinoda, T., & Putnam, F.W. (1969). Macroglobulin structure: homology of mu and gamma heavy chains of human immunoglobulins. Science 163, 75-78.

Wilhelm, E. & Lamm, M.E. (1966). Absence of allotype *b*4 in the heavy chains of rabbit γG-immunoglobulin. Nature 212, 846-847.

Wilkinson, J.M. (1969). Variation in the N-terminal sequence of heavy chains of immunoglobulin G from rabbits of different allotype. Biochem. J. 112, 173-185.

Williams, A.F. (1982). Surface molecules and cell interactions. J. Theor. Biol. 98, 221-234.

Williams, A.F. (1987). A year in the life of the immunoglobulin superfamily. Immunol. Today 8, 298-303.

Williams, A.F. & Barclay, A.N. (1988). The immunoglobulin superfamily—domains for cell surface recognition. Annu. Rev. Immunol. 6, 381-405.

Wilson, I.A. & Stanfield, R.L. (1993). Antibody-antigen interactions. Curr. Opin. Struct. Biol. 3, 113-118.

Wofsy, L., Kimura, J., & Truffa-Bachi, P. (1971). Cell separation of affinity columns: the preparation of pure populations of anti-hapten specific lymphocytes. J. Immunol. 107, 725-729.

Wofsy, L., Metzger, H., & Singer, S.J. (1962). Affinity labeling—a general method for labeling the active sites of antibody and enzyme molecules. Biochemistry 1, 1031-1039.

Wu, T.T. & Kabat, E.A. (1970). An analysis of the sequences of the variable regions of Bence Jones proteins and myeloma light chains and their implications for antibody complementarity. J. Exp. Med. 132, 211-250.

AUTHOR INDEX

SUBJECT INDEX

www.ingramcontent.com/pod-product-compliance
Lightning Source LLC
LaVergne TN
LVHW011909200926
848484LV00002B/570
* 9 7 8 0 7 6 2 3 0 0 7 8 5 *